M

This b

World Survey of Climatology Volume 9

CLIMATES OF SOUTHERN AND WESTERN ASIA

World Survey of Climatology

World Survey of Climatology Volume 9

Climates of Southern and Western Asia

edited by

K. TAKAHASHI

6-26-15, Seijo
Setagaya-ku
Tokyo 157 (Japan)

and

H. ARAKAWA

Daita 3-5-27
Setagaya-ku
Tokyo 157 (Japan)

ELSEVIER SCIENTIFIC PUBLISHING COMPANY
Amsterdam-Oxford-New York 1981

ELSEVIER SCIENTIFIC PUBLISHING COMPANY
Molenwerf 1, 1014 AG
P.O. Box 211, 1000 AE Amsterdam, The Netherlands

ELSEVIER/NORTH-HOLLAND, INC.
52 Vanderbilt Avenue
New York, New York 10017

Library of Congress Cataloging in Publication Data
Main entry under title:

Climates of southern and western Asia.

   (World survey of climatology ; v. 9)
   Includes bibliographies and index.
   1. Asia, Southeastern--Climate. 2. Near East--
Climate. I. Landsberg, Helmut Erich, 1906-   .
II. Arakawa, Hidetoshi, 1907-   . III. Takahashi,
Kōichirō, 1913-   . IV. Series.
QC980.15.W67 vol. 9 [QC990.A7542]   551.69s   81-1046
ISBN 0-444-41861-X   [551.6956]   AACR2

Library of Congress Card Number 81-1046
ISBN 0-444-41861-x (vol. 9)   √
ISBN 0-444-40734-0 (series)
With 50 illustrations and 180 tables

Printed in The Netherlands

World Survey of Climatology

List of Contributors to this Volume

S. A. HARB
Ministry of Civil Aviation
The Meteorological Authority
Koubry El-Quobba
Cairo (Egypt)

M. K. NAGIB
Ministry of Civil Aviation
The Meteorological Authority
Koubry El-Quobba
Cairo (Egypt)

S. NIEUWOLT
c/o MARDI
Bag Berkunci 202
Universiti Pertanian P.O.
Serdang, Selangor
(Malaysia)

K. N. RAO
India Meteorological Department
Meteorological Office
Ganeshkind Road
Poona—5 (India)

Y. P. RAO
B3/33 Azad Apartments
Aurobindo Marg
New Delhi—110016 (India)

M. F. TAHA
Ministry of Civil Aviation
The Meteorological Authority
Koubry El-Quobba
Cairo (Egypt)

A. H. TANTAWY
Ministry of Civil Aviation
The Meteorological Authority
Koubry El-Quobba
Cairo (Egypt)

Contents

Chapter 1. THE CLIMATES OF CONTINENTAL SOUTHEAST ASIA
by S. Nieuwolt

Chapter 2. THE CLIMATE OF THE INDIAN SUBCONTINENT
by Y. P. Rao

Chapter 3. THE CLIMATE OF THE NEAR EAST
by M. F. Taha, S. A. Harb, M. K. Nagib and A. H. Tantawy

Contents

Contents

Chapter 4. TROPICAL CYCLONES OF THE INDIAN SEAS
by K. N. Rao

XII

Contents

The Climates of Continental Southeast Asia

S. NIEUWOLT

Introduction

Continental Southeast Asia comprises the territories of the following states: Vietnam, Laos, Kampuchea, Thailand, Burma, Western Malaysia (Malaya) and Singapore. It is a large and highly differentiated region, stretching over 27 degrees of latitude, from 1°20′ (Singapore) to 28°N (northern Burma). It includes the narrow Malayan Peninsula, where maritime conditions prevail, and areas of northern Burma and Thailand, almost 1,000 km from the nearest coast. Extensive lowlands and high mountain ranges are found in this area, which will be referred to as "the region".

In spite of these large differences, the climates of Continental Southeast Asia have one factor in common: they are all controlled to a very large extent by the system of the Asian monsoons. These monsoons have their origin outside Continental Southeast Asia, and their effects are essentially the same over the whole area. A description of these climates should therefore start with the seasons, which are mainly caused by the monsoons.

The northeast monsoon season

This season lasts approximately from November to March. Fig.1 illustrates the surface conditions during January, a fairly typical month of this season. Conditions in the atmosphere, both at the surface and aloft, are quite stable during this season. The average conditions, as shown on the maps and described below, are therefore similar to actual conditions on a large number of days (RAMAGE, 1952, p.407).

Pressure conditions in the low latitudes are always dominated by a regular diurnal oscillation, with maxima at 10h00 and 22h00, and minima at 04h00 and 16h00 local time. The diurnal range of pressure in Singapore is around 3.5 mbar and at Phu-Lien (near Hanoi) approximately 4.6 mbar (BRUZON et al., 1940, p.26; WATTS, 1955a, pp.13–15). However, these diurnal variations have no climatic consequences and they can therefore be disregarded here.

In the low latitudes, pressure differences are usually quite small compared to conditions in the higher latitudes. But during the northeast monsoon, there is a fairly strong pressure gradient from north to south over the region (Fig.1B).

The resultant surface wind is the northeast monsoon. Its velocity is relatively high for this part of the world, being generally around 4–6 m/sec, with somewhat higher speeds in the northern parts of the region, where the pressure gradient is larger, than in the

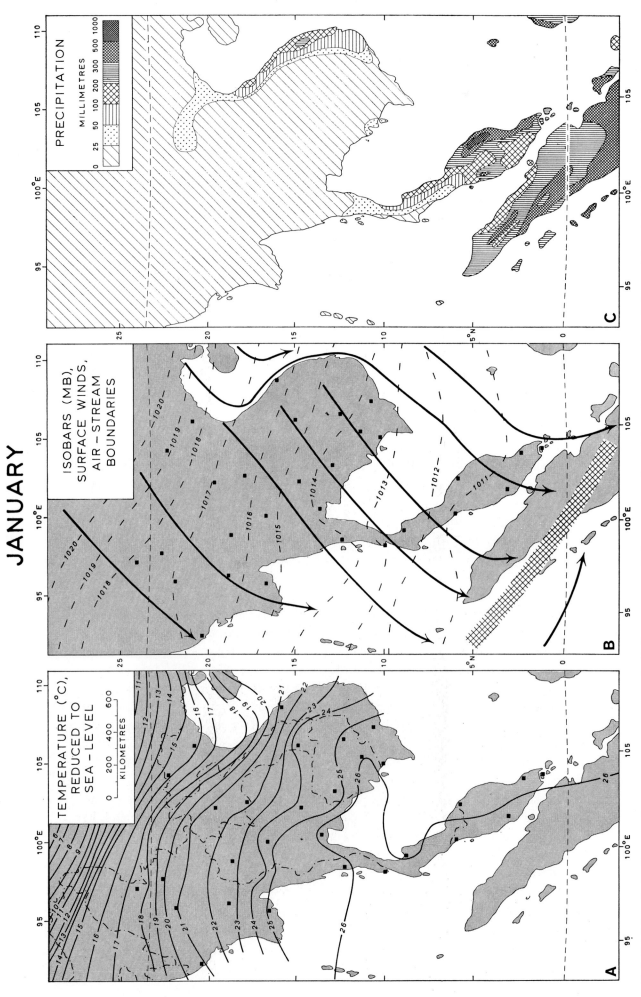

Fig.1. Conditions during January: A. mean temperatures at sea level; B. mean atmospheric pressure, surface winds and average position of air-stream boundaries; C. mean precipitation.

south. The monsoon winds are locally deflected by mountain ranges, as over the mountains of Annam, but over the sea they are very constant in direction, reaching a constancy figure of over 75%. This means that more than three-quarters of all observations were within 45° on the main wind direction (DALE, 1956, p.14).

The northeast monsoon brings two main air masses to Continental Southeast Asia.

(1) Air from a large high pressure cell over Siberia and Mongolia. This polar continental air is originally very cold, dry and stable. But when it reaches Continental Southeast Asia it is always thoroughly modified, especially if it has moved over the relatively warm China Sea. It is not a deep air mass: over southern China its thickness rarely exceeds 3,000 m, and on its way southwards it becomes shallower so that in Singapore it is usually not more than 1,350 m thick (JOHN, 1950, p.7). Occasionally a surge of the northeast monsoon brings this type of air to northern Vietnam via a land route, and in this case its modification is limited, with a consequent drop in temperature. Although these surges of the monsoon are not frequent, they explain the large temperature gradients over Vietnam, as shown in Fig.1A. On its way southwards over Vietnam the air is rapidly warmed.

(2) The second air mass has its origin over the Pacific Ocean, north of the Equator. It is tropical maritime air, carried by the northeastern trade winds, and it is generally warm and stable, as a consequence of the trade wind inversion. This air mass being warmer than the first type it generally moves over the polar air when the two air masses meet. On its way westwards this air mass is humidified over the China Sea. Normally this type of air occurs quite frequently over Continental Southeast Asia, especially in the southern part of the region. However, during surges of the northeast monsoon the trade wind system usually retreats far eastwards to the Pacific and the first air mass dominates over the area.

The boundary between these two air masses forms part of the polar front. Its position at sea level during January is around 25°N, but it frequently moves over large distances and can be as far south as 10°N (RAMAGE, 1952, p.406). At higher levels the front is further south: at 600 m its position is around 15°–20°N from southwest to northeast over the region (THOMPSON, 1951, p.572). Even over Singapore the front is still discernible at higher levels, though the temperature differences between the two air masses become very small near the Equator, as both are rapidly warmed from below.

Upper air conditions over Continental Southeast Asia are not very well known because of a scarcity of observations. Interpretation of the limited amount of information which is available is somewhat controversial, as it is not generally accepted that there are major differences between upper-air and sea-level conditions (THOMPSON, 1951; WATTS, 1955a). Until more observations are made, the following generalizations are possible. At about 3,000 m the position of the polar front is approximately the same as at 600 m. North of it upper westerlies prevail, and these increase in velocity with elevation. At approximately 12,000 m there is a strong westerly jet stream around 25°N, the result of the confluence of the two branches of the jet stream, north and south of the Himalayan–Tibetan massif (TREWARTHA, 1958, pp.208–210). South of the upper polar front, easterlies from the Pacific dominate the circulation at the 3,000 m level. At the extreme south of the region equatorial westerlies replace these easterlies (THOMPSON, 1951, p.572).

During the northeast monsoon season most disturbances associated with the polar front pass north of Continental Southeast Asia (RAMAGE, 1952, p.405). South of the jet stream strong subsidence prevails in thick layers of the atmosphere. This explains not only the rarity of invasions of polar air from the north to lower latitudes than about 15°N, but also the general lack of rainfall over most of the region during this season (Fig.1C).

There are only two areas with widespread precipitation during the northeast monsoon season: eastern and northern Vietnam, and the Malay Peninsula, south of about 10°N (Fig.1C).

The precipitation in Vietnam is largely caused by disturbances, which follow tracks near the polar front, situated not far to the north. Rainfall from these disturbances is locally increased by orographic lifting, as for instance in the mountains of Annam. But the importance of orographic lifting should not be overestimated, as there is no doubt that the main source of the precipitation are the disturbances. This is demonstrated by the intermittent and irregular character of the rainfall, by the occurrence of rain in the predominantly flat Red River Basin, and by the fact that there is no close correlation between the strength of the monsoon winds and the amounts of rainfall received (PÉDE-LABORDE, 1958, p.130). On the western slopes of the mountains the rain shadow effect reduces the rainfall to less than 25 mm per month.

Conditions in northern Vietnam, locally known as "crachin", are also related to disturbances of the polar front. This type of weather occurs in periods of 3–5 days. It is characterized by fogs or heavily overcast skies, and light drizzle, but only small amounts of actual precipitation. These conditions occur when a shallow depression is situated over the Gulf of Tonkin. The main season of "crachin" is from the end of January to April.

Rainfall in the Malay Peninsula during this season is also mainly caused by disturbances, but these are of a different origin. They travel with the monsoon current and are related to the air mass boundary between the two air masses of the northeast monsoon, or to convergence within the monsoon currents (WATTS, 1949, p.13). The amounts of rainfall in this region are higher than in Vietnam, due to the intensive humidification of the monsoon air masses over the warm South China Sea. Orographic lifting further increases the amounts of rainfall in eastern coastal areas. Convection over land, frequent in these low latitudes where temperatures are quite high, is the third cause of rainfall. There is, however, no doubt that the disturbances are the main factor, as is demonstrated by the intermittent and irregular character of the rainfall along the east coast of Malaya, and by the inverse correlation which exists between the strength of the monsoon and the amounts of rainfall received, so that stronger monsoon winds usually bring less rainfall than relatively weak winds (NIEUWOLT, 1966a, p.171).

From the coastal areas of eastern Malaya the amounts of rainfall decrease both towards the west and towards the north, (Fig.1C). The decrease to the west is due to the rain-shadow effect of the mountain ranges of the peninsula. The decrease to the north is caused by the shorter trip of the monsoonal air masses over the South China Sea, which limits their humidification to relatively shallow layers.

The final destination of the air masses of the northeast monsoon is in the Southern Hemisphere, where they become the northwest monsoon of Indonesia, reinforced by equatorial westerlies from the Indian Ocean (Fig.1B).

November–December

Conditions during January are representative for the whole of the northeast monsoon season, from November to March. The upper-air circulation shows no basic changes during these five months (THOMSON, 1951, pp.571–582). But the first two months show some interesting differences from the described conditions in January.

The northeast monsoon starts its move southwards over Continental Southeast Asia in October, and by the end of that month it normally reaches the south coast of Indo-China. It is usually well established in Singapore by the middle of December. But the southward movement is very irregular and often even temporarily reversed. During November and December equatorial westerlies still occur frequently over the southern part of the region, and they collide with the advancing northeast monsoon. The resulting air mass boundary zone is always accompanied by widespread and heavy rainfall. This is especially the case in the eastern coastal areas of the Malay Peninsula, where orographic lifting increases the amounts of precipitation considerably. November is therefore the wettest month of the year in the northern parts of the peninsula, as in Bandon and Kota Bharu. In the south, December is usually correlated with surges of the monsoon and close proximity of the air mass boundary. Disturbances play a much lesser role than during January (GAN TONG LIANG, n.d., pp.7, 8).

In Vietnam, November and December are normally considerably wetter than January. Here the typhoons continue to bring large amounts of precipitation until the end of the year, though their frequency of occurrence decreases gradually during these two months. Still, they bring the strongest winds of the northeast monsoon season (BRUZON et al., 1940, p.14). In Malaya, however, the strongest winds occur during January (NIEUWOLT, 1966a, pp.173, 177).

February–March

While the basic conditions remain the same as in January, the last two months of the northeast monsoon season bring some changes. The most important development is the general decrease in the velocity of the northeast monsoon over the whole region, and the gradual change to more easterly wind directions. This development is, of course, related to the slow weakening of the high pressure cell over continental Asia and the consequent attenuation of invasions of polar air. In Hongkong the thickness of the polar air shows a general decrease from October, when it reaches about 3,000 m, to about 2,000 m in January, to only 1,500 m in March (AIR MINISTRY, 1937, Vol.II, p.147). This decrease proceeds even more rapidly further south, and as the importance of the polar air mass diminishes, the trade wind air from the Pacific Ocean takes over, and increases its frequency of occurrence over most of the region. As this air mass is originally stable up to a high level, the result is a lessening in precipitation, compared to January, at the eastern coast of Vietnam.

A second change is the rapid rise in surface temperatures by 5°–7°C between January and March in the inland areas north of about 12°N. This is due to increased insolation, caused by longer days and higher elevations of the sun. The high daytime temperatures give rise to convective thunderstorms. These bring the "mango-rains" of Thailand and Burma, which are of great importance for agriculture, although the total amounts of rainfall received are not high. But they indicate the end of the dry season. These local disturbances may be connected with the eastward movement of a tropical trough at the 500 mbar level from southern India. This movement creates convergence at lower levels, which favours the development of thunderstorms (RAMAGE, 1955, p.257).

The temperatures over the mainland become sufficiently high to create a shallow surface low over the inland areas. This low causes a deflection of the main air current over the South China Sea, which shows a divergence starting from February (DALE, 1956, pp.15, 16). This divergence in turn results in subsiding movements and this largely prevents the formation of disturbances. This might be the main reason for the sharp decrease of precipitation in February, compared to January, in the eastern coastal areas of Malaya, where the force of the northeast monsoon is practically the same during these two months (NIEUWOLT, 1965a, fig.18; 1966a, pp.172–173).

The inter-monsoon period of April and May

During these two months the transition zone between the retreating northeast monsoon and the slowly advancing southwest monsoon moves over Continental Southeast Asia from south to north. In the southern parts of the region the transition period lasts about one month, which is usually April, but further north the transition takes longer and in northern Burma the southwest monsoon is usually not well established until the end of May. During this period the processes which started in February and March develop further. The continental high pressure area over Siberia and Mongolia has disappeared and the northeast monsoon consists almost entirely of extended trade winds from the north Pacific Ocean. Its velocities are now very low.

The map (Fig.2B) shows four different air streams over the region: (*1*) the retreating northeast monsoon, a part of which is now clearly diverted into the mainland by a surface low over Thailand; (*2*) the southeasterly trades, from the south Pacific and Australia, which in April just reach the Equator, but rarely cross into the Northern Hemisphere; (*3*) equatorial westerlies from the Indian Ocean; their origin is not well known, but might be in the Southern Hemisphere; as the season proceeds these westerlies increase in force; (*4*) westerlies from the northwest, the first signs of an extension of the Indian southwest monsoon.

However, it must be emphasized that all these air streams are very weak and highly variable in direction. Actually, during most of this period there is no recognizable general circulation and the weather maps are dominated by the broad air stream boundaries. These are areas of slowly converging air masses, characterized by a general instability of the lower atmosphere, caused by rising air movements. They bring much cloudiness and precipitation. Calms are prevalent in the areas where the air stream boundaries occur, and if there are winds they are of variable direction.

The air stream boundaries are very changeable, both in intensity and in location. Fig.2B shows the average position during April. On actual weather maps the boundaries are often not recognizable. They disappear frequently, and may move hundreds of miles in a few days. Usually the most intensive boundary is the one near the Equator.

During this period the general circulation over Continental Southeast Asia comes to an almost complete standstill. Light winds of very variable directions, called "doldrums", prevail over most of the region. The absence of a clear general circulation has some important consequences. Over inland areas strong surface heating takes place, as the heated air is only slowly removed by turbulence. In most places in the northern parts of the region April is the hottest month of the year. The low pressure area over central Thailand and Burma is directly connected with these high temperatures (Fig.2A, B).

APRIL

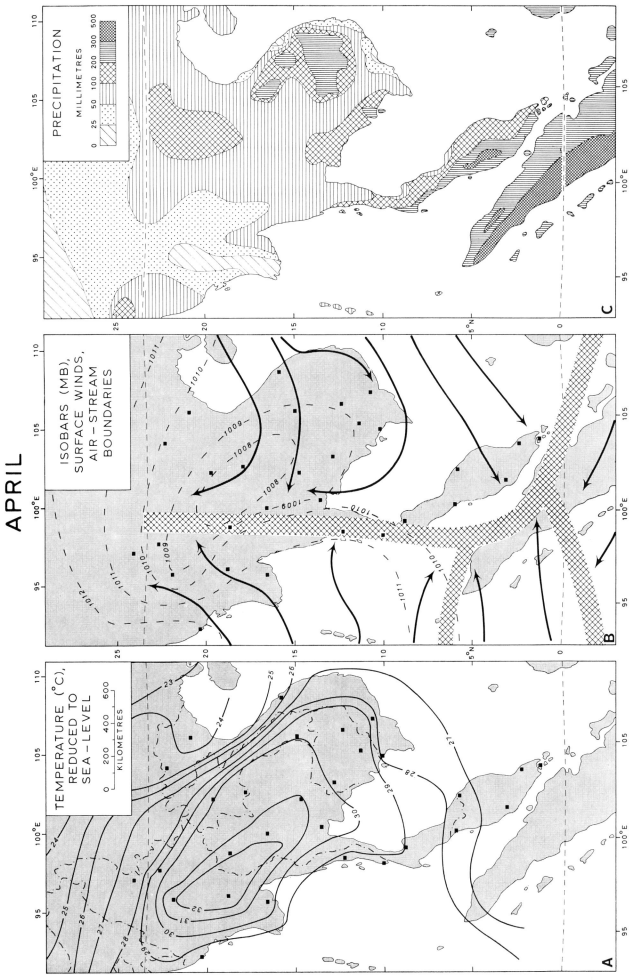

Fig.2. Conditions during April: A. mean temperatures at sea level; B. mean atmospheric pressure, surface winds and average position of air-stream boundaries; C. mean precipitation.

A second consequence is that local wind systems dominate the circulation. These are the land and sea breezes and the mountain and valley winds. The strong sea breezes along the southern coasts of Burma, Thailand and Campuchea become almost continuous because of the large temperature differences between land and sea.

Furthermore, local convection is rather active. The diurnal rainfall maximum is during the hot hours of the day, when thunderstorms occur. The frequency of these pre-monsoonal thundershowers is quite high, though the amount of rainfall they bring is relatively modest (Fig.2C). This is a consequence of conditions in the upper troposphere.

Upper air conditions are dominated by the southern branch of the jet stream which at this time is still south of the Himalayan–Tibetan massif. Therefore westerlies prevail over the region. In the northern parts these are extensions of the jet stream, in the southern parts equatorial westerlies. Easterlies from the Pacific Ocean do not enter the region, but may just reach the coast of the Tonkin Delta. The two systems of westerlies converge at a latitude of about 15°–20°N, in a southwest–northeast line (THOMPSON, 1951, p.583). This convergence causes subsidence and stability at lower levels, and this is the main reason for the relatively low precipitation during the transition period.

Areas of highest rainfall during April are those near the air mass boundary in the south, where the air masses are very humid and warm, so that even a slight lifting will result in precipitation. This is the case over Malaya, especially in the western parts. Other areas with relatively high rainfall are mountainous areas, where local convection is particularly effective, as in the mountains of Annam and northern Vietnam. The very small amounts of rainfall in lower Burma are related to the prevalent westerly winds, and the resultant rain-shadow effect of the mountain ranges (Fig.2C).

May

In the southern parts of Continental Southeast Asia May is part of the southwest monsoon season. Here the monsoon begins gradually and is usually well established by May 10 (WATTS, 1955a, p.9). But in the north May is still a month of transition. The beginning of the monsoon in the northern parts of the region is a much more irregular process, which may last, in the far north, until the first days of June. The southwest monsoon comes earlier to Burma than to India. This is caused by an upper air meridional trough at about 85°E, which accelerates southwesterly winds in Burma, but has the opposite effect further to the west, in India (YIN, 1949).

The start of the southwest monsoon in the north is related to conditions in the upper troposphere. The upper westerlies of the southern branch of the jet stream are slowly replaced by upper easterlies, which develop their own jet stream at the 15,000 m level. But this replacement is not a simple movement; it consists of a number of disappearances and re-appearances of the upper westerlies, each disappearance being related to a surge northwards of the monsoon (RAMAGE, 1952, p.405). Thus the beginning of the monsoon is usually accompanied by rapid changes of weather conditions. Still, it is much more gradual than the "burst" of the southwest monsoon in India.

The southwest monsoon season

This season lasts from about June to September, and the month of July (Fig.3) is quite

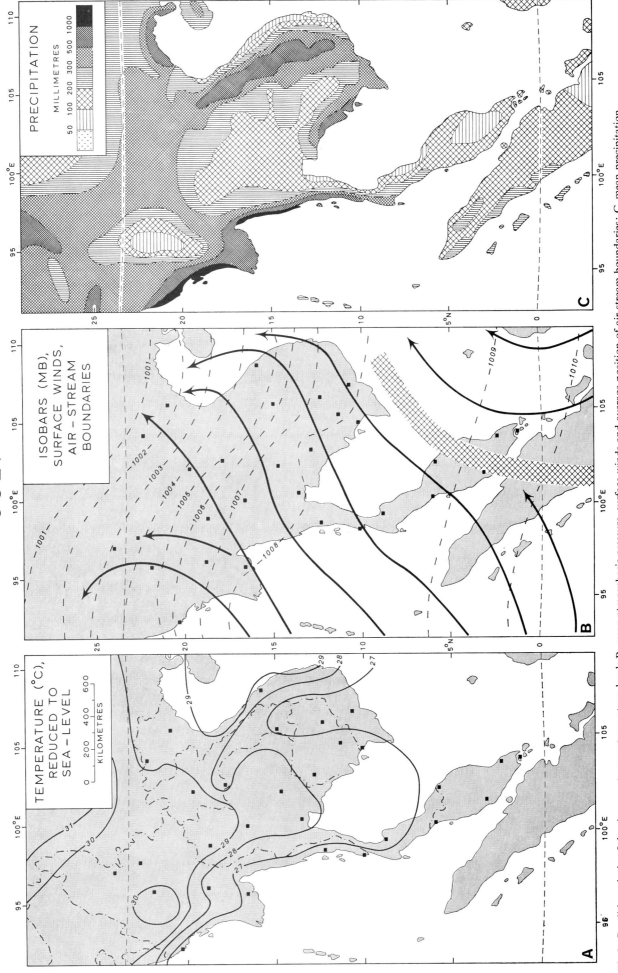

JULY

A TEMPERATURE (°C), REDUCED TO SEA-LEVEL

KILOMETRES
0   200   400   600

B ISOBARS (MB), SURFACE WINDS, AIR-STREAM BOUNDARIES

C PRECIPITATION
MILLIMETRES
50   100   200   300   500   1000

Fig.3. Conditions during July: A. mean temperatures at sea level; B. mean atmospheric pressure, surface winds and average position of air-stream boundaries; C. mean precipitation.

typical for conditions during this period, when the whole region is under the dominance of southwesterly winds.

Generally the southwest monsoon is weaker than the northeast monsoon, its average velocity being around 2 m/sec. (PÉDELABORDE, 1958, p.123). But the winds increase in speed towards the northern parts of the region, according to the pressure gradient (Fig.3B). In northern Vietnam the southwesterlies reach average velocities up to 5 m/sec and are stronger than the northeast monsoon winds (BRUZON et al., 1940, p.18).

The air masses of the southwest monsoon have two main origins.

(*1*) The southeasterly trades from Australia and the south Pacific bring originally very stable air to the equatorial latitudes. These air masses cause a dry season in the eastern parts of Indonesia, but on their way further westwards they are thoroughly modified over the seas and islands of Indonesia. When they reach Continental Southeast Asia they are very humid and unstable. These air masses are only of importance in the extreme south of the region (Fig.3B). They are relatively shallow, rarely reaching levels over 2,000 m. Their crossing into the Northern Hemisphere seems to be related to equatorial clockwise eddies (THOMPSON, 1951, p.594).

(*2*) The main air masses of the southeast monsoon come from the Indian Ocean. They can be equatorial in origin or come from further north, but it is usually impossible to distinguish between these two branches, both extensions of the Indian southwest monsoon. These air masses are quite deep, reaching up to 9,000 m along the Burma coast. Their thickness decreases as the season progresses, but they remain quite deep until September, when they measure about 3,000 m in the south and about twice that amount in the north.

Between these two main branches of the southwest monsoon exists an air mass boundary, which in this case is a zone of convergence. As the two air streams are very similar in temperature and humidity and their directions almost parallel, the convergence is not very intensive. And because the air masses from the southeast are rather shallow, the convergence zone is limited to low levels, usually below 3,000 m. Despite its shallowness and low intensity the boundary zone causes widespread but light rainfall in Malaya. Its position is highly variable and it is often not shown on daily weather maps because of its low intensity and its position over the sea.

Conditions at higher levels during the southwest monsoon season are not as stable as during the northeast monsoon. The westerlies prevail up to about 7,000 m. But at the 8,000-m level easterlies dominate over most of the region, and these increase in velocity with elevation. At an elevation of about 15,000 m an easterly jet stream exists during most of this season. Its position fluctuates widely between 10° and 25°N. There seems to be a correlation between the position of this easterly jet stream and disturbances at lower levels, which often bring heavy precipitation to the region (KOTESWARAM, 1958, p.47). As the whole region is under control of only one type of air mass, temperature differences during this season are relatively small (Fig.3A). It is striking that the areas with the highest temperatures, as for instance central Burma and central Thailand, are also the driest (Fig.3C). This is caused by rain shadow and foehn effects of the mountains further to the west.

In most of Continental Southeast Asia this is the main rainfall season of the year, as is illustrated by Table I and Fig.5 (p.18). Rainfall is always caused by one, or a combination of several of the following four processes.

TABLE I

MEAN RAINFALL DURING THE SOUTHWEST MONSOON (JUNE–SEPTEMBER), IN PERCENTAGES OF AVERAGE
ANNUAL TOTAL PRECIPITATION

Lashio	69	Chiang Mai	24	Hanoi	67
Mandalay	55	Phitsanulok	65	Vientiane	67
Akyab	83	Bangkok	57	Tourane	38
Rangoon	76	Bandon	33	Paksé	79
Mergui	73	Kota Bharu	21	Phnom Penh	51
Victoria Point	67	Kuala Lumpur	24	Ho Chi Minh City	63
Alor Star	40	Singapore	29	Ha-tien	56

(1) Cyclonic disturbances. These are not related to fronts, as in the higher latitudes, because temperature differences between the two air masses are too small for the creation of fronts. But as has been previously mentioned there are relations between conditions in the upper parts of the troposphere and the occurrence of disturbances at lower levels. The disturbances usually move with the monsoonal air streams. They are of greatest importance as a cause of precipitation at latitudes north of about 10°N. In these areas the Coriolis force is appreciable and cyclonic vorticity prevents the rapid filling up of the depressions. At lower latitudes this is not the case and cyclonic disturbances are generally short-lived and of low intensity. A special type of disturbances are the typhoons. They have their origin outside the region, but are of great importance in the coastal areas of Vietnam. They reach their maximum frequency in September and are discussed in the section dealing with conditions during that month.

(2) Convection. Convectional thunderstorms are frequent over the whole region, but especially at latitudes south of about 10°N. This is not only due to strong insolation, but also to the low wind velocities which prevail here, as strong winds usually impede convection. In these areas convection is the main process that brings precipitation, because disturbances are generally weak. Clouds of convectional thunderstorms can reach elevations of 12,000 m and more. Rainfall is usually very intense, but of short duration and strongly localized (WATTS, 1955b). In coastal areas the occurrence of convectional thunderstorms is closely related to sea and land breezes (RAMAGE, 1964, pp. 66–68). Rainfall caused by convection therefore shows a clear pattern in diurnal variation, depending on the prevailing winds and the distance from the coasts (NIEUWOLT, 1968a). In the northern parts of the region convection is limited to the warmer months of the year (BRUZON et al., 1940, p.132).

(3)Orographic lifting. Even at mountainous coasts facing the full impact of the monsoon winds, precipitation is rarely continuous but comes in spells of a few very wet days followed by some relatively dry days. This shows that orographic lifting is not the main origin of precipitation, but it has the effect of enormously increasing the amounts of rainfall caused by other factors. The wettest areas of the region are those where orographic lifting occurs frequently (compare Figs.6 and 7, pp.22, 25). The leeward (eastern) sides of the mountains are much drier, but here the effect is merely a reduction in the amounts of rainfall. Rain shadow areas, such as central Burma and central Thailand in July, still receive a mean rainfall of up to 100 mm during that month (Fig.3C).

(4) Convergence. Air mass boundaries are always areas of increased precipitation. The effectiveness of air mass boundaries as bringers of precipitation depends mainly on the thickness and humidity of the air masses involved, but other factors are the speed and

the angle of convergence. The general instability of the atmosphere and the low wind velocities which they bring favour convection. But many disturbances are also related to convergence zones and in many cases the three processes cannot be separated.

The annual rainfall maximum which over most of the region comes during the southwest monsoon season, is related to the character and thickness of the air masses which the monsoon brings. They have their origin mainly in equatorial regions, have travelled over warm seas for long distances, and are consequently very warm and humid. Even more important, these conditions prevail up to high elevations, much higher than during the northeast monsoon. Consequently, any lifting of these air masses, by convection, convergence, disturbances or by relief, will bring large amounts of rainfall, and all four processes outlined above are most effective during this period. Even though the number of disturbances is less than during the other periods of the year, the yield of precipitation per disturbance is much higher. Convection is also very efficient because of the generally high temperatures (Fig.3A). And orographic lifting increases the amounts of rainfall received at westerly coasts to very high amounts: Akyab receives a mean of 1,151 mm in June, 1,400 mm in July and 1,133 mm in August, but actual amounts may be as high as 2,000 mm per month.

The only process which is not showing a maximum during the southwest monsoon season is convergence. By itself the airmass boundary over Malaya and the South China Sea brings only limited amounts of rainfall. This is because the two air masses run on almost parallel courses and the convergence is therefore of low intensity. Malaya is relatively .dry during this season, the main reason being the rain shadow effect of the high mountain ranges of Sumatra (Fig.3C).

The characteristics of the air masses of the southwest monsoon are the main reason why Continental Southeast Asia is one of the wettest regions of the world.

September

During this month the circulation is still dominated by the southwest monsoon, but both speed and constancy of the monsoon flow are reduced and its direction tends to become more southerly.

September is the wettest month of the year in large parts of Vietnam, Kampuchea and Thailand. The main reason for the high precipitation is to be found in the higher parts of the troposphere, around the 3,000-m level, where upper westerlies are slowly and irregularly replaced by easterlies, which have their origin in the anticyclone of the north Pacific Ocean. This replacement progresses from north to south and in September the boundary between the two upper wind systems is around 19°N latitude (THOMPSON, 1951, p.596). The boundary zone is accompanied by a broad zone of doldrums, with winds of low velocity and highly variable directions. The general instability of the atmosphere favours convection.

A second reason for the high amounts of rainfall in these areas during September are the typhoons, which attain their maximum frequency in this month. This is related to the high surface water temperatures over the South China Sea. The typhoons are most important in the coastal areas of Vietnam, generally north of 15°N. But in September and October the typhoons often occur farther south and the whole of Vietnam might be influenced. They also show a greater tendency to move inland to the west during September, and often reach Kampuchea or even Thailand. Rain can fall as far as 150 km from the centre of a typhoon and the amounts can be enormous. In coastal areas a typhoon may bring as much as 400 mm in 24 h. Further inland the amounts of rain brought by typhoons decrease rapidly with distance from the sea. Similarly, wind velocities are very high near the coasts and over sea, and may reach values of 40 m/sec there, but further inland wind velocities decrease rapidly. Most destruction and floods by typhoons are therefore limited to the coastal areas of Vietnam (AIR MINISTRY, 1937, Vol.I, Part 2).

The inter-monsoon period of October

October is the principal month of the inter-monsoon season between the southwest and the northeast monsoon. In the northern parts of the region this period begins in September, but it continues well into November in the most southern areas. Fig.4 illustrates the average conditions during October.

The Asian continent begins to cool off in September and consequently the southwest monsoon weakens. It begins to retreat to lower latitudes in October while at the same time the northeast monsoon, still mainly consisting of trade winds from the northern Pacific Ocean, begins its advance southwards. But the wind velocities of both air streams are very low and the general circulation comes to an almost complete standstill. Local wind systems dominate the circulation, as in April and May. The only dominating feature on the weather maps for this month is the air stream boundary between the two monsoon systems (Fig.4B). The general movement of this boundary is southwards. In spite of the low wind velocities the intensity of convergence is relatively high, definitely higher than during April and May. The boundary zone is therefore accompanied by a broad zone of heavy precipitation.

The movement of the air mass boundary is related to conditions in higher levels of the troposphere. The easterly jet stream at about 15,000 m, which during the southwest monsoon was situated around 15°N, now retreats to lower latitudes. This is due to the re-appearance of the westerly jet stream over India. It now passes south of the Tibetan–Himalayan massif, at around 27°N at a level of about 12,000 m (KOTESWARAM and PARTHASARATHY, 1954; STAFF MEMBERS, ACADEMICA SINICA, 1957–58). An extension of this westerly jet stream forms between 15° and 20°N in a southwest–northeast direction over Continental Southeast Asia. Once this jet stream has been established it remains more or less fixed in this position by the various mountain ranges, and this explains why weather conditions during October are much less changeable than during the April–May intermonsoon period.

This westerly jet stream controls the position of the discontinuity between easterlies and westerlies at the 3,000 m level. This air stream boundary appears in September, as described above, but its effects during that month are of minor importance. During October it moves further southwards and its associated disturbances now bring large amounts of rainfall to the northern parts of the region (TREWARTHA, 1958, p.206).

October is quite a wet month (Fig.4C). In the coastal areas of Vietnam and as far west as Phnom Penh, it is the wettest month of the year. The precipitation in the northern parts of the region is the combined results of four factors: the disturbances mentioned above, the typhoons, orographic lifting, and convergence in the early northeast monsoon flow.

13

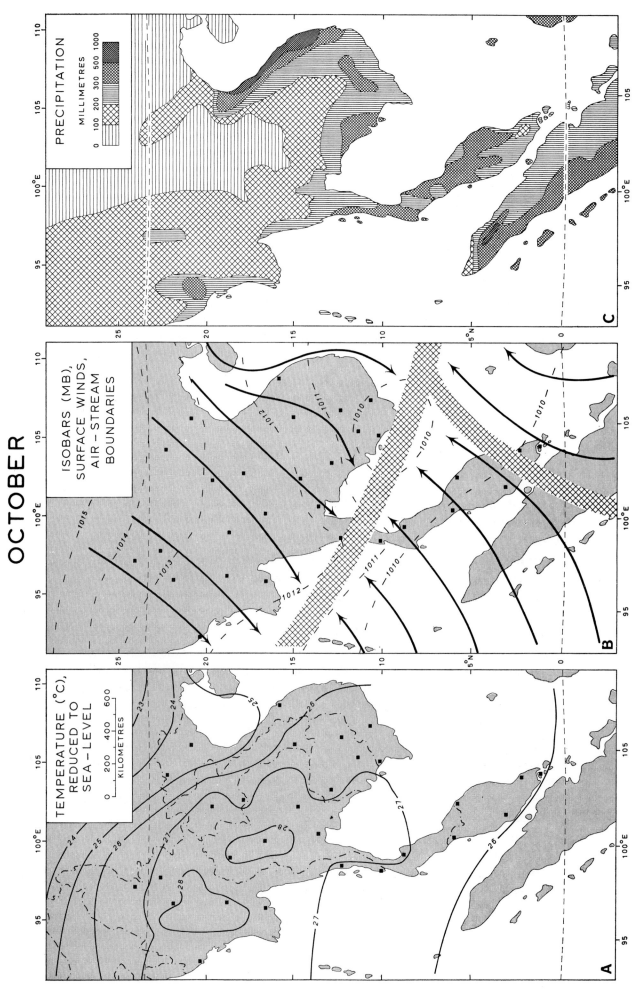

OCTOBER

A TEMPERATURE (°C), REDUCED TO SEA-LEVEL

KILOMETRES
0 200 400 600

B ISOBARS (MB), SURFACE WINDS, AIR-STREAM BOUNDARIES

C PRECIPITATION
MILLIMETRES
0 100 200 300 500 1000

Fig.4. Conditions during October: A. mean temperatures at sea level; B. mean atmospheric pressure, surface winds and average position of air-stream boundaries; C. mean precipitation.

Typhoons in Vietnam and further east are almost as frequent in October as in September, but they generally move slightly further to the south, thus affecting the whole of Vietnam. The South China Sea still has high surface water temperatures and the amount of rainfall brought by the typhoons is as large as during September.

Orographic lifting of the first air masses of the northeast monsoon to reach the mountains of Annam is not very rapid, as wind velocities are low. But the air masses have spent a relatively long time over the warm waters of the South China Sea and are humidified and warmed to high levels, and even a slow uplift therefore results in large amounts of rainfall.

In the monsoon flow of the northeast monsoon a convergence exists during October, caused by an upper air anticyclone over Thailand (FISHER, 1966, p.38). This convergence causes rising air movements and the resulting instability of the air masses increases the efficiency of the other factors bringing rainfall to parts of the region north of 10°N.

In the southern areas of the region the main factor responsible for the high rainfall during this month is the air mass boundary. In combination with convection and orographic lifting, widespread precipitation is caused. Because of the low wind velocities there is almost no rainshadow effect of mountain ranges, and it is difficult to detect differences between windward and leeward slopes during this month.

Climatic variations

The above description of the monsoonal seasons is, of course, a broad generalization based on long-term averages. But departures from these averages can be of great importance.

Near the Equator where conditions vary but little from the mean, average figures are considered to be fairly representative of the actual conditions. But even here important departures from the averages can occur. This is shown in the case of Singapore, where seasonal variations are very small, but where actual conditions show a number of other variations which are not indicated by the usual long-term averages (NIEUWOLT, 1968b). More important from a practical point of view are the departures from the averages from year to year, which can bring prolonged dry periods even in equatorial regions (NIEUWOLT, 1966a, pp.178–180). Variability in time of occurrence is especially important where most of the precipitation falls during one season only. In those parts of Continental Southeast Asia where a dry season occurs regularly, a large proportion of the rainfall during the wet season is caused by such extremely variable factors as the number of disturbances and the position and intensity of air mass boundaries. Variations in time of the beginning and end of the rains are therefore usually quite large. This is illustrated in Thailand, where during 20 years of observations the beginning of the wet season came as much as one month too early or too late, and the end of the rains differed by as much as 17 days from the average date. During those 20 years the length of the rainy season varied between 174 and 236 days, which means a variation of about 15% on either side of the average (CREDNER, 1935, p.75).

Amounts of rain also vary from year to year. Rainfall variability has been studied in a number of regions (CREDNER, 1935, pp.74–75; AIR MINISTRY, 1937, Vol. I, pp.128–129; BRUZON et al., 1940, p.81; WATTS, 1955a, pp.61–64; DALE, 1960, p.15–20; STERNSTEIN, 1962, pp.70–73; NIEUWOLT, 1965a). Unfortunately, it is impossible to compare these studies quantitatively, because not only the periods of observation, but also the methods of computation and statistical expression differ widely. Only a few general conclusions can be drawn from these sources. Considering the generally high amounts of precipitation in the region, the rainfall variability is large. It is especially high where an important part of the rainfall is derived from disturbances. Therefore it shows a general increase with latitude. Where orographic lifting is responsible for an appreciable proportion of the precipitation, the rainfall variability is usually lower. The general rule that the variability decreases with higher total rainfall holds true in most parts of the region. There are some exceptions to this rule, as for instance along the east coast of Malaya (DALE, 1960, pp.16–19). Still, in the same region the rainfall variability is usually much higher during relatively dry periods than in the rainy season.

Temperature variability has been studied much less, as it is of little practical importance in a region where high temperatures prevail. A few studies exist (AIR MINISTRY, 1937, Vol. I, p.105; BRUZON et al., 1940, pp.34–65; DALE, 1963, p.65). Generally, the temperature variability is small, especially in the lower latitudes, where all air masses are of similar temperature. In the northern parts of the region the variability of temperature is larger. During the winter, when cold air masses can reach the region, it is of some practical importance.

Microclimatic differences occur, as everywhere else. In other climates the relatively small microclimatic differences are usually overshadowed by regional or seasonal variations. But in an area where conditions are similar over large distances and long periods, microclimatic differences are felt keenly by the inhabitants, who are sensitive to small variations. Microclimatic differences are caused by local wind systems, such as mountain and valley winds or land and sea breezes. They can also be the result of landforms, such as valleys, basins or variations in the coastline, or they can be caused by urban–rural differences (WATTS, 1955a, pp.156–177; 1955b, pp.43–53; RAMAGE, 1964; NIEUWOLT, 1966b).

The effects of latitude

The climatic uniformity of Continental Southeast Asia, caused by the monsoons, is differentiated by three factors: latitude, relief and continentality. Because the region stretches over 27 degrees of latitude, the differences resulting from this factor are pronounced. They are partly the consequence of differences in solar radiation, of which both duration and intensity vary considerably with latitude.

The possible duration of solar radiation is indicated by the length of day. At the Equator all days of the year are equally long, but the difference between the longest and the shortest day of the year increases in the low latitudes by about seven minutes for each degree of latitude. Thus this difference is only 9 min in Singapore, but over 3 h in northern Burma.

Intensity of solar radiation is largely controlled by the elevation of the sun above the horizon. This is always indicated at its daily maximum, at noon meridional time. This factor also shows only a small variation during the course of the year near the Equator, but this variation increases with latitude. In Singapore the difference between minimum and maximum elevations of the sun at noon is 25°. But at Bhamo, 24°N, this difference is 48°.

Because the period of the longer days coincides with the season of the higher elevations of the sun, the two factors, duration and intensity of solar radiation, reinforce each other at higher latitudes.

The combined effect is shown in Table II. It illustrates that seasonal differences in solar radiation increase sharply with latitude. It also shows that the amounts of radiation received at the same date at various latitudes differ much more during the winter than during the summer.

TABLE II

TOTAL DAILY DIRECT SOLAR RADIATION REACHING THE GROUND (in cal. cm^{-2})*
(From LIST, 1963, p.421)

Latitude	Mar. 21	June 22	Sept. 23	Dec. 22
0°	447	372	440	397
10°N	436	428	430	323
20°N	404	465	398	237
30°N	350	481	345	152

* Assuming an atmospheric transmission coefficient of 0.6.

The effect of latitude on temperatures in Continental Southeast Asia is illustrated in Table III and Fig.5.

All stations in Table III are at coastal locations and near sea level, so that the influence of continentality and elevation has been excluded. As expected, the figures show an increase of annual ranges with latitude. Temperature differences between various latitudes are much larger in winter than in summer. But the effect of latitude is insignificant between 1° and 8°N. This is related to the prevalence of tropical marine air masses in these low latitudes throughout the year. These air masses are well known for their

TABLE III

AVERAGE MONTHLY TEMPERATURES AND ANNUAL RANGES

Station	Latitude (°N)	January (°C)	July (°C)	Annual range (°C)
West coast				
Singapore	1°21′	25.6	27.3	2.0
Alor Star	6°12′	26.0	26.7	1.6
Victoria Point	9°59′	26.6	26.2	2.7
Mergui	12°26′	25.6	25.9	3.1
Akyab	20°8′	21.0	26.9	7.8
Difference along west coast	1°–20°	4.6	1.4	
East coast				
Mersing	2°27′	25.5	25.8	1.1
Kota Bharu	6°10′	25.3	26.7	2.1
Songkhla	7°9′	26.8	28.0	2.2
Bandon	9°7′	25.9	27.8	3.4
Tourane	16°5′	21.2	28.5	7.8
Tanh Hoa	19°48′	15.9	28.5	13.5
Difference along east coast	1°–20°	9.7	1.2	

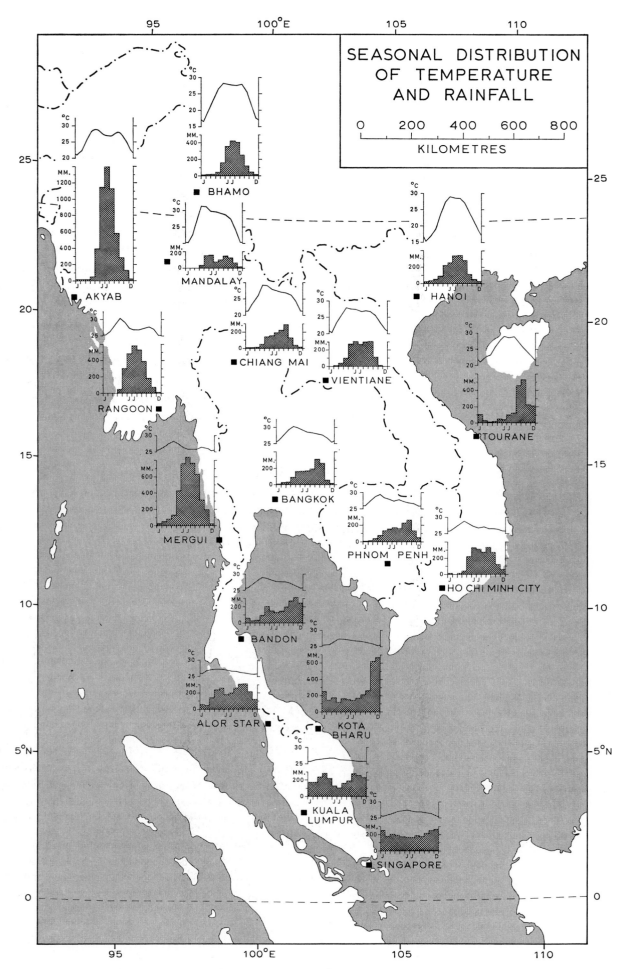

Fig.5. Monthly means of temperature and rainfall.

uniformly high temperatures at all seasons. The low winter temperatures in the northern parts of the region are largely the result of invasions of cold air masses. This is demonstrated by the differences between west and east coasts: while Burma is well protected from these invasions by high mountain ranges, northern Vietnam lies more open to them and the winter temperatures are considerably lower than at the same latitude in Burma. The effects of solar radiation are, of course, strongly influenced by cloudiness. This is illustrated by temperature conditions at latitudes 15°–22°N. In April and May temperatures are very high, the result of long and intensive insolation (Fig.2A). But in July the temperatures are lower, in spite of even longer days and higher elevation of the sun (Fig.3A). This is the result of increased cloudiness, brought by the humid air masses of the southwest monsoon.

For the same reason daily temperature ranges show almost no correlation with latitude. They are always higher during dry seasons, because of reduced cloudiness, than during wet periods. Almost everywhere in the northern parts of the region they are therefore larger in winter than in summer.

The conclusion is that while latitudinal differences in temperature are primarily caused by insolation conditions, the results are largely modified by air masses and their associated cloudiness.

Rainfall conditions also vary with latitude but in a more complicated manner, as four factors are involved. *Convection* is most closely related to differences in insolation. It occurs at all seasons near the Equator, but with increasing latitude becomes more and more limited to the warmer months of the year. *Disturbances* become more and more important as bringers of precipitation with increasing latitude (p.11). *Air masses* are warm and humid over the whole of the region during the summer, but during the northeast monsoon period the northern parts of the region are under the influence of stable and relatively cold air masses, which can bring only very small amounts of precipitation. *Air stream boundaries* move over the region twice a year, and the times of their occurrence vary according to latitude.

The combined result of these four factors is shown in Fig.5. Near the Equator rainfall is copious throughout the year. With increasing latitude a dry season becomes more and more pronounced. Stations exposed directly to the monsoon winds have usually one clear maximum. Stations which are more sheltered have a double maximum, separated by a slight minimum. They receive most rainfall during the transition periods when air stream boundaries are predominant. Such stations are Kuala Lumpur, Alor Star, Ho Chi Minh City, Phnom Penh, Vientiane and Mandalay. Usually the second maximum (Sept.–Oct.) is the strongest. At higher latitudes these two maxima are united into one, because the two transition periods are very close to each other, as is shown by the diagrams for Hanoi and Bhamo.

Continentality

Continentality is the effect of distance from the sea. In Continental Southeast Asia the seasonal reversal of the prevailing wind directions results in different degrees of continentality during the seasons. During the northeast monsoon period it is the distance from the eastern and northern coasts which is decisive, but during the southwest monsoon season the most important factor controlling continentality is the distance from the southern and western coasts. Moreover, the strongly marine air masses of the southwest monsoon carry maritime conditions far into the continent, but the originally continental air masses of the northeast monsoon lose their maritime characteristics, which are limited to relatively shallow surface layers, rather quickly. Consequently, in the northern parts of the region continentality is usually more pronounced during the winter than in the summer.

It is not only the distance from the sea, but also the relief which decides the degree of continentality, as mountain ranges increase continentality strongly over short distances (BRUZON et al., 1940, pp.60–61).

Normally, continentality is expressed by annual ranges of temperature (BRUZON et al., 1940, pp.57–60; BLÜTHGEN, 1966, pp.451–455). But in the low latitudes this is not a reliable indicator, because it is almost entirely controlled by elevation and latitude (GORCZYNSKI, 1920). Moreover, seasonal variations in continentality cannot be indicated by annual figures.

The degree of continentality in Continental Southeast Asia is therefore best measured by diurnal temperature ranges (Table IV). The stations in this table are grouped in west–east sections, to exclude all differences due to latitude. The increase of the diurnal ranges is not proportional to distance from the sea. This is the result of relief and differences in elevation. While in the northern parts of the region the diurnal ranges are larger in January than in July, seasonal differences in diurnal temperature ranges in Malaya are insignificant.

TABLE IV

DISTANCE FROM THE SEA AND TEMPERATURE RANGES

Station	Lat. (°N)	Long. (°E)	Elev. (m)	Distance from the sea (km)	Average daily range (°C)		Annual range (°C)
					Jan.	July	
A. Malaya:							
Kuala Selangor	3°21′	101°15′	12	0	8.5	8.5	0.8
Bentong	3°31′	101°55′	97	77(W); 160(E)	10.0	11.9	1.4
Temerloh	3°27′	102°26′	49	105(W); 124(E)	7.9	9.9	1.8
Kuala Pahang	3°32′	103°28′	3	0	4.7	7.1	1.3
B. Burma:							
Akyab	20°8′	92°55′	9	0	12.5	4.4	7.8
Minbu	20°11′	94°53′	50	155	14.7	7.3	10.7
Yamethin	20°26′	96°9′	196	295	16.1	7.9	9.8
C. Laos–Vietnam:							
Luang Prabang	19°51′	102°8′	290	380	14.0	8.8	8.4
Thanhoa	19°48	105°47′	5	15	6.9	8.3	13.5
D. Thailand–Kampuchea–Vietnam:							
Bangkok	13°44′	100°30′	12	30(S)	11.7	7.8	4.2
Battambang	13°11′	103°12′	16	140(S)	11.5	7.5	4.5
Kratié	12°29′	106°1′	24	340(SW); 360(E)	12.7	7.9	4.8
Nha Trang	12°15′	109°12′	6	0	7.3	8.6	4.2

The increase of diurnal ranges of temperature with distance from the sea is mainly due to higher maximum temperatures in inland regions. The minimum temperatures differ much less (DALE, 1963, pp.66–69). During the day sea breezes effectively cool coastal areas, while inland regions are heated. But at night air drainage and land breezes are equalizing factors and the warming effect of the sea is negligible during nights with a well developed land breeze, which prevails frequently.

Differences between marine and continental stations are larger in the northern parts of the region. This is not only due to the greater distances from the sea, but also to the dry season, when the lack of clouds frequently increases diurnal temperature ranges by strong heating during the day and radiation cooling at night.

A second result of continentality is a difference in the diurnal variation of rainfall (WATTS, n.d.; NIEUWOLT, 1968b). Coastal areas generally have a night or early morning maximum of rainfall. This is caused by radiational cooling from the tops of the clouds during the night, while surface layers are kept warm by the sea surface. Therefore, over sea the lapse rate is increased during the night and the resulting instability brings rainfall during the early hours of the day to the coastal areas. But inland areas usually have an afternoon maximum of rainfall, caused by convection over the land. At coastal locations important variations of this general scheme occur because of the interference of sea and land breezes with the prevailing monsoon winds (RAMAGE, 1964).

The influence of relief

Landforms have important effects on the differentiation of climates in Continental Southeast Asia. Their influence is especially important in mountainous areas, which form a large part of the region (Fig.6). However, conclusions about climatic conditions in these areas can only be drawn in a very general way, because of the very limited amount of climatic information available for the highland areas. Not only are these areas lightly settled in comparison with the lowlands, but in some cases the settlers in the highlands are also in political conflict with the central government of the state to which they belong. In these circumstances few recording stations exist and even fewer are able to supply reliable statistics. This low density of meteorological stations in the highlands is a great disadvantage, because microclimatic differences are pronounced, as in all mountainous areas, and spatial interpolation of data is therefore largely a matter of conjecture. For the highland regions of Continental Southeast Asia the climatic maps must be regarded as very generalized approximations.

Relief has two different influences on climate. The first one is on temperatures, which differ according to the elevation. The second influence is on air currents, and the results are mainly noticeable in winds and precipitation.

Temperature

Temperatures decrease with elevation, and in a region where high temperatures prevail this is of great practical significance. The rate of decrease of temperature with elevation in the region is generally around the normal value of 0.6°C per 100 m (HUKE, 1962). This is a high value in comparison with the small annual and daily ranges of temperature which prevail in the southern parts of Continental Southeast Asia. There, the only way to escape from the frequently oppressive conditions in the lowlands is to go up into the mountains, and some stations at higher elevations function as health resorts, for instance Fraser's Hills (1,300 m) and Tanah Rata (1,450 m) in Malaya.

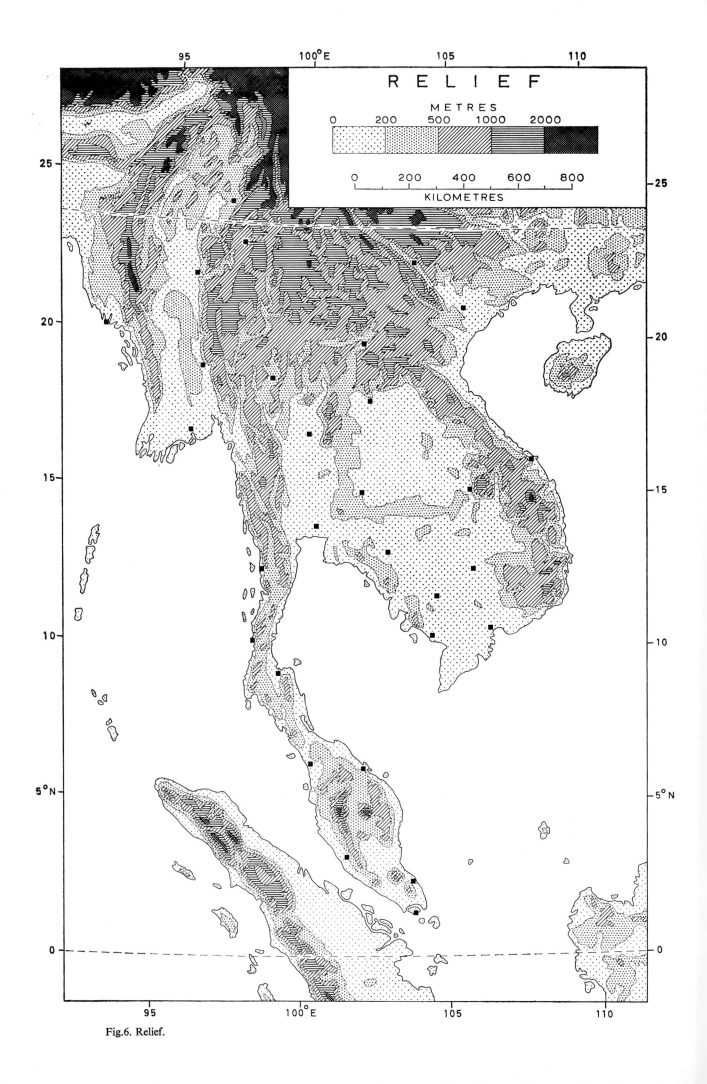

Fig.6. Relief.

The intensity of both solar and terrestrial radiation increases with elevation. Highland stations, except when they are located on isolated mountain tops, therefore usually have larger daily ranges of temperature than the lowland stations nearby. The difference in temperature with the lowlands is small during the day, when intense insolation heats up the highlands rapidly, but at night the lapse rate increases sharply, as rapid cooling takes place at the higher elevations (BLÜTHGEN, 1966, pp.474–475).

Conditions in Continental Southeast Asia, however, are different, as illustrated by Table V. Lapse rates are generally larger during the day than at night, and diurnal temperature ranges are smaller at highland stations than in the lowlands.

TABLE V

ELEVATION AND TEMPERATURE

Station	Elev. (m)	Distance (km)	January temp. lapse rate (°C/km) day[1]	night[2]	mean daily range (°C)	July temp. lapse rate (°C/km) day[1]	night[2]	mean daily range (°C)	Annual range (°C)
A. Malaya:									
Tanah Rata	1,448				8.3			9.2	1.4
		34	7.3	5.7		7.3	5.8		
Tapah	46				10.7			11.4	1.3
Fraser's Hills	1,301				4.8			4.3	2.3
		19	8.9	4.0		6.9	4.0		
Kuala Kubu Bharu	61				10.5			10.9	1.1
Penang Hill	731				6.0			4.8	1.4
		8	9.6	5.5		9.6	4.2		
Georgetown	5				9.1			8.7	1.2
B. Burma:									
Taunggyi	1,524				16.2			6.2	9.2
		100	5.5	5.5		6.0	4.7		
Yamethin	196				16.0			7.9	9.8
C. Laos:									
Paksong	1,200				12.7			4.7	4.2
		50	6.5	5.7		6.4	5.0		
Paksé	96				13.5			6.3	5.5

[1] Calculated on the basis of mean daily maximum temperature.
[2] Calculated on the basis of mean daily minimum temperature.

The explanation for this exception to the usual conditions lies in the heavy cloudiness over the mountains. Valley and upslope winds start early in the day and clouds form much earlier over the mountains than over the neighbouring lowlands, as can be observed on almost every clear morning. These clouds prevent a further increase of temperature over the mountains, while in the lowlands the diurnal rise of temperature frequently continues for much longer. The result is a large difference of temperature between highlands and lowlands during the day. At night the decrease of temperature in the mountains is relatively slow, because the thick cloud cover often reduces radiation cooling. The lowlands cool faster during clear nights, which occur more frequently than over mountains, and the result is a small lapse rate at night. Therefore, in Continental Southeast Asia diurnal ranges of temperature diminish with elevation in most areas (Table V).

That cloudiness is the main factor to explain these conditions is shown by the January values for Burma and Laos. During that dry month, when there is little cloudiness, the diurnal variation in the lapse rate is very small. Similarly, differences between lowland and highland in the diurnal temperature range are insignificant during this month (Table V).

Annual ranges of temperature are not affected by elevation.

In the low latitudes, where the elevation of the sun is high, even during the winter, exposure of the different slopes to the sun does not create any considerable differences of temperature. A differentiation of ubac and adret is therefore not possible. There is, however, a difference between east-facing slopes, which get the morning sun, and west-facing slopes, which have the sun in the evening. In wet periods the increase of cloudiness during the day reduces the number of actual sunshine hours on the latter, while the former have no such large reduction.

Winds

Because of reduced friction, wind velocities at higher levels are generally higher than in the lowlands (BRUZON et al., 1940, p.16). But compared to the mid-latitudes wind velocities are generally low in Continental Southeast Asia and deflection of winds by mountain ranges is rarely of more than local importance. The only significant case is the deflection of the northeast monsoon by the mountain ranges of Annam, but this is an extraordinarily strong wind for the region (Fig.1B).

For the same reason, namely the sluggish general circulation, the development of local winds, such as mountain and valley winds is usually strong, as local heating of the air is not disturbed by turbulence. This is especially the case during the inter-monsoon periods. The valley and upslope winds are essential to the creation of cloudiness and rainfall over the mountains.

Coastal landforms can be a factor in the development of local disturbances, such as the "Sumatras" of west Malaya (WATTS, 1955a, pp.159, 172).

Precipitation

The most important effect of relief on air currents is the forced rising of air masses, which often results in orographic precipitation on the windward side and a rain-shadow effect on the leeward side of a mountain range. Thus, if the rain-bringing winds come mainly from one direction, mountain ranges can become sharp climatic divides. This is shown by the large differences over short distances between west, central and east Burma, west and central Thailand and between the highlands and lowlands of Vietnam and Laos (compare Figs.6 and 7).

Because of the generally low wind velocities in the region, both cumuliform and stratiform clouds form on the windward sides of mountains. Orographic lifting mainly causes an increase in the amounts of precipitation produced by other factors. The diurnal maximum of precipitation in the highlands falls at an earlier hour than in the lowlands, because of the increased speed of lifting. In Malaya this diurnal maximum is around

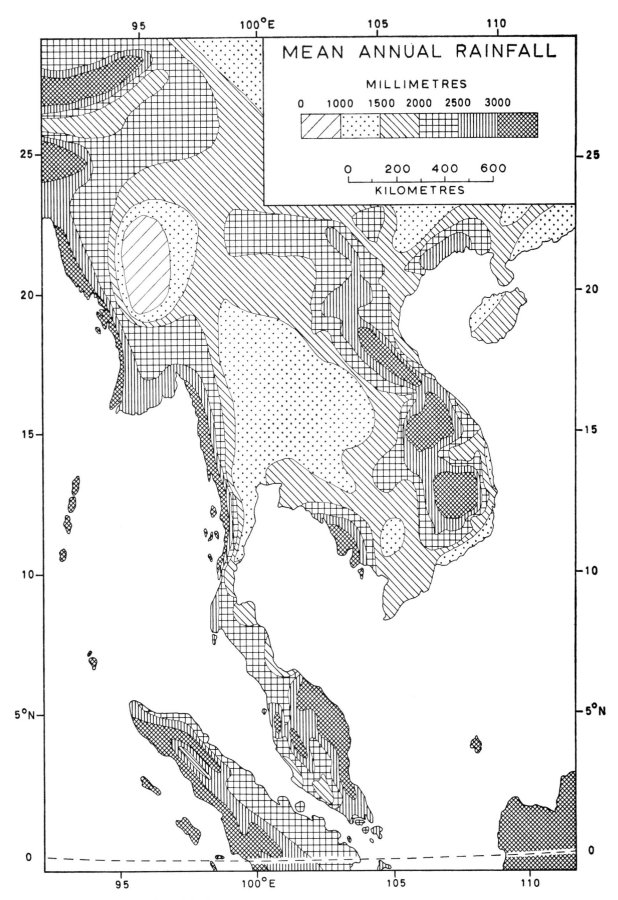

MEAN ANNUAL RAINFALL

MILLIMETRES

0   1000   1500   2000   2500   3000

0   200   400   600

KILOMETRES

Fig.7. Mean annual precipitation.

noon in the highlands, and about two hours later in the nearby lowlands (WOO KAM SENG, 1968, pp.15–16). With the increase of precipitation its variability is usually smaller in the highlands than in the lowlands.

It is generally agreed among climatologists that in the humid tropics the amounts of rainfall increase with elevation up to a certain level, above which a slow decrease with height takes place. But there is some doubt about the elevation of the zone of maximum rainfall in the mountains. This uncertainty is caused by the lack of reliable data in many parts of the tropics, by microclimatic differences which cannot always be recognized as such, and by seasonal differences. WEISCHET (1965), basing his conclusions largely on observations in Java and the Andes, puts it at 1,200–1,500 m, which agrees with estimates by BRAAK (1928, Vol.1, p.173) and DE BOER (1950) for Indonesia. In Malaya, where only a limited number of observations is available, the zone of maximum precipitation seems to be at a lower level. DALE (1959, p.27) has put it around 300 m, but WOO KAM SENG (1968, p.11) estimates it to vary between 150 and 1,200 m. Some figures for Continental Southeast Asia are given in Table VI. Except in Malaya, no comparison can be made of more than two stations at different elevations but in the same climatic region. This is due to the lack of observations in the highlands.

TABLE VI

ELEVATION AND RAINFALL

Station	Elevation (m)	Distance (km)	Mean precipitation (mm)				
			Jan.	Apr.	July	Oct.	Year
A. Malaya:							
Tanah Rata	1,448		168	307	121	329	2648
		34					
Tapah	46		315	405	166	413	3668
Fraser's Hills	1,301		303	284	110	258	2718
		19					
Kuala Kubu Bharu	61		189	303	159	382	2882
B. Burma:							
Taunggyi	1,524		0	33	288	173	1552
		100					
Yamethin	196		3	48	102	150	970
C. Laos							
Paksong	1,200		14	187	1125	243	3987
		50					
Paksé	96		1	83	505	103	2135
D. Thailand:							
Chiang Rai	400		12	75	309	120	1744
		150					
Chiang Mai	313		7	51	188	124	1246
		250					
Phitsanulok	50		7	75	204	132	1354
E. Kampuchea:							
Bokor	1,050		21	174	883	420	4644
		12					
Kampot	5		13	143	236	250	1993

The decrease of rainfall at higher elevations has not been observed at higher latitudes. Its explanation has created some controversy. It is certain that the following four factors play a role.

(*1*) The lower temperatures at higher levels. The difference in temperature between lowlands and highlands is approximately the same at all latitudes. But starting from the high temperatures in the tropics, the decrease in the capacity of the air to yield precipitation is more rapid than starting from relatively low temperatures, as prevail in the higher latitudes.

(*2*) The reduced speed of uplifting at higher levels, caused by gaps in the mountain ranges. The air can consequently move more sideways than over the mountains, but at lower levels this is not possible. This factor is certainly of great importance, but it should have the same effect at all latitudes.

(*3*) The clouds over the mid-slopes of tropical mountains are often thicker than near the top. This observation is related to the speed of uplifting (BRAAK, 1928, Vol.1, p.174), but it too, does not necessarily differ with latitude.

(*4*) Differences in humidity of the air at different levels (WEISCHET, 1965, p.10). The lower parts of many tropical and equatorial air masses have been humidified if they moved over tropical oceans, but the higher layers might still have remnants of the trade wind inversion. This explanation is certainly valid in many parts of the tropics. In Singapore the difference in humidity can be observed during the northeast monsoon season. The boundary between very humid and somewhat drier layers of the troposphere is then around 1,500 m (JOHN, 1950, p.7; NIEUWOLT, 1969, pp.25–27).

Information available for Continental Southeast Asia, limited though it is, seems to point clearly to the first and the last of these four factors as being of most importance.

Moisture conditions

Few statistics are available for the *vapour pressure* of the air over Continental Southeast Asia. Some observations were published for six stations in Indo-China (BRUZON et al., 1940, pp.89–90). They generally indicate higher values in summer than in winter, in accordance with the prevailing type of air masses.

Relative humidity

More figures are available for relative humidity. As in all tropical areas, the means of relative humidity are quite high. They show a general decrease with increasing distance from the sea.

The largest variation in mean relative humidity is seasonal, with high summer and low winter values. This seasonal variation is, of course, limited to the northern parts of the region. In the southern parts the mean relative humidity is high throughout the year. Comparatively high winter figures prevail in Tonkin and northern Vietnam (Hanoi, Lao Kay). This is related to the frequent periods of "crachin" weather conditions (see p.4) and disturbances over this area (BRUZON et al., 1940, pp.105–106).

In mountainous areas there is a large difference in relative humidity between windward and leeward sides, as illustrated in Table VII.

Owing to the large horizontal distances between the stations, the differences indicated

TABLE VII

MEAN RELATIVE HUMIDITY ON DIFFERENT SIDES OF MOUNTAIN RANGES

Month	Windward side	R.H. (%)	Leeward side	R.H. (%)	Distance (km)	Difference in R.H. (%)
January	Kota Bharu	84	Alor Star	78	200	6
	Tourane	86	Paksé	67	250	9
July	Akyab	91	Mandalay	70	400	21
	Rangoon	88	Phitsanulok	83	460	5

in the table are not entirely representative; they would, probably, be larger when measured nearer to the mountains. Extensive highlands and plateaux usually record lower relative humidities than the lowlands but exhibit the same seasonal and diurnal variations.

A smaller, but in its effects more important, variation of relative humidity is the diurnal cycle. Relative humidity changes inversely with temperature, and in Continental Southeast Asia the night maximum often reaches the saturation point. This has a number of consequences.

The first is the frequent occurrence of fogs. Radiation fogs in the early mornings are common in all parts of the region, except near the coasts, where land breezes create too much turbulence at night. Table VIII gives some figures for Malaya.

TABLE VIII

AVERAGE NUMBER OF DAYS WITH FOG PER YEAR*
(From WATTS, 1955a, p.35)

Singapore	0	Temerloh	78
Kota Bharu	9	Kluang	86
Kuala Lumpur	52	Kuala Lipis	236

* All stations, except Singapore, are in Malaya.

The fogs are rarely thicker than about 30 m. They occur mainly in valleys, where air drainage favours their formation, and near inundated rice fields, swamps and rivers, where there is a large supply of water vapour. Morning fogs are often combined with mild inversions which may reach as high as 200 m. The fogs usually dissipate within a few hours after sunrise.

Other forms of fogs are relatively rare. Sea fogs may occur over the South China Sea in March and April, when the sea surface temperatures are still comparatively low, as the result of a northerly cool current during the northeast monsoon season. Equatorial air masses moving over the sea with the first winds of the southwest monsoon might be cooled rapidly enough to produce sea fogs in these two months (WATTS, 1955a, p.37). Another form of fog is related to the "crachin" conditions in northern Vietnam.

A second consequence of the high relative humidity during the night is the common formation of dew. As the nights are often clear and calm, radiational cooling is rapid and dew is frequent in all parts of the region throughout the year. During the wet periods the relative humidity of the air is so high that even a slight cooling will produce dew, and it can often be observed within a few hours after sunset. During the dry seasons in the northern parts of the region the lower humidity of the air is compensated by more rapid cooling, caused by less cloudiness, and also by the longer nights of the winter season. At Phu-Lien, near Hanoi, the lowest temperature of the night can come as early as two hours before sunrise (BRUZON et al., 1940, p.54). This effect is due to the latent heat of condensation, set free by the foimation of dew, which prevents further cooling. This has not been observed in Malaya (DALE, 1963, pp.62–64) or Burma (HUKE, 1962, p.92). But a similar effect has been noted at about 10 cm above the ground in Singapore (R.D. Hill, University of Singapore, personal communication, 1968).

Evaporation

Evaporation figures are only available for Indo-China and Malaya. But these data are not comparable, because different types of instruments are used. In Indo-China Piche evaporimeters have been used, in Malaya evaporation pans. The first usually indicate values that are about twice those produced by pans under the same conditions (BRUZON et al., 1940, p.128). In spite of this, the published figures for Indo-China are around 1,000 mm per year, those in Malaya approximate 1,600 mm. Differences in temperature, cloud cover, relative humidity and sunshine hours are too small to explain this large difference. Wind velocities are higher in Indo-China, which would favour evaporation there.

Not surprisingly, evaporation figures in Malaya show almost no seasonal variation but in Indo-China there is a general maximum of evaporation during the dry season (NIEU-WOLT, 1965b).

Climatic regions

In spite of the general uniformity of climate caused by the monsoons, latitude, relief and continentality in their combined effects produce a number of clearly distinguishable climatic regions in Continental Southeast Asia.

The main subdivision is between "equatorial" and "tropical" types of climate. Equatorial climates are characterized by very small seasonal variations, both in rainfall and in temperature. Tropical climates, on the other hand, have clearly marked dry and wet seasons and, in some cases, also cold and warm periods.

The boundary between these two main types is given by the presence of a dry season. This is defined here as a period of at least two months during which the mean monthly rainfall is less than 50 mm. This limit has important effects on the vegetation. The boundary, so defined, runs across the Malayan Peninsula at about 6°–9° northern latitude (Fig.8). South of it the equatorial types (*P, Q, R*) are found; north of it the tropical climates (*A, B, C, D, E*) prevail.

Equatorial climates

The equatorial climates have three main characteristics. The first is *seasonal uniformity*. Not only is there (by definition) no dry season, but rainfall is copious throughout the year: in almost every region with this type of climate all months of the year have a mean rainfall over 100 mm. Seasonal differences of temperature are also small: the annual

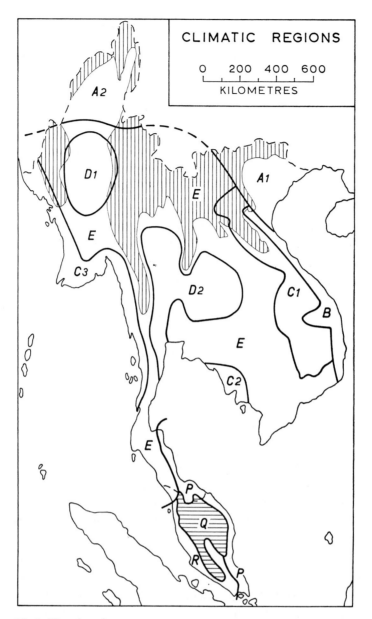

Fig.8. Climatic regions.

range is under 2.5°C, and only in the extreme north of the region is this value exceeded (Bandon: 3.4°C; but Songkhla: 2.2°C).

The second feature is that *wind velocities are generally low*, exceptions occurring locally only during thunderstorms. The only regularly strong winds are those of the northeast monsoon in eastern coastal regions during December and January.

The third characteristic of the equatorial climates is the *dominance of the diurnal cycle* in the general weather conditions. Diurnal ranges of temperature are up to five times the annual ranges, as is illustrated by the old adage: "The nights are the winter of the tropics", "tropics" here referring to the equatorial regions. The diurnal variation of rainfall is also of greater practical importance than the small seasonal variation. The only seasonal contrasts are in wind directions and in the diurnal variation of rainfall

(NIEUWOLT, 1968a, b). Equatorial climates occur over most of Indonesia and Malaysia. In Continental Southeast Asia they can be subdivided into three regions (Fig.8).

Region *P* is the area along the Malayan east coast, where the effects of the northeast monsoon are strongest. The annual rainfall maximum is here during the beginning of the monsoon season. The wettest month of the year is November in the north (Bandon, Kota Bharu), December in the south (Kuantan, Mersing, Singapore). The annual rainfall minimum is during the southwest monsoon season, when this region is on the leeward side of the central mountain ranges of Malaya, but a minimum also occurs during the last two months of the northeast monsoon (February–March) when the monsoon becomes more and more like an extended trade wind (NIEUWOLT, 1966a, p.172–173; 1969, pp.65–66). The diurnal rainfall maximum shows a clear seasonal variation: during the northeast monsoon it is early in the morning, but during the rest of the year it comes in the afternoon or evening (RAMAGE, 1964, pp.63–66; NIEUWOLT, 1968a).

Region *R* is along the Malayan west coast. Because this region is sheltered from the northeast monsoon by the central ranges of Malaya, and from the southwest monsoon by the mountain ranges of Sumatra, both monsoon periods are relatively dry. The maxima of rainfall come here during the two inter-monsoon seasons and usually the maximum of October–November is the highest. At coastal stations the diurnal variation of rainfall shows an opposite seasonal change from the east coast region. During the southwest monsoon the diurnal maximum comes early in the morning, during the rest of the year in the afternoon. Inland stations (Ipoh, Kuala Lampur) have an afternoon maximum of rainfall throughout the year (NIEUWOLT, 1968a).

Region *Q* comprises the central parts of Malaya, as far as they have an elevation over 500 m. This is the dominating factor in the climate of this region. It may seem that an elevation of 500 m is too low to have an appreciable effect on climate, but this is not the case in equatorial climates. At 500 m the temperature is about 4°C lower than in the lowlands (compare Table V), and to the inhabitants of Malaya, very sensitive as they are to small variations in temperature, this is a considerable difference. Another feature of this region is its high precipitation, but in this respect the difference with the lowlands is not of great significance. This is an inland region, and the other characteristics of its climate are caused by distance from the sea rather than elevation. The diurnal maximum of rainfall is in the afternoon, throughout the year. Annual rainfall maxima depend mainly on exposure: on western slopes and in sheltered valleys the maxima are in the inter-monsoon periods; on eastern slopes the maximum rainfall is at the beginning of the northeast monsoon season. This area is largely without settlements and only a few stations are available for statistical information.

Tropical climates

The tropical climates all have a dry season. The boundary with the equatorial climates was drawn where two months have less than 50 mm rainfall, but at most places the dry period is longer. Its length increases with latitude, but its intensity varies according to more local conditions (Table IX). Everywhere the dry season comes during the northeast monsoon.

The tropical climates occupy the largest part of Continental Southeast Asia (Fig.8). It is

TABLE IX

DURATION AND INTENSITY OF THE DRY SEASON

Station	Latitude (°N)	Number of dry months*	Average rainfall during dry months (mm/month)	Total average rainfall during 3 driest months (mm)
Bhamo	24	6	24	38
Mandalay	22	5	11	11
Hanoi	21	5	36	84
Chiang Mai	19	5	16	28
Vientiane	18	5	15	24
Rangoon	17	4	6	15
Bangkok	14	4	20	45
Phnom Penh	12	4	26	57
Ho Chi Minh City	11	4	18	32
Victoria Point	10	2	14	84
Bandon	9	2	29	121

* A dry month is a month with an average rainfall of less than 50 mm.

impossible to subdivide them according to one element of climate; different criteria must be used to delimit the various types.

The *shading* indicates all extensive areas with an elevation over 1,000 m. At this level elevation becomes an important factor, which modifies the corresponding lowland climate. The main modifications are lower temperatures, higher precipitation, more cloudiness and stronger winds. Moreover, local wind systems of the mountain and valley wind type develop almost every day when the general circulation is weak. In the equatorial climates the influence of elevation is so strong that a separate climatic region was delimited at levels above 500 m. But in the tropical climates the minimum elevation must be set around 1,000 m because at lower levels the seasonal differences still completely overshadow the effect of elevation. The shaded areas are therefore those where elevation is the main climatic factor, dominating all other regional differences. They form highland modifications of the lowland types of tropical climate, indicated by *A, B, C, D,* and *E* (Fig.8).

Regions *A1* and *2* denote the areas where the winters are cold. The limit has been drawn at the January sea-level isotherm of 18°C. Latitude is the most important factor for the occurrence of cold winters, and these areas are found in the extreme north of the region. They reach farther south in Vietnam, which is relatively open to cold air invasions from the north, than in Burma, sheltered by high mountain ranges further north. At sea level, temperatures during the winter are rarely very low, but in the elevated parts of these areas occasional frosts occur.

Region *B* includes the coastal areas and eastern mountain slopes of Vietnam. Here the annual maximum of rainfall comes from September to December, and it is followed by a relatively dry period from February to April. A similar rainfall regime is experienced by a narrow strip in the eastern coastal areas of the Malayan Peninsula, just to the north of the boundary with the equatorial climates. This region is too small to be represented on Fig.8.

Regions *C1*, *2* and *3* are the wet areas, with annual rainfall totals over 2,500 mm (see Fig.7). These are mountainous areas and coastal strips, markedly exposed to the southwest monsoon. They derive their heavy precipitation largely from orographic lifting, in combination with other factors. Rainfall during the wet season can be as high as 2,000 mm in a single month.

Regions *D1* and *2* are the relatively dry parts of the region. They have annual rainfall totals under 1,200 mm. These areas are in the rain shadow of high mountain ranges during the southwest monsoon season and receive most of their precipitation during the inter-monsoon periods.

Region *E* combines all other areas with tropical climates, with no special features to distinguish them from each other. In this large region there are, of course, many variations. But it is impossible to relate them to a common factor. The most important differences within this region are related to latitude. The southern parts have almost equatorial climates, with a short dry season and small annual ranges of temperature. The northern parts have cold and dry winters, hot and wet summers. But the transition between these two extreme types is gradational and no clear boundaries can be drawn. There are also many local variations, caused mainly by relief, which interfere with the normal sequence of climates according to latitude.

Acknowledgements

The author wishes to express his thanks to Mr. R. D. Hill, University of Hong Kong, and Mr. Chia Lin Sien, University of Singapore, for their assistance in finalizing this paper.

The Director of the Meteorological Service, Singapore, made available a number of unpublished data. His help, and that of his staff, is gratefully acknowledged.

References for the illustrations

For maps of temperature (Figs.1A, 2A, 3A, 4A)
BRAAK, 1928, Vol.I; maps following p.342.
BRUZON et al., 1940, p.52.
Atlas of Southeast Asia, New York, 1964, pp.4, 42, 47, 54.
BARTHOLOMEW, J. G. and HERBERTSON, A. J., 1899: *Atlas of Meteorology.* The Royal Geographical Society, London, Edinburgh, 95 pp., p.9.
THAI METEOROLOGICAL DEPARTMENT, 1965: *Mean Annual and Monthly Temperature over Thailand,* Bangkok.

For maps of pressure, winds and air stream boundaries (Figs.1B, 2B, 3B, 4B)
AIR MINISTRY, METEOROLOGICAL OFFICE, 1937, Vol.I.
BRAAK, 1928, Vol.I, pp.68, 136–141.
BRUZON et al., 1940, pp.10–11, 16–17.
CREDNER, 1935, p.67.
DALE, 1956, pp.11–22.
WATTS, 1955a, p.10.
Atlas of Southeast Asia, 1964, pp.4 34, 42, 47, 54.
BARTHOLOMEW and HERBERTSON, 1899, p.25.

For maps of precipitation (Figs.1C, 2C, 3C, 4C, 7)
BRUZON et al., 1940, maps following p.74.
DALE, 1959, pp.30–32.
WATTS, 1955a, pp.54–55.
THAI METEOROLOGICAL DEPARTMENT, 1964: *Mean Annual and Monthly Rainfall over Thailand*, Bangkok.
SERVICE MÉTÉOROLOGIQUE DU CAMBODGE, 1964: *Cartes d'Isohyètes*, Phnom Penh.
MALAYAN METEOROLOGICAL SERVICE: *Rainfall Maps*, Singapore, no date.

Statistical information

It proved impossible to obtain reliable and recent climatological data for the whole of Continental Southeast Asia for the same period of sufficient length, which is about 30 years. Political and military events during the last 40 years have frequently disrupted the continuity of observations in many parts of the region, but even where records were kept, their reliability is often questionable when they were obtained under conditions of war or armed insurrection. Even under peaceful circumstances, the accuracy of climatological observation is doubtful at many remote stations, where no trained personnel is employed. Comparability of records in the region is further reduced by its division into six different states, which use different types of instruments, times of observation and methods of publication. It was therefore decided to use, for each state in the region, the longest and most reliable series of data which was available, regardless of the period during which these were collected.

A comparison between stations in different states is impossible for data on evaporation, which are based on readings from evaporation pans in Malaya and on Piche evaporimeters in Vietnam, Laos and Kampuchea. Data on the number of days with precipitation are also not comparable, because the minimum amount of rainfall used was 1 mm in Burma and Thailand, 0.25 mm in Malaya, and "measurable rainfall", which is approximately 0.1 mm, in Vietnam, Laos and Kampuchea.

The stations in the following statistical tables were chosen primarily because they are considered representative for the climatic region in which they are situated (Fig.9). Preference was given to stations with long and reliable records. However, stations in mountainous areas are only typical for the valley or basin in which they are located; they must be considered as illustrations only for the conditions over wider areas, as microclimatic differences can be large.

Sources for Malaysia and Singapore

Climatological Summaries; Malayan Meteorol. Service, Singapore, no date.
  (*I*) Winds (1951–1960).
 (*II*) Precipitation, revised edition, 1965.
(*III*) Temperature, relative humidity and air pressure.
Unpublished data, supplied by the Director, Meteorological Service, Singapore, 1967: evaporation (1958–1966); sunshine hours (1951–1960); cloudiness (1951–1960).

Sources for Burma

NUTTONSON, M. Y., 1963. *The Physical Environment and Agriculture of Burma*. American Institute of Crop Ecology, Washington, D.C., pp.7–33.

Sources for Thailand

Rainfall Data 1931–1960. Thai Meteorol. Service, Bangkok, no date.
YOSHINO, M. M. and KOBAYASHI, M., 1977. *Climatic Records of Monsoon Asia*. Part Ia: Precipitation in Thailand, Malaysia and Singapore. *Climatological Notes*, Vol. 20, University of Tsukuba, pp.3–44.

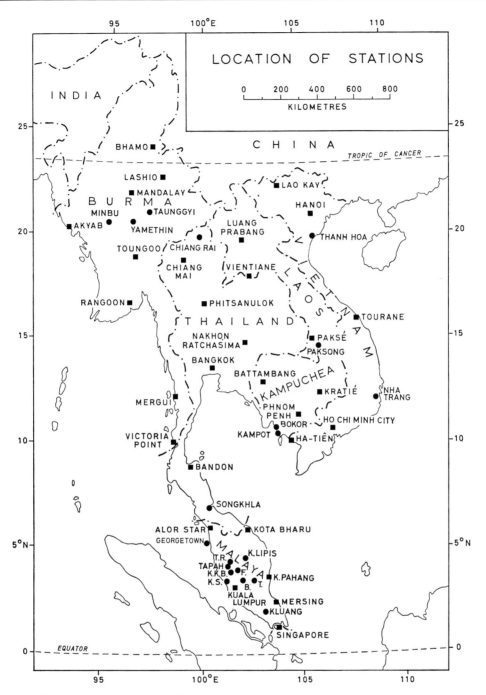

Fig.9. Location of stations. ■: Stations for which statistical information is supplied. ●: Stations mentioned in the text: *B* = Bentong; *F* = Fraser's Hills; *T* = Temerloh; *K.K.B.* = Kuala Kubu Bharu; *K.S.* = Kuala Selangor; *T.R.* = Tanah Rata.

Sources for Vietnam, Laos and Kampuchea

BRUZON et al., 1940, pp.27, 122, 130, 168–203.
KHIOU BONTHONN, 1965. *Le Climat du Cambodge*. Phnom Penh, pp.37, 40.
NUTTONSON, M. Y., 1963. *Climatic Data of Vietnam, Laos and Cambodia*, Washington, D.C. (max. precipitation in 24 h.)
DIRECTORATE OF METEOROLOGY, 1965. *Climatological Mean Values*. Saigon.

Sources for the whole region

World Weather Records, 1951–1960. U.S. Dept. of Commerce, Washington, D.C., 1967, pp.40–45, 537–543.
CLINO, Climatological Normals 1931–1960. W.M.O., 1962, pp.48042–48930.
Tables of Temperature, Relative Humidity, Precipitation for the World, V. Asia, 1937–1954, Meteorol. Office, London, 1958, pp.84–90.

References

AIR MINISTRY, METEOROLOGICAL OFFICE, 1937. *Weather in the China Sea and in the Western Part of the North Pacific Ocean*. H.M.S.O., London, Vol.I—255 pp., Vol.II—771 pp.

BLÜTHGEN, J., 1966. *Allgemeine Klimageographie*. De Gruyter, Berlin, 2nd ed., 720 pp.

BRAAK, C., 1928. *Het Klimaat van Nederlandsch Indië*. Kon. Magn. Meteorol. Observ., Batavia, Vol.I—528 pp., Vol.II—540 pp.

BRUZON, E., CARTON, P. and ROMER, A., 1940. *Le Climat de l'Indochine et les Typhons de la Mer de Chine*. Gouvernement Général de l'Indochine, Hanoi, 209 pp.

CREDNER, W., 1935. *Siam, das Land der Tai*. Engelhorn, Stuttgart, 422 pp.

DALE, W. L., 1956. Wind and drift currents in the south China Sea. *Malayan J. Tropical Geogr.*, 8: 1–31.

DALE, W. L., 1959. The rainfall of Malaya, I. *J. Tropical Geogr.*, 13: 23–87.

DALE, W. L., 1960. The rainfall of Malaya, II. *J. Tropical Geogr.*, 14: 11–28.

DALE, W. L., 1963. Surface temperatures in Malaya. *J. Tropical Geogr.*, 17: 46–56.

DALE, W. L., 1964. Sunshine in Malaya. *J. Tropical Geogr.*, 19: 20–26.

DE BOER, H. J., 1950. On the relation between rainfall and altitude in Java, Indonesia. *Chronica Naturae*, 106: 424–427.

FISHER, C. A., 1966. *Southeast Asia*. Methuen, London, 2nd ed., 831 pp.

GAN TONG LIANG, no date. *A Study of Some Heavy Rain-Spells on the East Coast of Malaya during the North East Monsoon Season*. Malayan Meteorol. Serv., Singapore, 20 pp.

GORCZYNSKI, L., 1920. Sur le calcul du degré du continentalisme et son application dans la climatologie. *Geogr. Ann.*, 2: 324–331.

HUKE, R. E., 1962. *Temperature Change with Elevation in Burma*. Indiana Univ. Found. Res. Division, Bloomington, Ind., 143 pp.

JOHN, I. G., 1950. The properties of the upper air over Singapore. *Malayan Meteorol. Serv. Mem.*, 4: 33 pp.

KHIOU-BONTHONN, 1965. *Le Climat du Cambodge*. Service Météorologique, Phnom Penh, 240 pp.

KOTESWARAM, P., 1958. The easterly jet stream in the tropics. *Tellus*, 10: 43–57.

KOTESWARAM, P. and PARTHASARATHY, S., 1954. The mean Jet Stream over India in the pre-monsoon, and post-monsoon seasons and vertical motions associated with subtropical jet streams. *Indian J. Meteorol. Geophys.*, 5: 138–156.

KOTESWARAM, P., RAMAN, C. R. V. and PARTHASARATHY, S., 1953. The mean jet stream over India and Burma in winter. *Indian J. Meteorol. Geophys.*, 4: 111–122.

LIST, R. J., 1963. *Smithsonian Meteorological Tables*. Smithsonian Institution, Washington, D.C., 7th ed., 537 pp.

LOCKWOOD, J. G., 1966. 700-mb. contour charts for Southeast Asia and neighbouring areas. *Weather*, 21: 325–334.

LOCKWOOD, J. G., 1967. Probable maximum 24-h precipitation over Malaya by statistical methods. *Meteorol. Mag.*, 96: 11–19.

LOCKWOOD, J. G., 1974. *World Climatology—An Environmental approach*. Edward Arnold, London, pp.144–206.

NAVAL INTELLIGENCE DIVISION, 1943. *Indo-China*. Geographical Handbook Series, B.R. 510, London, 535 pp.

NIEUWOLT, S., 1965a. Rainfall probability in Malaya. *Malayan Meteorol. Serv., Mem.*, 7: 24 pp.

NIEUWOLT, S., 1965b. Evaporation and Water Balance in Malaya, *J. Tropical Geogr.*, 20: 34–53.

NIEUWOLT, S., 1966a. A comparison of rainfall in the exceptionally dry year 1963 and average conditions in Malaya. *Erdkunde*, 20: 169–181.

NIEUWOLT, S., 1966b. The urban microclimate of Singapore. *J. Tropical Geogr.*, 22: 30–37.

NIEUWOLT, S., 1968a. Diurnal variation of rainfall in Malaya. *Ann. Assoc. Am. Geogr.*, 58: 313–326.

NIEUWOLT, S., 1968b. Uniformity and variation in an equatorial climate. *J. Tropical Geogr.*, 27: 23–39.

NIEUWOLT, S., 1969. *Klimageographie der malaiischer Halbinsel.* Mainz, 155 pp.

OOI JIN BEE and CHIA LIN SIEN, 1974. *The Climate of West Malaysia and Singapore,* Oxford Univ. Press, Singapore, 262 pp.

PDELABORDE, P., 1958. *Les Moussons.* Colin, Paris, 208 pp.

RAMAGE, C. S., 1952. Relationship of general circulation to normal weather over southern Asia and the western Pacific during the cool season. *J. Meteorol.,* 9: 403–408.

RAMAGE, C. S., 1955. The cool season tropical disturbances of Southeast Asia. *J. Meteorol.,* 12: 252–262.

RAWAGE, C. S., 1964. Diurnal variation of summer rainfall of Malaya. *J. Tropical Geogr.,* 19: 62–68.

RAMASWAMY, S., 1956. On the sub-tropical jet stream and its role in the development of large-scale convection. *Tellus,* 8: 26–60.

SCHMIDT, F. H., 1951. Streamline patterns in equatorial regions. *J. Meteorol.,* 8 :300–306.

STAFF MEMBERS, ACADEMICA SINICA, 1957–58. On the general circulation over eastern Asia. *Tellus,* 9: 432–446(I); 10: 58–75(II); 10: 299–312(III).

STERNSTEIN, L., 1962. *The Rainfall of Thailand.* Indiana University Foundation Research Division, Bloomington, Ind., 150 pp.

THOMPSON, B. W., 1951. An essay on the general circulation of the atmosphere over Southeast Asia and the West Pacific. *Q.J.R. Meteorol. Soc.,* 77: 569–597.

TREWARTHA, G. T., 1958. Climate as related to the jet stream in the Orient. *Erdkunde,* 12: 205–214.

TREWARTHA, G. T., 1961. *The Earth's Problem Climates.* University of Wisconsin Press, Madison, Wisc., 334 pp.

WATTS, I. E. M., 1955a. *Equatorial Weather, with Particular Reference to Southeast Asia.* University of London Press, London, 224 pp.

WATTS, I. E. M., 1955b. Rainfall of Singapore Island. *Malayan J. Tropical Geogr.,* 7: 1–68.

WATTS, I. E. M., 1949. The equatorial convergence lines of the Malayan–East Indies area. *Malayan Meteorol. Serv., Mem.,* 3: 30 pp.

WATTS, I. E. M., no date. *The Diurnal Variation of Frequency of Precipitation over Southeast Asia.* Royal Observatory, Hong Kong, 11 pp.

WEISCHET, W., 1965. Der tropisch-konvektive und der aussertropisch-advektive Typ der vertikalen Niederschlagsverteilung. *Erdkunde,* 19: 6–14.

WOO KAM SENG, 1968. *Rainfall in the Highlands of Malaya.* Unpublished Academic Exercise of the Department of Geography, University of Singapore, Singapore, 74 pp.

YIN, M. T., 1949. A synoptic-aerologic study of the onset of the summer monsoon over India and Burma. *J. Meteorol.,* 6: 393–400.

YOSHINO, M. M. and KOBAYASHI, M., 1977. Climatic records of Monsoon Asia, Part Ia. Precipitation in Thailand, Malaysia and Singapore. *Climatological Notes,* Vol. 20. Institute of Geoscience, University of Tsukuba, 88 pp.

Appendix—Climatic tables

TABLE X

CLIMATIC TABLE FOR BHAMO
Latitude 24°16′N, longitude 97°14′E, elevation 118 m

Month	Mean press. at sea level (mbar)	Temperature (°C)						Mean relative humidity (%)
		mean	mean daily max.	mean daily min.	mean daily range	abs. max.	abs. min.	
Jan.	—	16.7	24.4	9.1	13.3	30.6	2.8	84
Feb.	—	19.2	26.9	11.6	15.3	32.8	5.0	75
Mar.	—	23.2	31.1	15.3	15.8	37.8	7.8	64
Apr.	—	26.6	33.7	19.6	14.1	40.6	12.8	58
May	—	28.2	34.0	22.5	11.5	41.1	16.7	70
June	—	27.9	32.1	23.9	8.2	38.9	18.3	83
July	—	27.6	31.1	24.2	6.9	37.8	18.3	86
Aug.	—	27.6	31.0	24.2	6.8	36.7	21.1	86
Sept.	—	27.8	31.9	23.6	8.3	36.7	16.1	86
Oct.	—	25.7	30.6	21.2	9.4	36.7	13.3	84
Nov.	—	21.4	27.1	15.7	11.4	32.8	8.3	85
Dec.	—	17.5	24.1	11.1	13.0	30.0	3.9	88
Year	—	24.1	29.8	18.5	11.3	41.1	2.8	79
Years of records		50	50	29	29	50	50	50

Month	Precipitation (mm)				Mean number of days with precipitation	Mean evaporation (mm)	Mean cloud cover 0–10	Mean sunshine hours	Most freq. wind direct.
	mean	max.	min.	max. in 24 hours					
Jan.	10	—	—	76	1	—	—	—	—
Feb.	15	—	—	28	2	—	—	—	—
Mar.	18	—	—	46	2	—	—	—	—
Apr.	46	—	—	58	4	—	—	—	—
May	155	—	—	119	10	—	—	—	—
June	358	—	—	150	19	—	—	—	—
July	424	—	—	173	20	—	—	—	—
Aug.	409	—	—	221	19	—	—	—	—
Sept.	249	—	—	183	13	—	—	—	—
Oct.	117	—	—	102	7	—	—	—	—
Nov.	43	—	—	86	3	—	—	—	—
Dec.	13	—	—	53	1	—	—	—	—
Year	1857	—	—	221	101	—	—	—	—
Years of records	50	n.a.	n.a.	50	50	n.a.	—	—	—

Note. See "Statistical information" on p. 34. The climatic tables have been arranged according to latitude from north to south.

TABLE XI

CLIMATIC TABLE FOR LASHIO
Latitude 22°56′N, longitude 97°45′E, elevation 854 m

Month	Mean press. at sea level (mbar)	Temperature (°C)						Mean relative humidity (%)
		mean	mean daily max.	mean daily min.	mean daily range	abs. max.	abs. min.	
Jan.	—	15.5	23.2	8.0	15.2	30.6	−0.6	72
Feb.	—	17.5	25.4	9.7	15.7	32.8	2.8	62
Mar.	—	21.4	29.6	13.2	16.4	35.0	5.0	50
Apr.	—	24.2	31.4	16.7	14.7	40.0	9.4	54
May	—	24.9	30.4	19.3	11.1	38.3	12.2	71
June	—	24.9	28.7	21.2	7.5	35.0	19.4	81
July	—	24.6	28.1	21.3	6.8	33.3	17.8	84
Aug.	—	24.4	27.8	21.1	6.7	33.9	16.7	86
Sept.	—	24.3	28.4	20.3	8.1	32.8	13.9	84
Oct.	—	22.6	27.6	17.9	9.7	33.9	7.2	84
Nov.	—	19.3	25.0	13.8	11.2	31.7	5.0	85
Dec.	—	16.1	22.8	9.3	13.5	27.2	2.2	83
Year	—	21.7	27.4	16.0	11.4	40.0	−0.6	74
Years of records	n.a.	45	45	28	28	45	45	50

Month	Precipitation (mm)				Mean number of days with precipitation	Mean evaporation (mm)	Mean cloud cover 0–10	Mean sunshine hours	Most freq. wind direct.
	mean	max.	min.	max. in 24 hours					
Jan.	5	—	—	38	1	—	—	—	—
Feb.	10	—	—	28	1	—	—	—	—
Mar.	13	—	—	30	1	—	—	—	—
Apr.	56	—	—	56	5	—	—	—	—
May	170	—	—	76	11	—	—	—	—
June	257	—	—	160	16	—	—	—	—
July	297	—	—	170	18	—	—	—	—
Aug.	325	—	—	117	19	—	—	—	—
Sept.	201	—	—	107	13	—	—	—	—
Oct.	142	—	—	122	10	—	—	—	—
Nov.	74	—	—	102	5	—	—	—	—
Dec.	23	—	—	56	1	—	—	—	—
Year	1572	—	—	170	101	—	—	—	—
Years of records	45	n.a.	n.a.	45	45	n.a.	—	—	—

TABLE XII

CLIMATIC TABLE FOR LAO-KAY
Latitude 22°30′N, longitude 103°57′E, elevation 93 m

Month	Mean press. at sea level (mbar)	Temperature (°C)						Mean relative humidity (%)
		mean	mean daily max.	mean daily min.	mean daily range	abs. max.	abs. min.	
Jan.	—	15.4	20.7	13.2	7.5	31.1	2.2	90.0
Feb.	—	18.0	21.7	14.4	7.3	34.0	7.2	89.5
Mar.	—	20.2	25.0	17.2	7.8	37.0	6.8	88.8
Apr.	—	23.9	28.7	20.3	8.4	39.0	10.0	89.4
May	—	26.8	31.7	22.9	8.8	42.8	14.8	88.1
June	—	27.9	32.3	24.4	7.9	39.3	18.8	89.4
July	—	27.7	32.2	24.5	7.7	39.7	21.4	90.1
Aug.	—	27.6	32.2	24.2	8.0	38.3	20.4	88.3
Sept.	—	26.5	31.4	23.2	8.2	37.2	15.8	88.2
Oct.	—	23.8	28.5	20.5	8.0	36.5	10.7	88.9
Nov.	—	20.7	25.4	17.3	8.1	34.8	8.0	89.0
Dec.	—	18.0	22.6	14.2	8.4	32.8	3.6	88.6
Year	—	23.0	27.7	19.7	8.0	42.8	2.2	89.0
Years of records	n.a.	6	32	32	32	32	32	8

Month	Precipitation (mm)				Mean number of days with precipitation	Mean evaporation (mm)	Mean cloud cover 0–10	Mean sunshine hours	Most freq. wind direct.
	mean	max.	min.	max. in 24 hours					
Jan.	15	64	0	36	5	41	8.5	—	S
Feb.	38	99	1	41	7	48	8.3	—	S
Mar.	57	153	8	56	9	58	7.7	—	SE
Apr.	117	250	27	102	13	61	7.1	—	S
May	223	471	75	147	16	66	6.8	—	S
June	222	491	44	124	17	55	7.3	—	S
July	305	574	132	173	19	54	7.4	—	S
Aug.	345	810	140	173	19	53	7.0	—	S
Sept.	242	565	97	147	14	51	6.8	—	S
Oct.	118	288	9	107	11	53	7.3	—	S
Nov.	60	177	1	79	9	44	7.6	—	S
Dec.	25	105	2	58	6	44	7.5	—	S
Year	1767	2534	1179	173	145	628	7.5	—	—
Years of records	32	32	32	32	32	9	32	n.a.	n.a.

TABLE XIII

CLIMATIC TABLE FOR MANDALAY
Latitude 21°59′N, longitude 96°06′E, elevation 77 m

Month	Mean press. at sea level (mbar)	Temperature (°C)						Mean relative humidity (%)
		mean	mean daily max.	mean daily min.	mean daily range	abs. max.	abs. min.	
Jan.	1016.3	20.2	27.7	13.2	14.5	35.6	7.2	62
Feb.	1013.3	22.8	30.9	15.3	15.6	37.8	8.3	50
Mar.	1011.4	27.4	35.8	20.1	15.7	42.2	12.2	38
Apr.	1008.7	31.4	37.9	25.1	12.8	45.6	17.2	40
May	1006.3	31.2	36.6	26.0	10.6	45.0	20.0	60
June	1004.0	29.8	34.1	25.9	8.2	41.7	20.6	70
July	1004.5	29.6	33.7	25.9	7.8	41.1	22.2	70
Aug.	1005.0	29.1	33.1	25.6	7.5	39.4	21.1	76
Sept.	1007.6	28.7	32.9	25.2	7.7	39.4	20.6	78
Oct.	1011.6	27.4	31.7	23.7	8.0	38.9	16.7	76
Nov.	1014.6	24.1	29.3	19.7	9.6	36.7	13.3	74
Dec.	1016.9	20.3	26.9	14.8	12.1	33.9	6.7	64
Year	1010.0	26.8	32.6	21.7	10.9	45.6	6.7	64
Years of records	10	50	50	29	29	50	50	50

Month	Precipitation (mm)				Mean number of days with precipitation	Mean evaporation (mm)	Mean cloud cover 0–10	Mean sunshine hours	Most freq. wind direct.
	mean	max.	min.	max. in 24 hours					
Jan.	1	18	0	25	0	—	—	—	—
Feb.	5	53	0	25	0	—	—	—	—
Mar.	5	28	0	20	1	—	—	—	—
Apr.	36	99	5	79	3	—	—	—	—
May	150	343	13	140	8	—	—	—	—
June	152	262	13	145	7	—	—	—	—
July	74	386	5	132	6	—	—	—	—
Aug.	102	211	23	191	8	—	—	—	—
Sept.	147	251	23	119	9	—	—	—	—
Oct.	127	320	53	109	7	—	—	—	—
Nov.	64	208	1	142	3	—	—	—	—
Dec.	10	112	0	38	1	—	—	—	—
Year	871	n.a.	n.a.	191	53	—	—	—	—
Years of records	50	18	18	55	50	n.a.	n.a.	n.a.	n.a.

TABLE XIV

CLIMATIC TABLE FOR HANOI

Latitude 21°02′N, longitude 105°51′E, elevation 7 m

Month	Mean press. at sea level (mbar)	Temperature (°C)						Mean relative humidity (%)
		mean	mean daily max.	mean daily min.	mean daily range	abs. max.	abs. min.	
Jan.	1019.2	15.5	20.2	13.4	6.8	33.1	5.6	82.4
Feb.	1016.2	17.0	20.4	14.4	6.0	34.4	6.3	86.1
Mar.	1013.9	19.2	23.1	17.1	6.0	36.8	8.5	86.5
Apr.	1009.6	23.6	27.5	20.5	7.0	38.5	9.8	87.9
May	1006.5	27.4	32.0	23.6	8.4	42.8	15.4	84.4
June	1003.0	29.0	33.2	25.3	7.9	40.0	20.5	83.2
July	1001.8	28.6	32.9	25.4	7.5	40.0	21.6	85.3
Aug.	1002.8	28.6	32.4	25.3	7.1	38.1	20.9	86.2
Sept.	1007.9	27.1	31.1	24.3	6.8	37.1	17.1	85.9
Oct.	1013.9	24.4	29.0	21.4	7.6	35.7	13.9	82.3
Nov.	1017.2	21.4	25.6	17.8	7.8	36.0	6.8	81.4
Dec.	1017.8	18.4	22.3	14.9	7.4	31.9	6.7	82.5
Year	—	23.4	27.5	20.3	7.2	42.8	5.6	84.5
Years of records	n.a.	10	30	30	30	30	30	12

Month	Precipitation (mm)				Mean number of days with precipitation	Mean evaporation (mm)	Mean cloud cover 0–10	Mean sunshine hours	Most freq. wind direct.
	mean	max.	min.	max. in 24 hours					
Jan.	22	108	0	46	9	47	7.7	1.4	NE
Feb.	36	92	10	48	14	36	8.4	1.4	N, NE
Mar.	45	119	4	63	15	40	8.7	1.3	SE
Apr.	89	202	29	104	14	42	8.2	2.2	SE
May	216	493	64	155	14	62	7.5	4.2	SE
June	254	597	26	244	15	46	7.9	5.0	SE
July	335	564	104	206	16	63	7.7	4.8	SE
Aug.	339	905	80	262	16	57	7.7	4.2	Variable
Sept.	276	711	53	252	14	52	6.8	4.3	Variable, N
Oct.	115	321	5	157	10	65	6.0	4.2	N
Nov.	48	150	2	69	7	64	6.6	3.2	N
Dec.	27	105	0	30	7	51	7.0	2.1	N
Year	1802	2741	1275	262	151	625	7.5	3.1	—
Years of records	30	30	30	30	30	9	30	11	n.a.

TABLE XV

CLIMATIC TABLE FOR AKYAB
Latitude 20°08′N, longitude 92°55′E, elevation 9 m

Month	Mean press. at sea level (mbar)	Temperature (°C)						Mean relative humidity (%)
		mean	mean daily max.	mean daily min.	mean daily range	abs. max.	abs. min.	
Jan.	1017.6	21.0	27.1	14.6	12.5	34.4	8.3	72
Feb.	1015.5	22.5	28.9	15.8	13.1	35.0	9.4	70
Mar.	1013.6	25.6	31.2	19.6	11.8	37.8	12.2	74
Apr.	1012.4	28.4	32.7	23.7	9.0	37.2	16.7	74
May	1008.8	28.8	32.2	25.1	7.2	37.2	18.9	79
June	1005.6	27.5	29.8	24.8	4.9	36.7	20.0	88
July	1006.1	26.9	29.0	24.6	4.4	33.9	21.7	91
Aug.	1006.2	26.9	28.9	24.6	4.4	32.8	21.7	89
Sept.	1008.9	27.6	30.0	24.8	5.2	34.4	21.1	87
Oct.	1012.1	27.5	30.6	24.1	6.5	33.9	18.3	84
Nov.	1015.2	25.3	29.2	21.1	8.2	32.8	15.6	82
Dec.	1017.3	22.1	27.0	16.7	10.3	30.6	10.6	84
Year	1011.6	25.8	29.7	21.6	8.1	37.8	8.3	80
Years of records	10	60	60	29	29	60	60	50

Month	Precipitation (mm)				Mean number of days with precipitation	Mean evaporation (mm)	Mean cloud cover 0–10	Mean sunshine hours	Most freq. wind direct.
	mean	max.	min.	max. in 24 hours					
Jan.	3	15	0	25	0	—	—	—	—
Feb.	5	71	0	58	0	—	—	—	—
Mar.	10	18	0	71	1	—	—	—	—
Apr.	51	183	3	292	2	—	—	—	—
May	391	1138	94	358	11	—	—	—	—
June	1151	1897	615	381	24	—	—	—	—
July	1400	2101	696	343	28	—	—	—	—
Aug.	1133	2238	490	383	27	—	—	—	—
Sept.	577	836	256	249	19	—	—	—	—
Oct.	284	620	36	330	9	—	—	—	—
Nov.	130	218	3	465	4	—	—	—	—
Dec.	20	185	0	135	1	—	—	—	—
Year	5154	n.a.	n.a.	465	126	—	—	—	—
Years of records	60	n.a.	n.a.	60	60	—	—	—	—

TABLE XVI

CLIMATIC TABLE FOR LUANG PRABANG
Latitude 19°53′N, longitude 102°08′E, elevation 290 m

Month	Mean press. at sea level (mbar)	Temperature (°C)						Mean relative humidity (%)
		mean	mean daily max.	mean daily min.	mean daily range	abs. max.	abs. min.	
Jan.	—	20.5	27.5	13.5	14.0	39.2	0.8	82.7
Feb.	—	23.1	31.7	14.5	17.2	39.1	8.0	75.7
Mar.	—	25.6	34.1	17.1	17.0	41.0	9.8	71.8
Apr.	—	28.1	35.5	20.6	14.9	44.8	14.0	73.8
May	—	28.9	35.1	22.6	12.5	44.0	17.0	81.2
June	—	28.7	33.9	23.6	10.3	40.0	14.0	84.6
July	—	28.0	32.4	23.6	8.8	39.0	19.2	87.2
Aug.	—	27.7	32.2	23.2	9 0	39.8	14.0	87.4
Sept.	—	27.8	32.9	22.8	10.1	38.0	10.8	86.8
Oct.	—	26.2	31.7	20.7	11.0	38.6	12.9	84.8
Nov.	—	23.6	29.3	17.8	11.5	36.3	6.0	84.3
Dec.	—	21.1	27.0	15.2	11.8	33.0	4.3	85.1
Year	—	25.8	31.9	19.6	12.3	44.8	0.8	82.1
Years of records	n.a.	27	27	27	27	27	27	9

Month	Precipitation (mm)				Mean number of days with precipitation	Mean evaporation (mm)	Mean cloud cover 0–10	Mean sunshine hours	Most freq. wind direct.
	mean	max.	min.	max. in 24 hours					
Jan.	16	140	0	109	2	49	5.2	—	NE, SW
Feb.	17	94	0	74	2	66	3.4	—	SW
Mar.	30	91	0	41	4	96	3.2	—	SW
Apr.	109	314	11	104	8	98	3.5	—	SW
May	163	383	21	79	13	78	5.1	—	SW
June	155	386	8	102	13	59	5.8	—	SW
July	230	485	68	127	17	46	6.8	—	SW
Aug.	299	541	72	89	19	41	6.8	—	SW
Sept.	166	366	24	81	12	46	5.3	—	NE, SW
Oct.	78	248	0	117	6	50	4.5	—	NE
Nov.	30	116	0	58	3	47	4.9	—	NE
Dec.	13	65	0	51	1	43	5.1	—	NE, SW
Year	1306	1879	510	127	100	719	5.0	—	—
Years of records	30	30	30	30	30	10	27	n.a.	n.a.

TABLE XVII

CLIMATIC TABLE FOR TOUNGOO
Latitude 18°56′N, longitude 96°26′E, elevation 48 m

Month	Mean press. at sea level (mbar)	Temperature (°C)						Mean relative humidity (%)
		mean	mean daily max.	mean daily min.	mean daily range	abs. max.	abs. min.	
Jan.	—	21.9	29.5	14.7	14.8	35.0	8.3	68
Feb.	—	24.3	32.5	16.0	16.5	38.2	8.3	58
Mar.	—	28.3	36.2	20.5	15.7	40.0	12.2	50
Apr.	—	31.1	37.7	24.5	13.2	42.2	16.1	54
May	—	30.0	35.2	24.8	10.4	42.2	18.9	72
June	—	27.7	31.4	24.1	7.3	38.9	18.3	86
July	—	27.1	30.2	23.9	6.3	37.2	21.1	88
Aug.	—	27.1	30.2	23.9	6.3	35.6	21.1	88
Sept.	—	27.8	31.5	24.1	7.4	37.2	18.9	86
Oct.	—	27.8	32.1	23.4	8.7	37.2	17.8	82
Nov.	—	25.8	30.7	20.9	9.8	35.0	12.8	80
Dec.	—	22.4	28.6	16.4	12.2	33.9	9.4	81
Year	—	26.8	32.1	21.4	10.7	42.2	8.3	74
Years of records	n.a.	60	60	29	29	60	60	50

Month	Precipitation (mm)				Mean number of days with precipitation	Mean evaporation (mm)	Mean cloud cover 0–10	Mean sunshine hours	Most freq. wind direct.
	mean	max.	min.	max. in 24 hours					
Jan.	5	—	—	69	0	—	—	—	—
Feb.	5	—	—	43	0	—	—	—	—
Mar.	8	—	—	107	1	—	—	—	—
Apr.	53	—	—	130	3	—	—	—	—
May	203	—	—	142	11	—	—	—	—
June	366	—	—	91	20	—	—	—	—
July	455	—	—	109	25	—	—	—	—
Aug.	480	—	—	168	24	—	—	—	—
Sept.	297	—	—	114	17	—	—	—	—
Oct.	183	—	—	163	9	—	—	—	—
Nov.	48	—	—	97	3	—	—	—	—
Dec.	10	—	—	76	1	—	—	—	—
Year	2113	—	—	168	114	—	—	—	—
Years of records	60	—	—	60	60	—	—	—	—

TABLE XVIII

CLIMATIC TABLE FOR CHIANG MAI
Latitude 18°47′N, longitude 98°59′E, elevation 313 m

Month	Mean press. at sea level (mbar)	Temperature (°C)						Mean relative humidity (%)
		mean	mean daily max.	mean daily min.	mean daily range	abs. max.	abs. min.	
Jan.	1016.5	21.3	28.9	13.3	15.6	36.1	6.1	75
Feb.	1014.1	23.1	31.7	14.4	17.3	36.1	9.4	67
Mar.	1012.3	26.0	34.4	17.2	17.2	38.9	12.8	60
Apr.	1010.1	29.0	36.1	21.7	14.4	40.6	15.0	60
May	1008.2	28.8	34.4	22.8	11.6	41.1	19.4	74
June	1006.3	27.9	32.2	23.3	8.9	37.8	20.6	80
July	1006.7	27.4	31.1	23.3	7.8	37.2	18.9	82
Aug.	1006.3	27.0	31.1	23.3	7.8	37.2	21.1	85
Sept.	1008.6	26.8	31.1	22.8	8.3	35.6	18.3	85
Oct.	1012.7	26.2	30.6	21.1	9.4	35.6	15.6	83
Nov.	1015.1	24.4	30.0	18.9	11.1	37.2	12.2	81
Dec.	1017.1	21.4	28.3	15.0	13.3	36.1	6.1	77
Year	1011.2	25.8	31.7	20.0	11.7	41.1	6.1	76
Years of records	10	10	13	13	13	13	13	10

Month	Precipitation (mm)				Mean number of days with precipitation	Mean evaporation (mm)	Mean cloud cover 0–10	Mean sunshine hours	Most freq. wind direct.
	mean	max.	min.	max. in 24 hours					
Jan.	7	59	0	18	1	—	—	—	S
Feb.	11	89	0	25	1	—	—	—	S
Mar.	15	68	0	51	2	—	—	—	S
Apr.	51	123	6	38	5	—	—	—	SE–SW
May	139	326	66	74	12	—	—	—	SW–S
June	154	347	39	56	16	—	—	—	SW–S
July	188	260	54	74	18	—	—	—	SW
Aug.	220	302	163	71	21	—	—	—	S, SW
Sept.	292	474	212	104	18	—	—	—	S
Oct.	124	291	20	107	10	—	—	—	Variable
Nov.	38	83	0	117	4	—	—	—	N
Dec.	10	85	0	18	1	—	—	—	N
Year	1246	2032	959	117	109	—	—	—	—
Years of records	30	10	10	8	30	n.a.	n.a.	n.a.	n.a.

TABLE XIX

CLIMATIC TABLE FOR VIENTIANE
Latitude 17°58′N, longitude 102°36′E, elevation 162 m

Month	Mean press. at sea level (mbar)	Temperature (°C)						Mean relative humidity (%)
		mean	mean daily max.	mean daily min.	mean daily range	abs. max.	abs. min.	
Jan.	1014.7	20.3	28.1	14.0	14.1	34.9	3.9	76.5
Feb.	1011.7	23.1	30.2	17.1	13.1	36.6	7.6	74.5
Mar.	1009.6	25.6	32.5	19.4	13.1	39.8	12.1	70.9
Apr.	1007.6	27.9	34.0	22.6	11.4	39.2	17.1	74.0
May	1006.3	27.4	32.4	23.6	8.8	38.9	20.8	81.8
June	1004.6	27.3	31.5	24.1	7.4	35.6	21.1	85.2
July	1005.0	26.8	30.5	24.0	6.5	34.3	21.2	86.5
Aug.	1004.4	27.0	30.9	24.1	6.8	36.6	21.1	85.7
Sept.	1006.7	26.8	30.8	23.9	6.9	35.2	21.2	86.0
Oct.	1010.9	25.5	30.3	21.2	9.1	34.2	12.9	81.6
Nov.	1013.1	23.4	29.5	18.2	11.3	34.4	10.3	79.2
Dec.	1015.2	21.2	28.6	15.4	13.2	33.4	5.0	78.0
Year	1009.2	25.2	30.8	20.6	10.2	39.8	3.9	80.0
Years of records	10	9	11	11	11	11	11	9

Month	Precipitation (mm)				Mean number of days with precipitation	Mean evaporation (mm)	Mean cloud cover 0–10	Mean sunshine hours	Most freq. wind direct.
	mean	max.	min.	max. in 24 hours					
Jan.	6	65	0	36	1	92	3.4	7.5	E–SE
Feb.	15	62	0	66	2	102	4.0	6.1	ESE
Mar.	37	134	0	53	4	136	3.5	5.9	ESE
Apr.	99	329	7	155	7	119	4.8	6.2	SE
May	268	439	79	140	15	81	7.3	4.9	Variable
June	302	499	102	163	17	63	8.1	4.4	WSW
July	267	515	116	193	18	56	8.8	2.6	SW
Aug.	292	506	50	132	18	56	8.4	3.0	WSW
Sept.	303	777	99	117	16	50	7.9	3.6	Variable
Oct.	108	329	0	130	7	64	5.2	6.3	E, ESE
Nov.	15	99	0	69	1	72	4.8	7.0	ESE, E
Dec.	3	24	0	23	1	81	4.0	7.3	ESE, NW
Year	1715	2138	1373	193	107	972	5.9	5.4	—
Years of records	27	27	27	27	27	10	11	7	n.a.

TABLE XX

CLIMATIC TABLE FOR PHITSANULOK
Latitude 16°50′N, longitude 100°16′E, elevation 50 m

Month	Mean press. at sea level (mbar)	Temperature (°C)						Mean relative humidity (%)
		mean	mean daily max.	mean daily min.	mean daily range	abs. max.	abs. min.	
Jan.	1014.8	24.7	31.6	17.1	14.5	35.7	7.5	73
Feb.	1012.6	26.7	33.7	19.7	14.0	38.3	11.8	70
Mar.	1011.4	29.0	35.8	22.5	13.3	39.8	15.0	68
Apr.	1009.8	30.9	36.7	24.4	12.3	41.8	17.0	66
May	1008.1	30.3	35.3	24.8	10.5	41.0	21.8	75
June	1006.8	29.1	33.4	24.6	8.8	38.8	22.0	81
July	1007.1	28.5	32.8	24.4	8.4	38.0	21.8	83
Aug.	1006.6	28.3	32.4	24.4	8.0	36.6	21.8	84
Sept.	1008.4	28.3	32.3	24.5	7.8	37.0	22.1	85
Oct.	1011.9	28.1	32.4	23.7	8.7	36.1	18.0	82
Nov.	1014.0	26.6	32.0	21.5	10.5	35.9	10.5	78
Dec.	1015.6	24.3	30.8	18.0	12.8	35.5	9.9	74
Year	1010.6	27.9	33.3	22.5	10.8	41.8	7.5	76
Years of records	10	10	22	22	22	22	22	10

Month	Precipitation (mm)				Mean number of days with precipitation	Mean evaporation (mm)	Mean cloud cover 0–10	Mean sunshine hours	Most freq. wind direct.
	mean	max.	min.	max. in 24 hours					
Jan.	7	50	0	—	1	—	—	—	Variable
Feb.	24	83	0	—	3	—	—	—	S
Mar.	39	142	15	—	5	—	—	—	S
Apr.	75	135	2	—	5	—	—	—	S
May	159	350	61	—	12	—	—	—	S
June	178	281	88	—	15	—	—	—	S
July	204	446	84	—	16	—	—	—	S
Aug.	227	432	60	—	18	—	—	—	S
Sept.	272	355	189	—	17	—	—	—	S
Oct.	132	179	51	—	10	—	—	—	Variable
Nov.	37	69	0	—	3	—	—	—	Variable
Dec.	1	17	0	—	0	—	—	—	N
Year	1354	1630	1026	—	105	—	—	—	—
Years of records	30	10	10	n.a.	30	n.a.	n.a.	n.a.	n.a.

TABLE XXI

CLIMATIC TABLE FOR RANGOON
Latitude 16°47′N, longitude 96°13′E, elevation 5 m

Month	Mean press. at sea level (mbar)	Temperature (°C)						Mean relative humidity (%)
		mean	mean daily max.	mean daily min.	mean daily range	abs. max.	abs. min.	
Jan.	1015.6	25.2	31.8	18.8	13.0	37.8	12.8	61
Feb.	1014.0	26.5	33.5	19.8	13.7	38.3	13.3	62
Mar.	1013.0	28.6	35.3	22.2	13.1	39.4	16.1	64
Apr.	1011.3	30.4	36.4	24.4	12.0	41.1	20.0	68
May	1008.9	29.2	33.2	25.1	8.1	40.6	20.6	80
June	1006.9	27.4	30.2	24.6	5.6	36.7	21.7	86
July	1007.9	26.9	29.6	24.3	5.3	33.9	21.1	88
Aug.	1007.9	26.9	29.5	24.3	5.2	33.9	20.0	89
Sept.	1009.4	27.3	30.1	24.6	5.5	34.4	22.2	86
Oct.	1012.2	27.8	31.1	24.6	6.6	35.0	21.7	80
Nov.	1014.1	27.1	31.2	23.1	8.1	35.0	16.1	76
Dec.	1015.8	25.3	30.9	20.1	10.8	35.6	12.8	68
Year	1011.4	27.4	31.9	23.0	8.9	41.1	12.8	76
Years of records	10	60	60	29	29	60	60	50

Month	Precipitation (mm)				Mean number of days with precipitation	Mean evaporation (mm)	Mean cloud cover 0–10	Mean sunshine hours	Most freq. wind direct.
	mean	max.	min.	max. in 24 hours					
Jan.	2	48	0	74	0	—	—	—	—
Feb.	5	86	0	48	0	—	—	—	—
Mar.	8	51	0	41	1	—	—	—	—
Apr.	51	244	0	361	2	—	—	—	—
May	307	691	28	229	14	—	—	—	—
June	483	823	255	152	23	—	—	—	—
July	582	907	188	137	26	—	—	—	—
Aug.	528	782	241	135	25	—	—	—	—
Sept.	394	658	201	132	20	—	—	—	—
Oct.	180	414	41	135	10	—	—	—	—
Nov.	69	389	0	150	3	—	—	—	—
Dec.	10	23	0	102	1	—	—	—	—
Year	2619	n.a.	n.a.	361	125	—	—	—	—
Years of records	60	60	60	65	60	n.a.	n.a.	n.a.	—

TABLE XXII

CLIMATIC TABLE FOR TOURANE
Latitude 16°5′N, longitude 108°13′E, elevation 3 m

Month	Mean press. at sea level (mbar)	Temperature (°C)						Mean relative humidity (%)
		mean	mean daily max.	mean daily min.	mean daily range	abs. max.	abs. min.	
Jan.	1016.5	21.2	24.1	18.9	5.2	30.5	11.0	85.9
Feb.	1014.6	22.4	25.8	19.9	5.9	36.8	14.3	85.9
Mar.	1012.4	23.3	27.0	20.8	6.2	36.3	15.4	85.9
Apr.	1010.0	25.8	30.0	22.9	7.1	39.8	17.6	84.6
May	1007.1	27.8	32.7	24.3	8.4	39.0	21.8	81.2
June	1004.6	29.0	34.4	25.0	9.4	40.0	22.6	76.8
July	1004.1	28.5	33.5	24.9	8.6	38.0	21.8	78.1
Aug.	1004.3	28.7	34.0	24.6	9.4	39.0	21.4	76.8
Sept.	1006.6	27.1	31.1	23.9	7.2	36.7	20.9	84.4
Oct.	1011.2	25.3	28.2	22.7	5.5	33.5	17.4	85.1
Nov.	1013.8	23.9	26.7	21.7	5.0	31.0	15.2	85.8
Dec.	1016.0	22.2	24.9	20.1	4.8	30.8	13.3	86.1
Year	1010.1	25.4	29.4	22.5	6.9	40.0	11.0	83.1
Years of records	20	8	8	8	8	8	8	10

Month	Precipitation (mm)				Mean number of days with precipitation	Mean evaporation (mm)	Mean cloud cover 0–10	Mean sunshine hours	Most freq. wind direct.
	mean	max.	min.	max. in 24 hours					
Jan.	102	219	49	257	15	60	7.7	4.4	NNW
Feb.	31	74	1	61	7	60	6.5	5.4	NNW
Mar.	12	34	1	137	4	75	6.0	5.8	NNW–ENE
Apr.	18	70	1	30	4	85	5.1	7.0	E–ENE
May	47	116	0	122	8	109	5.9	8.2	N–E
June	42	177	3	333	7	129	5.8	8.0	NNW–E
July	99	252	9	89	11	116	7.2	7.8	NNW–E
Aug.	117	293	11	140	12	125	6.6	7.0	NNW
Sept.	447	1176	156	325	17	76	7.5	5.8	NNW
Oct.	530	1256	67	279	21	77	7.7	5.1	NNW
Nov.	221	620	106	272	21	65	8.0	4.6	NNW–ENE
Dec.	209	450	40	320	20	64	8.0	4.0	NNW–ENE
Year	1875	2167	1501	333	147	1041	6.8	6.1	—
Years of records	10	10	10	10	10	8	10	13	n.a.

TABLE XXIII

CLIMATIC TABLE FOR PAKSÉ
Latitude 15°47′N, longitude 105°47′E, elevation 96 m

Month	Mean press. at sea level (mbar)	Temperature (°C)						Mean relative humidity (%)
		mean	mean daily max.	mean daily min.	mean daily range	abs. max.	abs. min.	
Jan.	—	23.9	31.1	17.6	13.5	36.8	8.2	66.5
Feb.	—	26.4	32.6	21.0	11.6	37.9	13.2	65.5
Mar.	—	28.3	34.3	22.7	11.6	38.0	12.6	62.7
Apr.	—	29.4	34.8	24.7	10.1	39.4	19.8	67.3
May	—	28.0	33.0	24.3	8.7	38.5	21.8	77.8
June	—	27.2	31.1	24.1	7.0	34.8	21.5	82.7
July	—	26.2	29.9	23.6	6.3	34.1	21.4	86.3
Aug.	—	26.6	30.4	23.8	6.6	34.4	21.5	85.5
Sept.	—	26.1	29.9	23.5	6.4	34.5	19.8	86.7
Oct.	—	25.6	30.5	22.1	8.4	35.0	16.7	80.4
Nov.	—	24.9	30.7	20.2	10.5	34.2	14.0	75.2
Dec.	—	23.9	30.7	18.5	12.2	35.0	9.0	70.7
Year	—	26.4	31.6	22.2	9.4	39.4	8.2	75.6
Years of records	n.a.	9	11	11	11	11	11	9

Month	Precipitation (mm)				Mean number of days with precipitation	Mean evaporation (mm)	Mean cloud cover 0–10	Mean sunshine hours	Most freq. wind direct.
	mean	max.	min.	max. in 24 hours					
Jan.	1	6	0	5	0	119	4.0	—	ESE, NNE–NNW
Feb.	12	86	0	71	1	101	4.1	—	ESE
Mar.	12	43	0	43	2	160	4.5	—	ESE
Apr.	83	188	34	122	7	139	6.1	—	ESE
May	230	363	111	91	16	106	7.3	—	ESE
June	314	421	82	114	21	60	8.2	—	ESE
July	505	736	308	122	27	43	8.8	—	E–SE
Aug.	450	832	243	137	26	45	8.6	—	E–SE
Sept.	404	752	151	160	22	42	8.6	—	E–SE
Oct.	103	230	11	117	11	26	6.8	—	N–NW
Nov.	17	38	0	36	4	89	5.9	—	NNW–NNE
Dec.	4	14	0	8	1	101	4.9	—	NW–NNE
Year	2135	2910	1441	160	138	1031	6.5	—	—
Years of records	11	11	11	11	11	11	11	n.a.	n.a.

TABLE XXIV

CLIMATIC TABLE FOR NAKHON RATCHASIMA
Latitude 14°58′N, longitude 102°07′E, elevation 181 m

Month	Mean press. at sea level (mbar)	Temperature (°C)						Mean relative humidity (%)
		mean	mean daily max.	mean daily min.	mean daily range	abs. max.	abs. min.	
Jan.	1015.9	23.4	32.8	15.6	17.2	37.8	5.0	70
Feb.	1013.3	26.5	35.0	18.9	16.1	40.6	11.1	67
Mar.	1012.0	28.8	36.7	21.7	15.0	41.1	15.0	68
Apr.	1010.2	30.0	36.7	23.3	13.3	42.2	15.6	69
May	1008.6	29.5	34.4	23.9	10.6	40.6	20.6	77
June	1007.3	28.7	33.3	23.9	9.4	38.9	21.1	78
July	1007.6	28.2	33.3	23.3	10.0	37.2	20.6	79
Aug.	1007.0	27.9	32.8	23.3	9.5	38.3	21.1	80
Sept.	1008.6	27.4	32.2	23.3	8.9	38.3	20.6	85
Oct.	1012.0	26.2	31.7	21.7	10.0	37.2	15.0	84
Nov.	1014.4	24.5	31.1	19.4	11.7	37.8	10.6	79
Dec.	1016.4	22.5	30.6	15.6	15.0	37.2	7.2	75
Year	1011.1	27.0	33.3	21.1	12.2	42.2	5.0	76
Years of records	10	10	14	14	14	12	12	10

Month	Precipitation (mm)				Mean number of days with precipitation	Mean evaporation (mm)	Mean cloud cover 0–10	Mean sunshine hours	Most freq. wind direct.
	mean	max.	min.	max. in 24 hours					
Jan.	7	28	0	51	1	—	—	—	NE
Feb.	33	107	0	71	3	—	—	—	NE
Mar.	48	125	13	71	5	—	—	—	Variable
Apr.	83	145	5	76	8	—	—	—	SW
May	157	350	54	89	14	—	—	—	W–S
June	111	242	46	38	13	—	—	—	SW, W
July	132	263	76	53	14	—	—	—	W, SW
Aug.	139	241	59	56	15	—	—	—	W, SW
Sept.	244	566	129	58	18	—	—	—	W, SW
Oct.	171	347	58	66	11	—	—	—	NE
Nov.	37	174	0	104	3	—	—	—	NE
Dec.	3	4	0	10	1	—	—	—	NE
Year	1162	1401	1006	104	106	—	—	—	—
Years of records	30	10	10	5	30	n.a.	n.a.	n.a.	n.a.

TABLE XXV

CLIMATIC TABLE FOR BANGKOK
Latitude 13°44′N, longitude 100°30′E, elevation 12 m

Month	Mean press. at sea level (mbar)	Temperature (°C)						Mean relative humidity (%)
		mean	mean daily max.	mean daily min.	mean daily range	abs. max.	abs. min.	
Jan.	1013.6	26.1	31.7	20.0	11.7	37.8	12.8	74
Feb.	1011.9	27.6	32.8	22.2	10.6	41.1	13.3	77
Mar.	1011.1	29.2	33.9	23.9	10.0	40.0	16.7	77
Apr.	1009.6	30.3	35.0	25.0	10.0	41.1	19.4	77
May	1008.1	29.8	33.9	25.0	8.9	41.1	21.7	80
June	1007.1	29.0	32.8	24.4	8.3	37.8	21.1	81
July	1007.4	28.4	32.2	24.4	7.8	38.3	21.7	82
Aug.	1007.0	28.2	32.2	24.4	7.8	37.2	22.2	83
Sept.	1008.2	27.9	31.7	24.4	7.2	36.7	20.6	85
Oct.	1010.6	27.6	31.1	23.9	7.2	37.8	17.8	85
Nov.	1012.3	26.7	30.6	22.2	8.3	37.2	13.3	82
Dec.	1013.9	25.4	30.6	20.0	10.6	37.8	11.1	76
Year	1010.1	28.0	32.2	23.3	8.9	41.1	11.1	80
Years of records	10	10	17	17	17	17	16	10

Month	Precipitation (mm)				Mean number of days with precipitation	Mean evaporation (mm)	Mean cloud cover 0–10	Mean sunshine hours	Most freq. wind direct.
	mean	max.	min.	max. in 24 hours					
Jan.	9	36	0	58	2	—	—	—	NE
Feb.	29	53	0	64	3	—	—	—	S
Mar.	34	51	0	79	3	—	—	—	S
Apr.	89	163	2	117	6	—	—	—	S
May	166	219	36	114	14	—	—	—	S, SW
June	171	244	79	89	16	—	—	—	S, SW
July	178	366	127	114	18	—	—	—	SW
Aug.	191	317	103	79	18	—	—	—	W–S
Sept.	306	500	210	175	21	—	—	—	W–S
Oct.	255	490	107	112	16	—	—	—	N–E
Nov.	57	115	2	142	6	—	—	—	N, NE
Dec.	7	10	0	107	1	—	—	—	N, NE
Year	1492	1957	1278	175	124	—	—	—	—
Years of records	30	10	10	17	30	n.a.	n.a.	n.a.	n.a.

TABLE XXVI

CLIMATIC TABLE FOR BATTAMBANG
Latitude 13°11′N, longitude 103°12′E, elevation 16 m

Month	Mean press. at sea level (mbar)	Temperature (°C)						Mean relative humidity (%)
		mean	mean daily max.	mean daily min.	mean daily range	abs. max.	abs. min.	
Jan.	1013.7	24.7	30.8	19.3	11.5	37.7	10.4	74.1
Feb.	1012.8	26.7	33.3	21.3	12.0	38.2	14.4	73.0
Mar.	1011.1	28.6	35.1	23.0	12.1	39.9	16.1	70.5
Apr.	1009.4	29.2	35.5	24.1	11.4	39.8	21.0	74.9
May	1008.4	28.1	33.7	24.5	9.2	39.0	22.0	80.5
June	1007.6	27.9	33.0	24.5	8.5	37.9	21.7	80.7
July	1008.1	27.2	31.8	24.3	7.5	35.3	21.2	83.6
Aug.	1007.5	27.1	31.9	24.1	7.8	35.1	22.1	84.2
Sept.	1008.7	26.7	31.1	24.1	7.0	34.4	21.9	86.6
Oct.	1011.3	26.4	30.1	23.6	6.5	34.1	19.6	84.6
Nov.	1013.2	25.7	30.0	22.1	7.9	33.4	13.1	81.8
Dec.	1014.2	24.8	30.1	20.2	9.9	33.6	10.7	78.6
Year	1010.5	26.9	32.2	22.9	9.3	39.9	10.4	79.4
Years of records	7	9	11	11	11	11	11	9

Month	Precipitation (mm)				Mean number of days with precipitation	Mean evaporation (mm)	Mean cloud cover 0–10	Mean sunshine hours	Most freq. wind direct.
	mean	max.	min.	max. in 24 hours					
Jan.	6	54	0	53	1	113	3.9	—	E
Feb.	17	113	0	46	2	119	4.1	—	E
Mar.	47	204	0	104	4	149	4.2	—	E
Apr.	87	266	5	84	7	127	5.2	—	WSW
May	155	340	41	89	13	100	6.6	—	SW–SSE
June	147	276	23	84	3	113	7.2	—	W–SW
July	155	284	71	79	16	89	7.9	—	SW
Aug.	158	289	70	74	16	80	7.7	—	W–SW
Sept.	262	609	79	150	16	62	7.9	—	W–SW
Oct.	228	524	79	132	15	63	6.9	—	Variable
Nov.	78	282	1	109	7	67	6.1	—	N, E
Dec.	25	132	0	46	3	87	5.2	—	N, E
Year	1365	1702	919	150	113	1169	6.1	—	—
Years of records	28	28	28	28	28	8	11	n.a.	n.a.

TABLE XXVII

CLIMATIC TABLE FOR KRATIÉ
Latitude 12°29′N, longitude 106°01′E, elevation 24 m

Month	Mean press. at sea level (mbar)	Temperature (°C)						Mean relative humidity (%)
		mean	mean daily max.	mean daily min.	mean daily range	abs. max.	abs. min.	
Jan.	—	25.5	32.2	19.5	12.7	35.6	11.4	66.6
Feb.	—	27.5	33.7	21.6	12.1	37.2	16.3	68.3
Mar.	—	28.9	35.3	23.1	12.2	38.3	18.4	67.0
Apr.	—	29.7	35.9	24.5	11.4	39.6	21.3	70.9
May	—	28.2	33.6	24.0	9.6	38.1	20.7	81.9
June	—	27.5	32.8	23.9	8.9	37.7	21.7	84.0
July	—	26.7	31.4	23.5	7.9	34.8	21.4	86.4
Aug.	—	27.1	31.6	23.9	7.7	36.0	21.0	85.6
Sept.	—	26.4	30.8	23.8	7.0	34.1	21.4	87.0
Oct.	—	26.2	31.0	22.8	8.2	34.5	17.0	83.9
Nov.	—	25.6	31.0	21.2	9.8	33.8	16.4	80.7
Dec.	—	24.9	31.0	19.8	11.2	35.0	9.5	75.1
Year	—	27.0	32.5	22.6	9.9	39.6	9.5	78.2
Years of records	n.a.	8	9	9	9	9	9	9

Month	Precipitation (mm)				Mean number of days with precipitation	Mean evaporation (mm)	Mean cloud cover 0–10	Mean sunshine hours	Most freq. wind direct.
	mean	max.	min.	max. in 24 hours					
Jan.	9	74	0	16	1	120	4.6	7.6	NE–E
Feb.	13	76	0	25	1	117	5.1	7.3	Variable
Mar.	23	113	0	52	2	144	5.1	8.6	E–SSW
Apr.	108	302	4	162	6	128	6.1	7.5	S
May	242	509	76	151	14	72	7.2	5.8	SSE–SSW
June	242	465	93	62	15	59	7.4	6.5	SSE–SSW
July	343	760	67	100	19	50	8.1	3.6	SSE–SSW
Aug.	256	617	84	72	18	51	7.9	4.8	SSE–SSW
Sept.	346	587	99	74	19	48	8.4	3.6	SSW
Oct.	175	387	54	70	12	57	7.1	6.0	NNE, E
Nov.	75	331	0	48	6	68	6.1	6.0	NNE, E
Dec.	26	163	0	16	2	85	5.5	7.2	NNE, E
Year	1858	2373	1402	162	115	999	6.5	6.2	—
Years of records	20	20	20	20	20	8	9	5	n.a.

TABLE XXVIII

CLIMATIC TABLE FOR MERGUI
Latitude 12°26′N, longitude 98°36′E, elevation 20 m

Month	Mean press. at sea level (mbar)	Temperature (°C)						Mean relative humidity (%)
		mean	mean daily max.	mean daily min.	mean daily range	abs. max.	abs. min.	
Jan.	1014.2	25.6	30.8	20.8	10.0	35.0	11.7	72
Feb.	1013.3	26.5	31.6	21.9	9.7	36.1	15.6	72
Mar.	1012.7	27.6	32.4	23.0	9.4	37.2	17.2	72
Apr.	1011.4	28.4	33.1	24.1	9.0	36.7	20.0	72
May	1009.8	27.6	31.6	24.0	7.6	36.7	19.4	82
June	1009.1	26.4	29.6	23.4	6.2	36.7	19.4	88
July	1009.6	25.9	28.9	23.2	5.7	33.9	18.9	90
Aug.	1009.5	25.9	28.9	23.2	5.7	33.9	18.9	90
Sept.	1010.5	25.9	29.1	23.1	6.0	32.8	18.9	89
Oct.	1012.0	26.5	30.2	23.0	7.2	33.9	17.2	83
Nov.	1012.8	26.2	30.6	22.4	8.2	34.4	15.0	76
Dec.	1014.2	25.3	30.3	21.0	9.3	34.4	12.8	72
Year	1011.6	26.5	30.6	22.8	7.8	37.2	11.7	80
Years of records	10	60	60	29	29	60	60	50

Month	Precipitation (mm)				Mean number of days with precipitation	Mean evaporation (mm)	Mean cloud cover 0–10	Mean sunshine hours	Most freq. wind direct.
	mean	max.	min.	max. in 24 hours					
Jan.	25	99	0	56	1	—	—	—	—
Feb.	51	142	0	76	3	—	—	—	—
Mar.	79	254	0	94	5	—	—	—	—
Apr.	127	586	15	137	7	—	—	—	—
May	424	925	66	155	18	—	—	—	—
June	762	1151	391	244	25	—	—	—	—
July	836	1359	333	239	26	—	—	—	—
Aug.	762	1176	493	178	26	—	—	—	—
Sept.	632	975	396	165	23	—	—	—	—
Oct.	307	569	135	119	16	—	—	—	—
Nov.	97	191	0	132	6	—	—	—	—
Dec.	20	58	0	137	2	—	—	—	—
Year	4122	n.a.	n.a.	244	158	—	—	—	—
Years of records	60	33	33	65	60	—	—	—	—

TABLE XXIX

CLIMATIC TABLE FOR PHNOM PENH
Latitude 11°33′N, longitude 104°55′E, elevation 12 m

Month	Mean press. at sea level (mbar)	Temperature (°C)						Mean relative humidity (%)
		mean	mean daily max.	mean daily min.	mean daily range	abs. max.	abs. min.	
Jan.	1012.4	26.0	30.7	21.3	9.4	35.3	14.0	71.1
Feb.	1011.4	27.4	32.1	22.0	10.1	36.7	15.2	71.5
Mar.	1010.2	28.7	33.6	23.2	10.4	39.0	19.0	70.1
Apr.	1009.0	29.4	34.6	24.3	10.3	40.5	20.0	72.5
May	1007.8	28.3	33.5	24.3	9.2	38.0	20.6	81.5
June	1007.3	27.9	32.7	24.3	8.4	38.4	21.2	81.0
July	1007.2	27.3	31.6	24.1	7.5	36.6	20.1	83.2
Aug.	1007.2	27.7	31.7	24.7	7.0	36.0	22.0	83.3
Sept.	1008.0	27.2	30.9	24.7	6.2	35.5	22.0	84.9
Oct.	1009.7	26.9	30.4	24.4	6.0	33.8	21.0	83.3
Nov.	1010.8	26.6	30.1	23.3	6.8	34.0	18.0	78.5
Dec.	1012.0	25.8	30.0	21.8	8.2	34.8	14.4	74.2
Year	1009.4	27.4	31.8	23.5	8.3	40.5	14.0	77.9
Years of records	19	9	22	22	22	22	22	9

Month	Precipitation (mm)				Mean number of days with precipitation	Mean evaporation (mm)	Mean cloud cover 0–10	Mean sunshine hours	Most freq. wind direct.
	mean	max.	min.	max. in 24 hours					
Jan.	7	51	0	28	1	100	3.5	8.4	N–NE
Feb.	10	127	0	71	1	94	3.6	8.0	S–SE
Mar.	40	193	0	142	3	123	3.6	8.6	S–SE
Apr.	77	359	0	104	6	113	4.5	8.0	S–SE
May	134	318	30	99	14	79	5.7	6.5	SE–SW
June	155	393	27	94	15	78	6.0	6.4	SW
July	171	359	55	99	16	75	7.0	4.6	SW
Aug.	160	380	66	140	16	74	6.4	5.6	SW
Sept.	224	443	93	109	19	56	7.1	4.3	SW
Oct.	257	650	63	157	17	62	6.2	6.5	N–NE
Nov.	127	298	2	109	9	74	5.7	7.1	N–NE
Dec.	45	176	0	64	4	88	4.6	7.8	N–NE
Year	1407	2310	969	157	121	1016	5.3	6.8	—
Years of records	32	32	32	32	32	10	9	5	n.a.

TABLE XXX

CLIMATIC TABLE FOR HO CHI MINH CITY
Latitude 10°47′N, longitude 106°42′E, elevation 11 m

Month	Mean press. at sea level (mbar)	Temperature (°C)						Mean relative humidity (%)
		mean	mean daily max.	mean daily min.	mean daily range	abs. max.	abs. min.	
Jan.	1012.2	25.8	31.5	21.0	10.5	36.4	13.8	76.5
Feb.	1012.0	26.7	32.9	21.8	11.1	38.7	16.2	74.5
Mar.	1010.6	27.8	34.0	23.3	10.7	39.4	17.5	73.5
Apr.	1009.5	28.9	34.8	24.7	10.1	40.0	20.0	76.0
May	1008.6	28.0	33.3	24.5	8.8	39.0	21.1	83.2
June	1008.3	27.2	31.9	23.8	8.1	37.5	20.4	85.9
July	1007.9	26.7	31.0	23.7	7.3	34.6	19.4	86.9
Aug.	1008.4	27.0	31.3	23.8	7.5	34.9	20.0	86.2
Sept.	1008.8	26.6	31.1	23.6	7.5	35.3	20.8	87.7
Oct.	1009.9	26.5	30.9	23.4	7.5	34.6	19.8	86.7
Nov.	1010.9	26.3	30.8	22.7	8.1	35.0	17.8	83.6
Dec.	1011.3	25.8	30.6	21.6	9.0	36.3	13.9	80.6
Year	1009.1	26.9	32.0	23.2	8.8	40.0	13.8	81.8
Years of records	n.a.	10	30	30	30	30	30	14

Month	Precipitation (mm)				Mean number of days with precipitation	Mean evapo-ration (mm)	Mean cloud cover 0–10	Mean sunshine hours	Most freq. wind direct.
	mean	max.	min.	max. in 24 hours					
Jan.	16	111	0	69	2	94	5.3	6.3	E
Feb.	3	10	0	10	1	106	4.6	7.1	SE
Mar.	13	129	0	104	2	135	4.9	6.8	SE
Apr.	42	178	0	89	5	126	5.8	6.7	SE
May	220	561	49	104	17	78	7.2	5.1	SW
June	331	522	180	137	22	66	7.9	5.0	SW–W
July	314	595	98	150	23	64	8.2	3.9	SW
Aug.	269	499	118	178	21	67	7.9	5.0	SW
Sept.	336	507	204	132	22	54	8.1	4.0	SW
Oct.	269	603	82	114	20	52	7.4	4.5	SW
Nov.	115	286	3	132	11	59	6.7	5.2	Variable
Dec.	56	173	13	71	7	76	6.3	5.7	N–E
Year	1984	2718	1552	178	153	977	6.7	5.4	—
Years of records	32	32	32	32	32	9	32	8	n.a.

TABLE XXXI

CLIMATIC TABLE FOR HA-TIEN
Latitude 10°23′N, longitude 104°29′E, elevation 4 m

Month	Mean press. at sea level (mbar)	Temperature (°C)						Mean relative humidity (%)
		mean	mean daily max.	mean daily min.	mean daily range	abs. max.	abs. min.	
Jan.	—	25.8	30.0	21.6	8.4	32.5	15.4	79.1
Feb.	—	26.5	30.4	22.6	7.8	34.1	19.1	79.7
Mar.	—	27.1	30.8	23.4	7.4	34.8	19.4	79.1
Apr.	—	27.9	31.2	24.6	6.6	33.1	19.6	80.6
May	—	28.0	31.1	24.9	6.2	34.5	21.0	83.6
June	—	27.6	30.4	24.8	5.6	32.8	20.2	83.4
July	—	27.1	29.7	24.4	5.3	32.1	20.9	84.5
Aug.	—	27.1	29.8	24.3	5.5	32.7	20.7	84.5
Sept.	—	26.9	29.6	24.3	5.3	32.6	20.9	85.3
Oct.	—	26.7	29.5	23.9	5.6	32.0	20.1	84.3
Nov.	—	26.4	29.7	23.1	6.6	32.1	17.1	83.1
Dec.	—	25.7	29.5	22.0	7.5	32.0	17.1	81.6
Year	—	26.9	30.1	23.7	6.4	34.8	15.4	82.4
Years of records	n.a.	11	32	32	32	32	32	9

Month	Precipitation (mm)				Mean number of days with precipitation	Mean evaporation (mm)	Mean cloud cover 0–10	Mean sunshine hours	Most freq. wind direct.
	mean	max.	min.	max. in 24 hours					
Jan.	12	138	0	56	2	100	5.5	—	NE
Feb.	14	97	0	38	1	83	5.6	—	NE, SE
Mar.	50	421	0	234	3	100	6.4	—	SE, SW
Apr.	136	345	0	99	8	87	7.4	—	SW, SE
May	231	504	46	104	14	74	8.0	—	SW
June	240	415	50	180	14	70	8.2	—	W, SW
July	311	613	84	185	16	64	8.4	—	W, SW
Aug.	274	549	40	173	15	68	8.1	—	W, SW
Sept.	254	521	33	112	15	62	8.5	—	SW, W
Oct.	240	432	26	109	14	71	7.9	—	NE
Nov.	129	370	0	130	9	83	7.1	—	NE
Dec.	47	261	0	89	4	93	6.5	—	NE
Year	1938	2929	1069	234	115	955	7.3	—	—
Years of records	32	32	32	32	32	12	11	n.a.	n.a.

TABLE XXXII

CLIMATIC TABLE FOR VICTORIA POINT
Latitude 9°59′N, longitude 98°33′E, elevation 48 m

Month	Mean press. at sea level (mbar)	Temperature (°C)						Mean relative humidity (%)
		mean	mean daily max.	mean daily min.	mean daily range	abs. max.	abs. min.	
Jan.	1013.8	26.6	30.2	23.0	7.2	32.2	17.8	68
Feb.	1013.1	27.4	31.3	23.5	7.8	35.6	18.9	66
Mar.	1012.7	28.2	31.9	24.5	7.3	33.9	21.7	68
Apr.	1011.2	28.6	31.9	25.2	6.7	34.4	21.7	73
May	1010.4	27.3	30.1	24.6	5.5	33.9	21.7	82
June	1010.1	26.4	28.9	24.1	4.8	31.1	21.7	85
July	1010.4	26.2	28.4	23.9	4.6	30.0	21.1	86
Aug.	1010.4	26.2	28.5	23.9	4.6	31.1	21.1	86
Sept.	1011.4	25.9	28.2	23.5	4.7	30.6	21.7	86
Oct.	1012.4	26.1	28.6	23.6	5.0	31.7	21.1	83
Nov.	1012.6	26.4	29.2	23.6	5.6	31.1	20.6	79
Dec.	1013.6	26.2	29.3	23.0	6.3	31.1	19.4	71
Year	1011.8	26.8	29.7	23.8	5.9	35.6	17.8	78
Years of records	10	30	30	28	28	30	30	30

Month	Precipitation (mm)				Mean number of days with precipitation	Mean evapo-ration (mm)	Mean cloud cover 0–10	Mean sunshine hours	Most freq. wind direct.
	mean	max.	min.	max. in 24 hours					
Jan.	18	—	—	31	1	—	—	—	—
Feb.	10	—	—	51	1	—	—	—	—
Mar.	56	—	—	84	4	—	—	—	—
Apr.	130	—	—	231	8	—	—	—	—
May	503	—	—	254	19	—	—	—	—
June	719	—	—	173	22	—	—	—	—
July	732	—	—	175	24	—	—	—	—
Aug.	663	—	—	158	22	—	—	—	—
Sept.	711	—	—	213	22	—	—	—	—
Oct.	447	—	—	178	18	—	—	—	—
Nov.	160	—	—	102	11	—	—	—	—
Dec.	58	—	—	119	4	—	—	—	—
Year	4206	—	—	254	156	—	—	—	—
Years of records	30	n.a.	n.a.	35	30	—	—	—	—

TABLE XXXIII

CLIMATIC TABLE FOR BANDON
Latitude 9°07′N, longitude 99°17′E, elevation 3 m

Month	Mean press. at sea level (mbar)	Temperature (°C)						Mean relative humidity (%)
		mean	mean daily max.	mean daily min.	mean daily range	abs. max.	abs. min.	
Jan.	—	25.9	31.3	20.5	10.8	36.0	12.4	81.7
Feb.	—	26.9	33.4	20.4	13.0	37.1	16.2	78.2
Mar.	—	28.1	34.9	21.2	13.7	39.0	15.5	75.3
Apr.	—	29.0	35.2	22.8	12.4	39.5	19.6	79.1
May	—	28.7	33.8	23.4	10.4	38.5	19.8	83.5
June	—	28.1	32.6	23.4	9.2	36.4	20.9	82.5
July	—	27.8	32.3	23.2	9.1	36.5	20.6	82.1
Aug.	—	27.8	32.5	23.1	9.4	36.7	20.8	82.2
Sept.	—	27.6	32.2	22.8	9.4	37.4	20.6	84.2
Oct.	—	27.1	31.5	22.7	8.8	35.5	20.6	86.5
Nov.	—	26.3	30.3	22.4	7.9	35.2	16.3	86.8
Dec.	—	25.6	29.9	21.8	8.1	34.2	16.4	85.4
Year	—	27.4	32.5	22.3	10.2	39.5	12.4	82.3
Years of records	n.a.	20	20	20	20	20	20	19

Month	Precipitation (mm)				Mean number of days with precipitation	Mean evaporation (mm)	Mean cloud cover 0–10	Mean sunshine hours	Most freq. wind direct.
	mean	max.	min.	max. in 24 hours					
Jan.	63	—	—	168	7	—	—	—	NE
Feb.	28	—	—	15	4	—	—	—	N–E
Mar.	30	—	—	61	4	—	—	—	N
Apr.	95	—	—	74	10	—	—	—	N
May	204	—	—	107	16	—	—	—	SW
June	145	—	—	53	16	—	—	—	SW
July	131	—	—	66	15	—	—	—	SW
Aug.	144	—	—	71	17	—	—	—	SW
Sept.	197	—	—	79	19	—	—	—	SW
Oct.	269	—	—	135	20	—	—	—	SW
Nov.	312	—	—	171	18	—	—	—	N, NE
Dec.	240	—	—	168	13	—	—	—	N, NE
Year	1858	—	—	171	159	—	—	—	—
Years of records	30	n.a.	n.a.	13	30	n.a.	n.a.	n.a.	n.a.

TABLE XXXIV

CLIMATIC TABLE FOR ALOR STAR
Latitude 6°12'N, longitude 100°25'E, elevation 4 m

Month	Mean press. at sea level (mbar)	Temperature (°C)						Mean relative humidity (%)
		mean	mean daily max.	mean daily min.	mean daily range	abs. max.	abs. min.	
Jan.	1010.3	26.0	31.7	21.4	10.3	35.6	16.7	77.7
Feb.	1009.9	26.8	33.4	21.4	11.9	37.2	16.7	75.3
Mar.	1009.1	27.2	33.8	22.4	11.4	37.2	18.3	77.7
Apr.	1008.7	27.3	32.9	23.4	9.6	37.2	20.0	83.1
May	1008.4	27.3	31.8	23.9	7.9	35.6	22.2	85.0
June	1008.6	27.0	31.5	23.6	7.9	34.4	21.1	84.5
July	1008.6	26.7	31.2	23.3	7.9	34.4	21.1	84.5
Aug.	1008.9	26.6	30.9	23.3	7.6	34.4	21.1	85.1
Sept.	1009.6	26.3	30.6	23.4	7.2	33.9	21.1	86.1
Oct.	1009.8	26.1	30.5	23.2	7.3	34.4	21.1	87.7
Nov.	1009.6	25.8	30.6	22.7	7.8	33.9	17.8	86.9
Dec.	1010.2	25.7	30.9	21.8	9.1	34.4	17.2	82.6
Year	1009.3	26.6	31.7	22.8	8.8	37.2	16.7	83.0
Years of records	15	23	24	24	24	24	24	23

Month	Precipitation (mm)				Mean number of days with precipitation	Mean evaporation (mm)	Mean cloud cover 0–10	Mean sunshine hours	Most freq. wind direct.
	mean	max.	min.	max. in 24 hours					
Jan.	58	171	1	—	7	—	6.5	8.0	NE, E
Feb.	52	159	0	—	5	—	6.4	8.6	NE
Mar.	143	417	9	—	12	—	6.2	8.3	NE
Apr.	238	491	73	—	18	—	7.5	8.0	Variable
May	258	483	56	—	18	—	7.8	7.0	W
June	184	480	76	—	15	—	8.0	6.5	W
July	198	551	52	—	17	—	8.0	6.5	W
Aug.	256	470	76	—	17	—	8.3	6.4	W
Sept.	309	588	84	—	19	—	8.4	5.7	W
Oct.	310	602	144	—	23	—	8.3	5.6	Variable
Nov.	216	457	69	—	19	—	8.1	5.7	NE
Dec.	120	412	16	—	11	—	6.7	6.8	NE
Year	2342	3379	1608	166	181	—	7.6	6.9	—
Years of records	42	42	42	22	19	n.a.	10	23	10

TABLE XXXV

CLIMATIC TABLE FOR KOTA BHARU
Latitude 6°10′N, longitude 102°17′E, elevation 4 m

Month	Mean press. at sea level (mbar)	Temperature (°C)						Mean relative humidity (%)
		mean	mean daily max.	mean daily min.	mean daily range	abs. max.	abs. min.	
Jan.	1011.6	25.3	28.8	22.4	6.4	31.1	16.7	84.2
Feb.	1011.0	25.5	29.8	21.9	7.9	32.8	17.8	83.6
Mar.	1010.0	26.4	31.2	22.5	8.7	34.4	17.8	82.8
Apr.	1009.0	27.3	32.3	23.4	8.9	35.6	18.3	82.4
May	1008.2	27.3	32.4	23.8	8.6	35.6	21.7	82.8
June	1008.3	27.0	32.1	23.4	8.7	34.4	21.1	83.1
July	1008.4	26.7	31.8	23.2	8.6	34.4	20.6	83.5
Aug.	1008.7	26.6	31.8	23.1	8.7	34.4	21.1	83.5
Sept.	1009.2	26.4	31.5	23.1	8.4	34.4	20.6	84.1
Oct.	1010.3	26.1	30.6	23.2	7.4	33.3	20.6	86.4
Nov.	1010.0	25.6	29.1	22.9	6.2	33.3	18.9	88.1
Dec.	1011.0	25.2	28.4	22.6	5.9	32.2	18.9	86.6
Year	1009.6	26.3	30.8	22.9	7.9	35.6	16.7	84.3
Years of records	19	22	21	22	21	20	22	22

Month	Precipitation (mm)				Mean number of days with precipitation	Mean evaporation (mm)	Mean cloud cover 0–10	Mean sunshine hours	Most freq. wind direct.
	mean	max.	min.	max. in 24 hours					
Jan.	251	1020	21	—	17	118	7.9	6.1	E, NE
Feb.	139	786	0	—	9	135	6.5	8.1	E
Mar.	172	749	0	—	10	165	6.5	8.4	E, NE
Apr.	116	858	4	—	9	175	6.4	8.7	E, NE
May	160	338	12	—	13	158	7.6	7.4	SW, S
June	152	342	48	—	12	145	7.9	7.0	SW, S
July	142	322	43	—	14	150	7.7	7.0	SW, S
Aug.	165	352	52	—	14	150	8.0	7.1	SW, S
Sept.	213	346	82	—	17	145	8.1	6.4	SW, S
Oct.	259	398	105	—	20	118	8.4	5.8	SW
Nov.	615	1230	199	—	21	91	8.4	4.8	E, NE
Dec.	659	1528	66	—	22	94	8.1	5.0	E, NE
Year	3053	4934	2155	358	179	1644	7.6	6.8	—
Years of records	44	44	44	20	18	9	10	22	5

TABLE XXXVI

CLIMATIC TABLE FOR KUALA LUMPUR
Latitude 3°07′N, longitude 101°42′E, elevation 34 m

Month	Mean press. at sea level (mbar)	Temperature (°C)						Mean relative humidity (%)
		mean	mean daily max.	mean daily min.	mean daily range	abs. max.	abs. min.	
Jan.	1010.3	25.8	31.9	22 1	9.8	35.6	17.8	82.4
Feb.	1010.0	26.3	32.9	22.2	10.7	36.1	18.9	81.3
Mar.	1009.5	26.5	33.3	22.7	10.6	36.1	20.0	83.0
Apr.	1009.0	26.6	33.0	23.1	9.9	35.6	20.6	85.3
May	1008.6	26.8	32.7	23.2	9.5	36.1	20.6	85.1
June	1009.0	26.7	32.7	22.8	9.9	35.6	20.0	82.9
July	1009.1	26.3	32.3	22.2	10.1	36.7	18.3	82.2
Aug.	1009.2	26.3	32.3	23.4	8.9	36.7	19.4	82.6
Sept.	1009.7	26.1	32.1	22.3	9.8	35.6	19.4	83.8
Oct.	1009.9	25.9	31.9	22.5	9.4	35.6	20.6	85.4
Nov.	1009.7	25.8	31.7	22.6	9.2	35.0	20.6	86.4
Dec.	1009.6	25.8	31.7	22.3	9.4	35.0	19.4	84.4
Year	1009.5	26.2	32.4	22.5	9.9	36.7	17.8	83.8
Years of records	20	24	24	24	24	24	24	23

Month	Precipitation (mm)				Mean number of days with precipitation	Mean evaporation (mm)	Mean cloud cover 0–10	Mean sunshine hours	Most freq. wind direct.
	mean	max.	min.	max. in 24 hours					
Jan.	171	440	18	—	16	126	8.0	6.2	NE
Feb.	169	432	20	—	15	128	7.6	7.4	NE
Mar.	237	451	91	—	17	142	8.0	6.5	Variable
Apr.	279	509	110	—	21	137	8.4	6.3	Variable
May	216	479	67	—	17	142	8.2	6.3	Variable
June	126	273	11	—	13	122	8.1	6.6	S
July	102	312	5	—	11	126	8.1	6.5	S
Aug.	157	370	33	—	14	118	8.6	6.3	S
Sept.	188	403	47	—	17	130	8.6	5.6	S
Oct.	275	595	75	—	21	118	8.9	5.3	Variable
Nov.	259	567	90	—	22	114	8.9	4.9	W
Dec.	230	488	66	—	18	118	8.4	5.4	NE
Year	2409	3149	1663	130	204	1521	8.4	6.1	—
Years of records	69	69	69	19	24	9	10	17	4

TABLE XXXVII

CLIMATIC TABLE FOR MERSING
Latitude 2°27′N, longitude 103°50′E, elevation 46 m

Month	Mean press. at sea level (mbar)	Temperature (°C)						Mean relative humidity (%)
		mean	mean daily max.	mean daily min.	mean daily range	abs. max.	abs. min.	
Jan.	1010.5	25.5	27.7	22.4	5.3	31.7	17.8	85.5
Feb.	1010.5	25.8	28.7	22.9	5.8	31.1	18.3	84.3
Mar.	1009.7	26.2	29.6	22.9	6.7	32.8	19.4	85.1
Apr.	1009.2	26.4	30.9	22.8	8.1	34.4	20.6	86.4
May	1008.7	26.3	31.3	22.9	8.4	35.0	20.0	87.4
June	1009.7	26.1	31.1	22.6	8.5	34.4	20.6	87.3
July	1009.1	25.8	30.8	22.3	8.6	33.9	19.4	86.6
Aug.	1009.4	25.6	30.7	22.2	8.6	33.9	20.0	87.4
Sept.	1009.9	25.6	30.8	22.1	8.7	33.9	19.4	87.4
Oct.	1010.0	25.6	30.6	22.3	8.3	34.4	20.0	87.7
Nov.	1009.7	25.3	29.4	22.4	7.0	33.9	20.6	88.4
Dec.	1010.4	25.4	28.0	22.8	5.2	33.9	20.0	86.9
Year	1009.7	25.8	29.9	22.6	7.3	35.0	17.8	86.7
Years of records	15	25	25	25	25	25	25	24

Month	Precipitation (mm)				Mean number of days with precipitation	Mean evaporation (mm)	Mean cloud cover 0–10	Mean sunshine hours	Most freq. wind direct.
	mean	max.	min.	max. in 24 hours					
Jan.	388	820	171	—	19	—	8.0	5.3	N, NE
Feb.	229	642	2	—	12	—	7.3	7.0	N, NE
Mar.	204	557	18	—	15	—	6.7	7.3	N, NE
Apr.	138	392	15	—	15	—	6.6	7.3	W
May	155	313	60	—	16	—	7.0	6.9	W, SW
June	148	277	62	—	14	—	7.1	6.6	SW, W
July	163	335	50	—	15	—	7.4	6.6	SW, W
Aug.	168	276	44	—	15	—	7.5	6.5	SW, W
Sept.	158	275	61	—	16	—	7.2	6.0	SW, W
Oct.	220	615	93	—	18	—	7.4	5.6	W, SW
Nov.	335	622	147	—	22	—	8.1	4.8	W
Dec.	516	869	143	—	22	—	7.8	4.7	N, NE
Year	2821	4012	1917	290	198	—	7.4	6.2	—
Years of records	32	32	32	21	20	n.a.	10	24	10

TABLE XXXVIII

CLIMATIC TABLE FOR SINGAPORE
Latitude 1°21′N, longitude 103°54′E, elevation 8 m

Month	Mean press. at sea level (mbar)	Temperature (°C)						Mean relative humidity (%)
		mean	mean daily max.	mean daily min.	mean daily range	abs. max.	abs. min.	
Jan.	1010.2	25.6	29.9	22.9	6.9	33.3	19.4	85.1
Feb.	1010.2	26.1	30.8	23.1	7.7	34.4	19.4	83.3
Mar.	1009.4	26.5	31.1	23.6	7.5	34.4	20.0	84.2
Apr.	1009.0	27.1	31.2	24.1	7.1	34.4	21.1	84.6
May	1008.7	27.3	30.9	24.5	6.4	35.0	21.7	84.4
June	1008.9	27.6	30.8	24.7	6.1	33.9	20.6	82.5
July	1009.1	27.3	30.6	24.7	5.9	33.9	21.1	82.0
Aug.	1009.3	27.1	30.5	24.4	6.1	33.9	20.0	82.5
Sept.	1009.7	27.0	30.4	24.2	6.2	33.3	20.6	83.1
Oct.	1009.8	26.7	30.4	23.9	6.4	33.3	20.6	84.1
Nov.	1009.5	26.2	30.1	23.7	6.4	33.3	21.1	86.0
Dec.	1010.0	25.7	29.9	23.2	6.7	33.9	20.6	86.0
Year	1009.5	26.7	30.6	23.9	6.7	35.0	19.4	83.9
Years of records	18	23	23	23	23	23	23	22

Month	Precipitation (mm)				Mean number of days with precipitation	Mean evaporation (mm)	Mean cloud cover 0–10	Mean sunshine hours	Most freq. wind direct.
	mean	max.	min.	max. in 24 hours					
Jan.	252	819	51	—	17	150	8.2	5.1	NE, N
Feb.	175	567	11	—	12	157	7.9	6.5	NE, N
Mar.	200	528	19	—	14	165	7.5	6.1	NE, N
Apr.	196	455	38	—	15	145	7.7	5.9	Variable
May	174	394	59	—	15	150	7.8	5.9	S
June	171	379	57	—	13	137	7.9	6.2	SE, SW
July	167	527	26	—	13	150	7.6	6.4	S, SE
Aug.	191	527	18	—	15	150	8.4	6.0	S, SE
Sept.	179	427	61	—	14	152	8.4	5.6	S, SE
Oct.	208	497	32	—	16	142	8.4	5.3	S
Nov.	251	521	91	—	19	122	8.5	4.6	Variable
Dec.	266	681	63	—	19	118	8.4	4.7	N, NE
Year	2430	3452	1606	277	183	1738	8.0	5.7	—
Years of records	86	86	86	13	71	9	10	23	10

Chapter 2

The Climate of the Indian Subcontinent

Y. P. RAO

Introduction

India, Pakistan, Bangladesh, Nepal, Bhutan and Sri Lanka form the Indian subcontinent. Nowhere else is the geographical separation of a vast region from the rest of the continent so complete. The lofty Himalayas to the north, the less high mountains to the northeast and northwest and the surrounding seas to the south isolate the region. The same geographical features modify the climate to a characteristic pattern.

The monsoon, the seasonal reversal of winds and associated rains, develops most prominently over the Indian subcontinent and adjacent seas. The average annual rainfall of about 100 cm over the plains is high for a region lying mostly in the subtropical high-pressure belt. Contrasts in climate are very striking. In almost the same latitude occur the heaviest rainfall in the world at Cherrapunji and an amount as low as 8 cm per year in the Thar Desert. Every type of climate, from equatorial to Alpine, is found in some part or the other. The Alpine climate occurs in the Himalayas within 400 km of the Tropic of Cancer.

The northern plains of the subcontinent experience higher temperatures than other land areas in the same latitude. The searing heat is mostly not in mid-summer but in spring. Perhaps this distortion of seasons made the ancient Indians classify the climate into six seasons. The mean annual range of temperature is 44°C at Bikaner and 65°C at Dras (3,066 m). Mean daily range varies from 4°C at Cochin in August to 18°C at Khanpur in May. While during the southwest monsoon relative humidity is over 80% over a large area, air is very dry at the height of summer in the northwest.

Physiography

Perhaps in no other part of the world has physiography determined the climate as in the Indian subcontinent. The significant features are the very high mountains in the north with broad plains below them, the tapering peninsula to the south, hill ranges to the northwest and northeast and a lower range running along the west coast of the peninsula. Fig.2 shows a schematic representation of the physiography.

The Great Himalaya runs unbroken from about 35°N and 74°E, southeastwards to 28°N and 87°E and thence to 29.5°N and 95°E. The height is generally above 4 km, frequently above 6 km. Within about 150 km in the western and central parts, there are two other parallel ranges successively to the south, the Lesser Himalaya, the Mahabharat

67

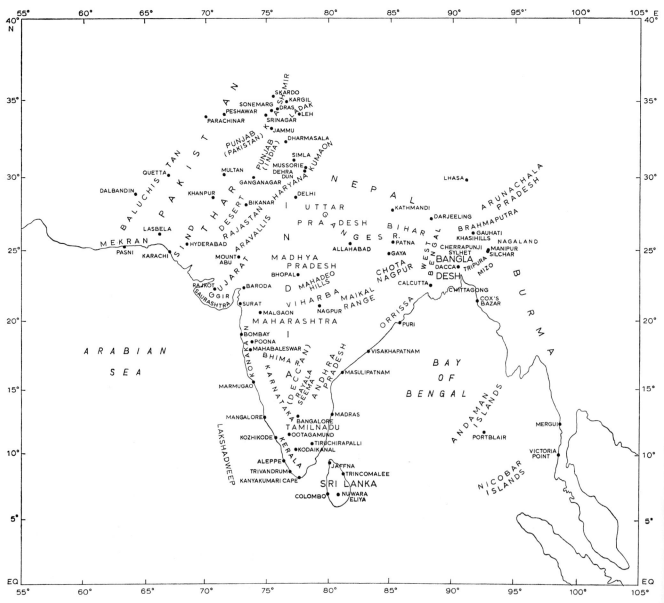

Fig.1. Location map.

and the Siwalik ranges. The heights range between 2 and 4 km in the former while the Siwaliks are not more than 2 km.

On the northern side of the Great Himalaya, the Karakoram and Kailas ranges run from 74°E to 84°E, at a distance of 200 km. The heights are upto 6 km. The parallel Ladakh and Zaskar ranges are in between, from 76° to 79°E and reach 4–6 km in height. The Hindu Kush and Kunlun mountains form another barrier between 67° and 85°E along the 36°N latitude.

Between the Karakoram to the west, Great Himalaya to the south, Kunlun Mountains to the north and Nan Shan to the northeast, lies the highest and largest plateau of the world, above 5 km. Lying between 30° and 40°N and 75° and 105°E, it makes an unmistakable imprint on the climate of the Indian subcontinent and Asia. Where the Hindu Kush and Kunlun mountains stop at about 36°N, the Tien Shan Mountains extend the

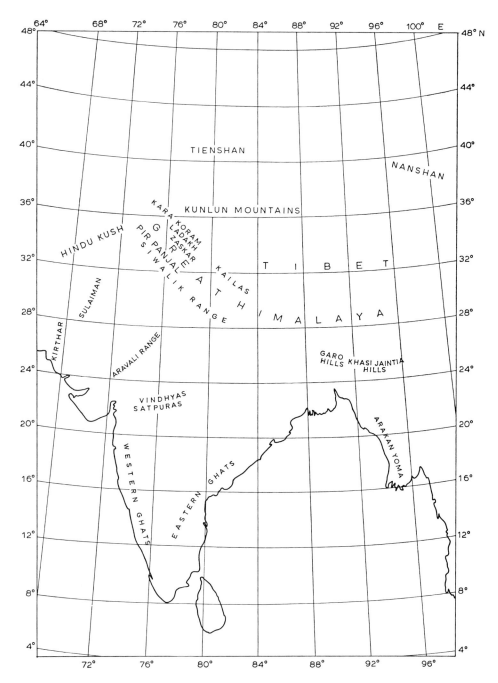

Fig.2. Schematic representation of the physiography in and around the Indian subcontinent.

barrier further north upto 43°N. Thus between 27°/35°N and 40°/43°N, from 70° to 105°E, there is a mountain wall preventing atmospheric exchange in the lower troposphere between the Indian subcontinent and the higher latitudes.

From the Hindu Kush near 36°N 72°E, a series of ranges run south-southwest almost upto the coast, near Karachi. Of these the Sulaiman and Kirthar ranges are the easternmost. These and the plateau to their west form a continuous barrier at least up to 1.5 km. From 28°N 97°E (near the eastern end of the Great Himalaya) the Patkai Dum, Naga

Hills and the Arakan Yoma (in Burma) form the eastern barrier, about 2 km in height, up to 18°N.

The northern plain of the Indian subcontinent is thus flanked by mountain and hill ranges on two or three sides. SIMPSON (1921) compared this to a box with two sides.

The triangularly shaped peninsula, 19° broad near the Tropic of Cancer, has as its vertex in the south Cape Kanniyakumari (8°N 78°E). This shape of the peninsula seems very important in allowing maritime air mass to reach the northern plains. Another important physical feature, the Western Ghats, the hill ranges within 100 km of the west coast of the peninsula, running from 21°N 74°E to almost the southern tip of India, and rising to heights of 1 km, all along its length, may have an overall influence on the climate of the subcontinent.

The less continuous Eastern Ghats within 200 km of the east coast have only local effect. So also are the Aravalli Range in the western part of the northern plains, the Garo and Khasi–Jaintia Hills which link up with the Naga Hills in the east, and the parallel ranges of Vindhya and Satpura which form the natural northern boundary of the peninsula.

Seasons

Before describing the distribution of climatic elements over the country, it is necessary to know the rather peculiar division of the year into seasons. Over most parts, the southwest monsoon rainfall is one order of magnitude more than in the rest of the year, so much so that both the annual and monsoon seasonal isohyets are very similar. Seasonal reversal of pressure gradients and winds makes patterns of annual averages misleading. The gradual rise in temperatures through spring to summer does not happen as, with the onset of the southwest monsoon, temperatures drop sharply in June or July. The monsoon period of four months from June to September is out of step with the three-monthly seasons. The usual classification into spring, summer, autumn and winter is, therefore, not adopted. January–February is called the winter period. The hot weather period is from March to May. The four months of June–September are called the southwest monsoon period and October–December form the post-monsoon period. In the northern and western parts of the subcontinent, December is like winter but in the extreme south the "northeast monsoon rains" still persist. The suffix "period" is intended to bring out the distinction from normal seasons.

Sea-level pressure

The distribution of sea-level pressure over the Indian subcontinent undergoes a complete reversal from January to July. January, the month of the northeast monsoon, has pressure decreasing to south. The gradients become strong and are directed towards the north in July, the southwest monsoon month. In the transition months of April and October, the pressure gradient over the subcontinent is flat.

January

In this month, the Indian subcontinent is at the periphery of the Siberian high centred at 45°N and 105°E. The Siberian high is the result of accumulation of cold continental air over the east central parts of Asia and the effect of subtropical high pressure belt which becomes prominent over land. The Himalayas obstruct the spreading of cold air from central Asia into northern India. The Siberian high is marked in December, January and February. The pressure gradient round the Siberian high is strong to the north of the Himalayas but weak over India.

The position of the equatorial low pressure area in the Indian Ocean in this month is between 10° and 15°S.

In the Indian area (Fig.3) a weak ridge runs from north Pakistan to Bihar with a weak

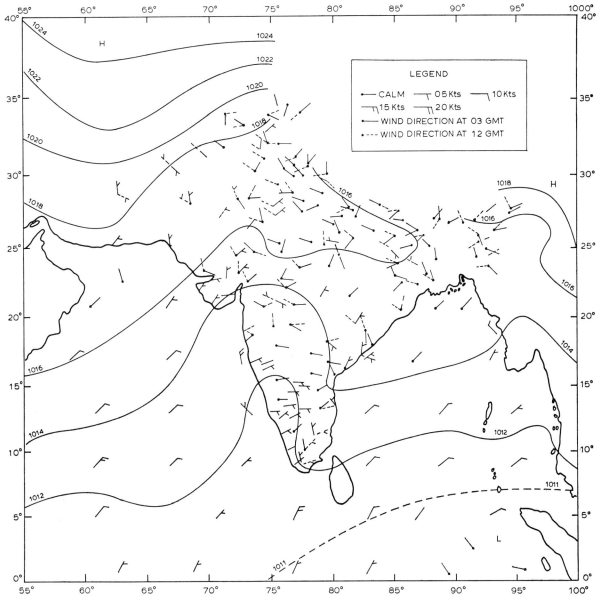

Fig.3. Mean pressure (mbar) and surface wind distribution, January.

71

trough to its north along the foot of the Himalayas. A marked trough extends from Kerala to Sind and a similar one from off the Tenasserim coast to central Burma. The ridge between them is prominent along the east coast of the peninsula.

April

Development of low pressure due to increased heating over land starts for India in March, when the whole area has a flat pressure distribution with slightly higher pressures over the Arabian Sea and the Bay of Bengal. By April the land lows have begun establishing themselves along about 10°N in north Africa and about the Tropic of Cancer in the Indian region and Burma. Peninsular India south of 20°N comes under considerable maritime influence. This makes the heating over land more marked to the north of

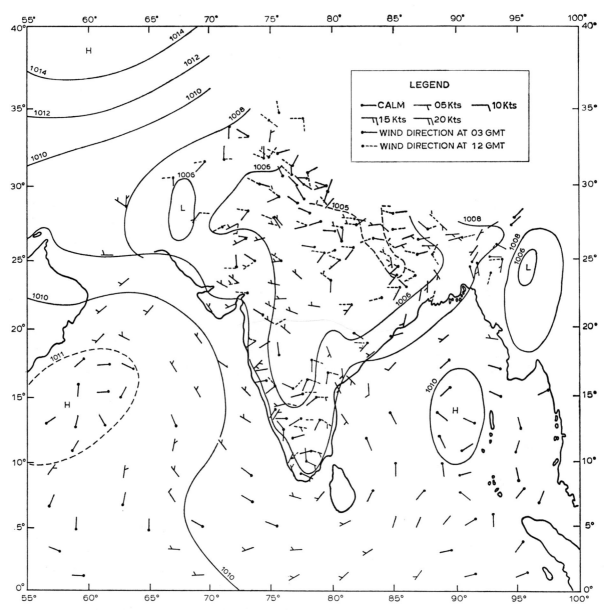

Fig.4. Mean pressure (mbar) and surface wind distribution, April.

20°N and the low is at a more northerly latitude over India than over Africa (Fig.4). There is a weak low over upper Sind and another over Bihar and east Uttar Pradesh. Over the peninsula a trough forms with the axis along longitude 78°E.

A ridge runs from Arabia into the west Arabian Sea where a weak high circulation is found around 14°N and 60°E. A similar circulation is also present over the Bay of Bengal around 13°N and 90°E.

In the Indian Ocean a weak trough is present at 2°–3°N and a better defined one at 8°S. By May, the summer continental low pressure area completely dominates the scene. Its main central area is over Sind and West Rajasthan and the low pressure extends as a trough to Orissa. This heat low becomes still more marked in June with the central area over Pakistan. In May the trough over the peninsula is along 79°E, the Madras coast. In these months the Indian Ocean trough is at about 3°–5°S.

July

The low pressure area extending from North Africa to Northeast Siberia is most intense. The lowest pressure is around upper Sind and surroundings (Fig.5). The axis of this low pressure belt runs from southwest to northeast. A trough lies over north India with its axis from Ganganager to the northern part of the Bay of Bengal which is referred to as the monsoon trough. The position of this monsoon trough at any time becomes an index of the activity of the monsoon over the subcontinent. The Indian Ocean high is strong and centred at 30°S and about 60°E. Pressure steadily decreases over the Indian Ocean northwards of this high pressure belt except for a weak trough at 2°S. Weak ridges are present in the Arabian Sea off the west coast of India and in the Bay of Bengal off the Tenasserim coast and over Burma. The weak trough along the east coast of the south peninsula persists throughout the monsoon months and is more pronounced in September.

October

Rather abruptly a trough develops over the Bay of Bengal with its axis along 13°/14°N and the pressure field is flat over the country (Fig.6). The trough in the south of the bay persists in November but disappears in December. The trough near 2°S continues in October.

Surface winds

January

Surface winds (Fig.2) are from between north and east (mainly northeast) to the south of about 25°N both over land and sea. In the central parts of the peninsula some south-easterlies also occur. North of about 25°N the wind direction is mainly west to northwest except for easterlies over Assam. Mean wind speed is generally low over the land and about 10 knots over the sea, getting a little stronger over the southwest Arabian Sea off the African coast. Light westerlies prevail very near the Equator to the east of 90°E.

April

A striking feature is the replacement of earlier northeasterlies by westerlies. Only in southwest Arabian Sea do some winds blow between north and east. With the heat low over Uttar Pradesh, east Madhya Pradesh and Bihar and a trough running from it southwards, winds are west to northwest to the west of the trough line and south to southwest to the east (Fig.3). Over lower Sind and west Rajasthan winds are west to southwest, on account of the trough over Sind. Easterlies continue over Assam, extending in the morning hours into north Bihar. Winds blow anticyclonically in the Arabian Sea around a centre at 14°N and 60°E. In the Bay, there is a similar circulation around the centre at 13°N and 90°E. Around the peninsular trough, winds are northwest in the east

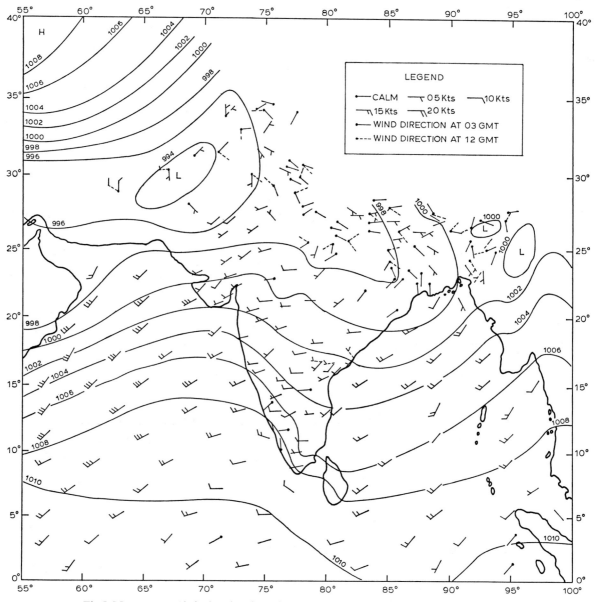

Fig.5. Mean pressure (mbar) and surface wind distribution, July.

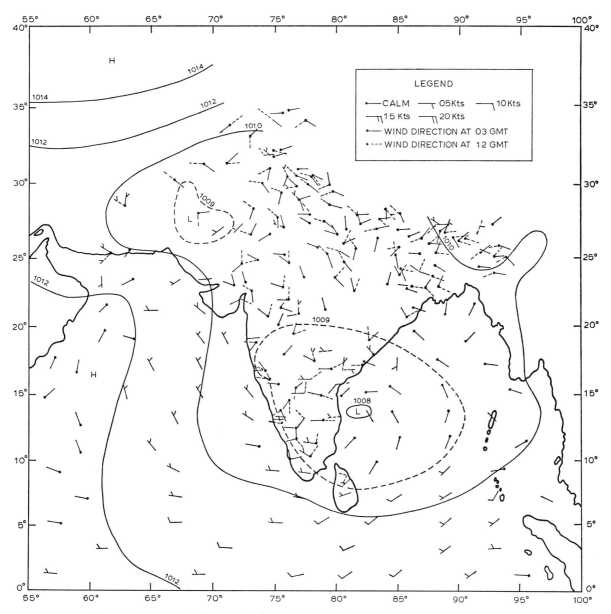

Fig.6. Mean pressure (mbar) and surface wind distribution, October.

Arabian Sea and south-southwest in the west Bay of Bengal. Between the Equator and 5°N, westerly winds prevail. Mean speeds are nowhere greater than 10 knots.

The features to be noted in May are west-southwestern winds near the Equator, southwest in the Bay, and west in the Arabian Sea.

July

Easterlies extend from Assam to north Pakistan to the north of the line through Multan, Ganganager, Gaya and Silchar, north of the monsoon trough (Fig.5). To the south of this line, southerlies occur over West Bengal and southwesterlies to westerlies elsewhere including the Bay and the Arabian Sea. East winds also blow in the central parts of

Pakistan. Strongest winds are in the southwest Arabian Sea. Nearer the west coast of the peninsula the direction is more westerly except along the Kerala coast where north-westerlies are observed. Mean speeds over land are not more than 10 knots. They are stronger upto 20 knots along the west coast. In the Arabian Sea west of 68°E and between 10° and 20°N, it is over 25 knots. Over most of the Bay and the rest of the Arabian Sea, the speed is about 15 knots. Winds are somewhat weaker between the Equator and 5°N. In August and September, winds are somewhat weaker over the seas.

October

Easterlies blow in the Brahmaputra Valley and along the foot of the Siwaliks. Otherwise, north of 25°N latitude winds are mainly from the west. They turn to northeasterlies in the upper parts of the peninsula, becoming variable south of 15°N. In the Arabian Sea, south of 8°N winds are mainly westerly and northerly between 8°N and 18°N changing to westerlies in the north Arabian Sea. There is an anticlockwise circulation of winds with its centre at about 17°N 58°E. In the Bay, winds are mainly west-southwest to the south of the trough line along 13°N and about southeasterly to the north of it. Speeds are about 10 knots or less.

In November, the east–west elongated trough in the Bay is near 7°N with northeasterly winds to the north of the trough. Weak westerlies still prevail south of 5°N both in the Arabian Sea and the Bay of Bengal. Winds are from north-northeast to northeast in the Arabian Sea north of 5°N becoming north-northwest in Lakshadweep area.

The winds over land described above refer to the morning hours. The effect of sea breezes is noticed along the coast in the afternoons. Where the prevailing wind is on-shore, the sea breeze penetrates even 200 km inland.

Table I gives the frequency of wind speeds at a few stations.

TABLE I

ANNUAL FREQUENCY (%) OF WIND SPEEDS (knots)

Station	Calm	1–5	6–10	11–17	>17
Karachi	0.2	21.6	35.0	32.4	10.8
Lahore	5.1	83.4	10.8	0.7	0.0
Madras	2.6	54.8	40.8	1.8	0.0
Bombay	0.2	48.6	44.7	6.3	0.2
Belgaum	2.2	27.5	35.5	22.9	11.9
New Delhi	13.0	46.9	29.0	10.1	1.0
Calcutta	15.0	72.9	11.5	0.6	0.0

Surface temperature

The mean temperature of the air at any place depends on many factors, of which altitude, latitude, proximity to the sea and temperature of the sea are important. The mean temperature falls by about 5.5°C per kilometre of ascent, but only 3/4°C for every degree of latitude, even in the middle latitudes. These values differ in different parts of the world and in different seasons. The altitude effect is thus dominant. A description of the actual temperatures over the subcontinent would, therefore, describe only the contour configurations. Temperatures reduced to a constant level would make them comparable. After a statistical analysis of the variation with altitude of the mean temperatures, a uniform reduction factor of 6°C per kilometre was considered to be the most suitable for reducing the temperatures to sea level. The few stations above 1 km are not considered.

The temperature charts discussed hereafter are based on the data for the period 1931–1960 for a large majority of stations, but for the rest data are available up to 1940. Significant differences in the reduced temperatures may not arise on this account. Air temperatures over sea have been taken from the *Monthly Meteorological Charts of the Indian Ocean* published by the LONDON METEOROLOGICAL OFFICE (1949).

Mean daily temperature

Fig.7 shows the distribution of mean daily temperature (reduced to sea level) in the four representative months of January, April, July and October. Mean daily temperature is the average of the mean daily maximum and the mean daily minimum.

January

The mean isotherms run more or less parallel to the latitudes except near the coasts. The temperature increases towards the south, markedly between 27° and 15°N. The gradient is about 0.9°C per degree of latitude between 20°N and 30°N along the longitude of Delhi. Temperatures are more even in the rest of the peninsula. Lower values occur on the east coast south of 17°N. The spatial range is about 13°, from 14° to 27°C.

April

Land gets progressively heated after January. By April, temperatures of the order of 33°–35°C occur over the peninsula. Temperatures along the coasts are between 28° and 30°C. The gradient of temperature is steep normal to the west coast, and reaches about 5°C per degree of longitude at some places, while it does not exceed 2.5°C per degree longitude towards the east coast. The gradient to the north is also gradual. The spatial range of temperature is less than in January. The highest temperatures occur around 20°N.

July

The southwest monsoon is established over India by July. The cloudiness is heavy: (1) between 17°N and 24°N in the central region: (2) west of 77°E and south of 17°N in the peninsula; and (3) to the east of approximately 85°E in the northeast. Temperatures are uniform in these regions, generally between 28° and 29°C; but on the west coast temperatures are of the order of 26°–27°C which are lower than the mean air temperatures over the open sea. A similar feature is noticed along the Burma coast. To the east of 77°E and south of 17°N temperatures are 30°–31°C. This large difference of 4°C between the west coast and areas further to the east is due to the foehn effect and lesser

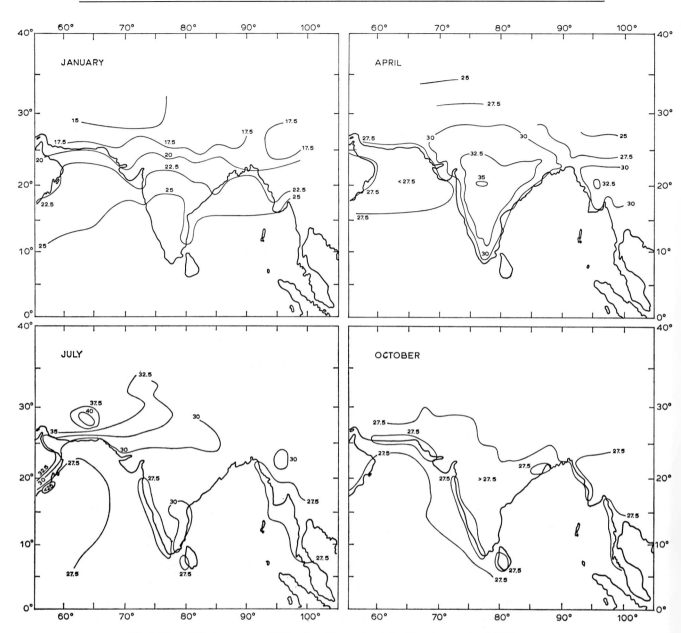

Fig.7. Mean daily temperature (°C) reduced to sea level at ad-hoc lapse rate of 6°C/km.

cloudiness. The hottest areas are West Rajasthan and Pakistan with temperatures exceeding 40°C in Baluchistan. The spatial range is about 13°C.

Very low temperatures along the Arabian coast are due to upwelling.

October

Practically throughout the subcontinent temperatures in October are generally within the range of 27°–29°C. Spatial range is only 3°C. This is the month of the most equable distribution of temperature in India; northern parts cooled down by the monsoon are nearing winter and the south is experiencing the northeast monsoon rains. The temperatures along the west coast of the peninsula and Burma are still lower than out at sea. It may be noted that along the north Orissa coast and the Sind–Mekran coasts, air temperature is lower than over the neighbouring sea.

Annual

The annual mean temperatures are the highest over Rayalaseema, north interior Karnataka and adjoining areas, being about 30°C. The lowest temperatures of about 25°C or less are experienced in the northern parts of the Gangetic plains and northwest India.

Mean daily maximum temperature

Fig.8 gives the distribution of mean daily maximum temperature reduced to sea level. The isotherms are more or less similar to the mean daily temperature charts.

January

The temperature is highest (about 33°C) over a region just to the east of the Western Ghats, which extends up to longitude 78°E between latitudes 11° and 20°N. North of 20°N the isotherms run along latitudes. On the west coast temperatures are about 30°C but on the east coast about 28°C. In this season, the influence of the sea breeze is greater on the east coast both in India and Sri Lanka due to a favourable pressure gradient. The spatial range of temperature is about 13°C.

April

A large part of the subcontinent between 72° and 85°E and 14° and 25°N has a uniformly high temperature between 40° and 42°C. The large contrast in the maximum temperature between air over land and sea in this month is seen as a packing of isotherms along the coasts; it also extends inland on account of the sea breeze effect. Temperatures are lower over a good part of the west coast than on the east coast, as the prevailing wind is more favourable for sea breeze on the west coast. However, the Western Ghats seem to obstruct the spreading and mixing of the sea breeze much beyond and the 40°C isotherm is much closer to the west than to the east coast. Rather low temperatures occur along the coast between Visakhapatnam and Puri. Northeastern parts of the subcontinent have relatively lower maxima due to the maritime airmass that prevails and greater cloudiness. The rest of the northern parts has mostly a temperature between 37° and 40°C. The spatial range of temperature is about 13°C, mainly between the area of continental and maritime influences.

July

With the spreading of the maritime airmass of the southwest monsoon over most of the subcontinent, temperature is nearly uniform (about 33°C), except over the northwest and along the west coast. The highest temperatures are over Baluchistan. Over the northwestern parts, continental airmass often prevails and cloudiness is least, leading to

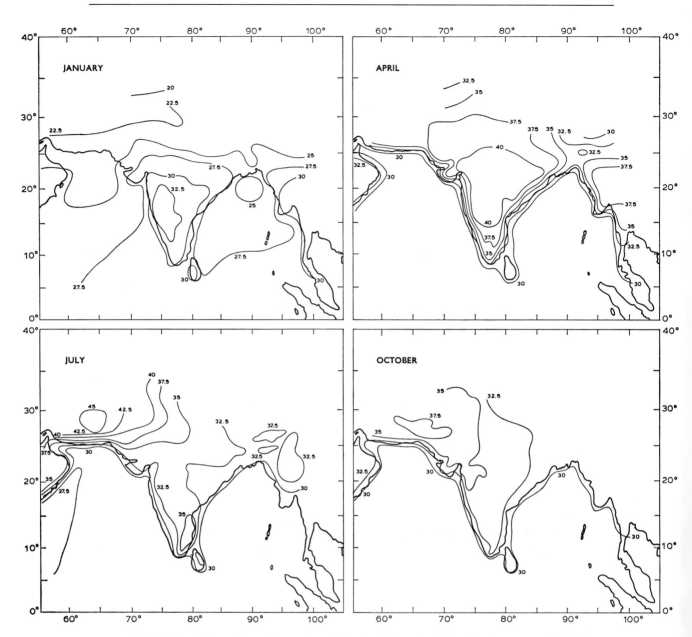

Fig.8. Mean daily maximum temperature (°C) reduced to sea level at ad-hoc lapse rate of 6°C/km.

high temperatures. From Pakistan to West Bengal temperature decreases and the shapes of the isotherms resemble closely the isobars. Assam is a little warmer than West Bengal. Compared to the west coast, temperatures are warmer by about 4°C within 100 miles to the east of the Western Ghats, both due to the foehn effect and decrease in the thickness and amount of cloud. The warmest zone in the peninsula is between 10° and 15°N and 78°E and the east coast where low-cloud amount is small. In Sri Lanka the east coast is warmer than the west coast. The spatial range of temperature is 16°C between Baluchistan and the west coast of the peninsula. The strongest temperature gradient is from the Mekran coast to central Baluchistan.

October

Temperatures fall substantially in the northwest. The highest values are still over Baluchistan. Temperature gradients are small except near coasts. The maritime influence reaches farther inland on the east coast than on the west coast. Temperatures over the south peninsula are about the same as in the Gangetic plains in spite of the greater elevation of the sun in the south. This is due to the greater cloudiness over the south peninsula.

Mean daily minimum temperature

Mean minimum temperature is likely to be affected by micro-meteorological conditions. However, an analysis of the January chart shows that such influences are very limited. Fig. 9 gives the mean daily minimum temperature (reduced to sea level) in January, April, July and October.

January

This is the month of the steepest north–south temperature gradient. The mean minimum temperature has a large variation of about 16°C across the subcontinent. The gradient is marked between 20° and 25°N. The isotherms are practically parallel to the latitudes. The coasts of the peninsula have a higher minimum temperature than the interior. Lower minimum temperatures seem to occur in the northwest of the peninsula.

April

Night temperatures are highest in the area around Bidar and Anantapur. The run of the isotherms north of this area is more or less east–west; but the area from Sind to Gujarat and north Konkan has a lower minimum than further to the east.

July

Mean minimum temperature is practically uniform except in the northwest. Minima are high over Baluchistan like the maxima.

October

Minima are between 20° and 25°C over the whole subcontinent.

Diurnal range of temperature

The range at hill stations (above 1 km) is generally lower than that at the neighbouring plain stations. Coastal stations also have a lower range.

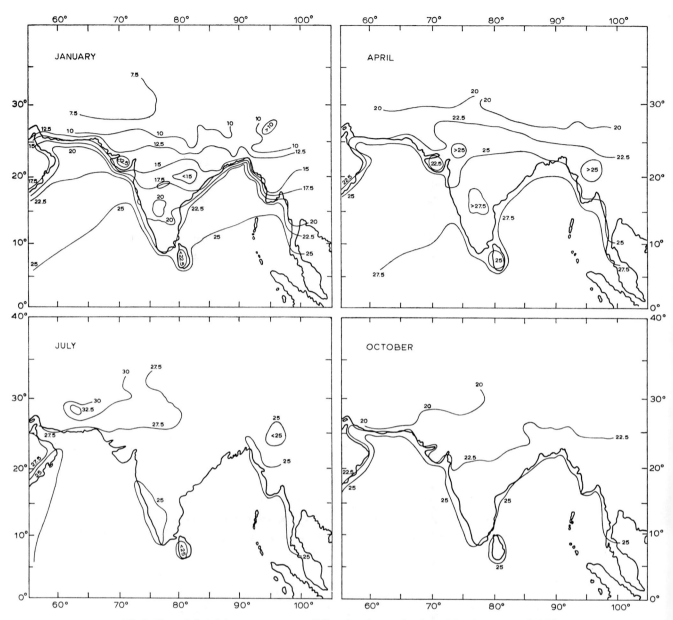

Fig.9. Mean daily minimum temperature (°C) reduced to sea level at ad-hoc lapse rate of 6°C/km.

January

Over most of the subcontinent the range is high, about 13°–15°C, due to the prevailing continental airmass. The highest range, about 19°C, is between Baroda, Malegaon, and Poona, where cloudiness is least. Along the coasts, it is generally less than 10°C due to maritime influence. The lowest values of about 5°C occur over southeastern Tamilnadu, where cloudiness is high and rains persist.

April

The highest values of the diurnal range in this month are about the same as in January, in spite of higher mean temperatures. Over most of the country north of 20°N, the range is higher than January but it has decreased over Upper Assam. The maximum range of about 18°C occurs in three regions, one covering parts of Gujarat, Saurashtra and Madhya Maharashtra, a second over north Madhya Pradesh, and a third over north Rajasthan and the adjoining Punjab. Along the west coast, with the prevailing wind becoming on-shore, the range comes down to about 6°C.

July

With the general increase of cloudiness, the mean range decreases all over the sub-continent. Over the extreme northwestern parts and over the peninsula upto about 200 km from the east coast, it is 8°–9°C. It is uniformly about 6°C over the rest of the country, except for the west coast and the Orissa–Bengal coast where it is about 5°C. It is worth mentioning that Cherrapunji with the heaviest rainfall has a range of 3.8°C. Perhaps 4°C is about the lowest range in these latitudes with the heavy clouds and rains of the month.

October

In this month the mean range increases all over the subcontinent. From about 6° to 7°C along the coasts, it increases northwestwards to 16°–17°C in the western-most regions.

Extremes of temperature

Highest temperatures

Most of the Indian subcontinent has recorded temperatures over 45°C, even about 47.5°C in the northwest and central parts, either in May or June. Temperatures are known to have gone above 50°C in the Great Indian Desert and Upper Sind. These values are about equal to the solar radiation in those areas.

Temperatures in the northeast of India and Bangladesh, where pre-monsoon showers are more marked, remain below 40°C. The west coast, south of Bombay, records upto 40°C in February or March. The east coast can reach 45°C when the marked sweep of continental air from the northwest cuts off the sea breeze in May or June.

Elevation does not always keep temperatures low, unless, in May–August, it is accompanied by cloudiness. In the plateau of Pakistan, temperatures are known to have reached 45°C or more in June or July, unless the height is above 1 or 1.5 km. In Shardu at 2,288 m, beyond the Great Himalaya, 43°C is recorded in July and August when there is little monsoon cloudiness. At 2,202 m at Simla temperatures do not rise above 31°C, due to heavy cloudiness on the southern slopes. Interestingly the highest temperature of 36°C at Colombo is in February and March.

Lowest temperatures

In the case of the lowest temperatures recorded, there is a strong south–north gradient over the peninsula and also towards the northwest. This is in contrast to the highest temperatures which show uniformity over a large area. In the northwest of the sub-continent the lowest temperatures recorded in the plains are around −4°C. Cold air invades from the northwest which gets modified while spreading over the subcontinent. The circulation becomes anticyclonic and cold air reaches the northwestern parts of the peninsula more than the northeastern areas. The Great Himalaya is a barrier to outbreaks of cold air from the north. The Himalayan hill stations have not recorded temperatures as low as those on the western plateau of Pakistan. Simla's lowest temperature is only −8°C while at Quetta it is −19°C, though the latter is slightly at a lower latitude and lower elevation. The lowest temperatures occur either in late December, January or early February.

Other significant features

The maritime airmass brought by the monsoon over most of the country, and the associated increased cloudiness counteract the effect of the higher elevation of the sun, and day temperatures decrease in the monsoon months of July and August instead of increasing. In the western parts of the peninsula the lowest mean daily maximum temperature is in July and August, during the mid-southwest monsoon.

Over most of India day temperatures are highest in May, after that the effect of maritime airmass of the southwest monsoon brings down the temperatures. In some parts the prevailing circulation brings maritime influence earlier, putting an end to the temperature rise. In the western half of the peninsula to the south of 15°N day temperatures decrease after April. Thunderstorm frequency is very high in May in this area. North of 15°N along the west coast, May has the highest mean day temperatures but there is a narrow strip to the east of the Western Ghats where April is the hottest month. At Trivandrum on the west coast, March is the hottest month and on the east coast June. To the east of 87°E (except Assam) day temperatures decrease after April. Maritime airmass prevails in lower levels in May and thunderstorms are very frequent. In Assam, July or August is the hottest month in spite of the monsoon.

In Kashmir the highest temperatures are reached in July. The sun reaches its highest altitude in late June in these areas and temperature has usually a lag of a month. It is only at stations north of 34°N that highest temperatures are recorded in July, uninfluenced by the monsoon. So also in Baluchistan. In the coastal regions of Saurashtra and Sind June is the hottest month, while in the interior it is May. On the Andaman Islands day temperatures are highest in April.

North of 15°N, day temperatures rise again towards the end of the monsoon and there is a second maximum. Between 15°N and 23°N and to the west of 82°E the second maximum is in October; in the rest of India to the north and east (except Assam and the northern half of West Bengal) it is in September. Decrease of cloudiness and low turbidity account for this second rise in day temperatures. From Bombay to Marmugao this is in November, and at Mangalore even in December–January. Port Blair shows this feature of a second temperature maximum in October–November. The second maximum is not apparent in a belt about 200 km wide on the east coast, to the south of Masulipatnam and along the west coast south of Mangalore; nor does it occur in Assam and the Himalayan foot-hill regions.

South of about 22.5°N, except in the western parts of the Deccan, day temperatures are lowest in December and to the north in January. In Gujarat, January has the lowest maximum temperatures. South of 19°N along the west coast and to about 250 km into the interior, July or August has the lowest day temperatures. Places further east in this zone enjoy low day temperatures without the heavy rains of the coast. The lowest day temperatures in the year occur in this region while the sun is almost overhead but obscured by clouds. On the Andaman Islands September is the month with the lowest maximum temperature.

Humidity

Relative humidity recorded at 17h30 IST (12h00 GMT) is discussed as it will be more representative of the atmosphere than at 08h30 IST (03h00 GMT). In the southwest monsoon period, humidities are high except in the northwest. At other times, except along the coast, most of the subcontinent has humidities less than 60%.

January

Most of the subcontinent has a relative humidity between 30% and 50%. Values of about 70% occur along the east coast and between 60% and 70% along the west coast. They are about 50% along the Sind and Saurashtra coasts. Higher humidities prevail over a wider belt into the interior from the east coast than from the west coast, due to on-shore winds there. The lowest humidity of about 25–30% is over southwest Rajasthan, Gujarat and Sind. From the central parts, humidity increases towards the northern mountains. All hill stations record about 60–70%. The plateau of Pakistan has values between 35% and 45%. Assam records about 70%, Bangladesh 60% and Sri Lanka 75%. Inland stations may record 30–40% higher values at 08h30 IST. A few coastal stations record humidities lower by 7% in the morning, but most of them have values in the morning higher by 10%. Many hill stations are less humid in the morning, up to 15%; some behave like plain stations.

April

In the northeast the relative humidity is over 50% and along coasts over 60%. The rest of the country has values between 40% and 15%, the lowest values being in the central parts, western desert and adjoining regions. Humidities increase towards the northern mountains, the hill stations recording 40–55%, but 70% in the northeast. The plateau of Pakistan has values between 25% and 40%. Lakshadweep, the Andaman Islands and Sri Lanka record humidities as high as 75% and such values also occur at the hill stations in the south of the peninsula. The excess of relative humidity in the morning over the evening is generally less in April than in January, but still ranges from 10% to 40%. Over Gujarat and lower Sind and parts of Bangladesh and West Bengal, morning excess is more than elsewhere. At Rajkot relative humidity is 66% at 08h30 IST but 18% at 17h30 IST. At most hill stations relative humidity is higher in the morning than in the evening, though it is the reverse at Ootacamund, Kodaikanal and Nuwara Eliya.

July

Relative humidities are highest in this month. Except in the northwestern parts of the subcontinent and in the southeast of the peninsula humidities are over 60%. Northeast India, the west coast, the Orissa coast and the Himalayas, mostly areas of good monsoon rains, record humidities of more than 80%. Humidity is less than 40% in Sind and even lower than 30% in Baluchistan. Tiruchirapalli has the lowest value of 43% in the peninsula. Colombo's 79% and Trincomalle's 55% on the two coasts of Sri Lanka are a striking contrast.

In the rainy areas where humidity is high, the difference between morning and evening is less than 10%. In the drier areas, mornings are more moist by 20–30%. At hill stations the difference is less than 5%. (At Lasbela in Pakistan humidity in the morning is higher than in the evening in all months, but with a maximum (60%) in the monsoon period.)

October

Lowest humidity of 30% occurs over the central parts of Pakistan and Rajasthan and values increase to the east to over 80% in Bangladesh and northeast India and to 60–70% towards the Himalayas. There is a sharp increase to 75–80% along the west coast. The whole east coast has a humidity of 75%. In the northern parts of the peninsula humidity increases from 40% in Gujarat to 60–70% to the east and south. Relative humidities of over 60% are recorded in the south of the peninsula.

Along the coast and in the northeastern parts of the subcontinent the morning humidity is higher than the evening humidity by 10–15%. The largest difference of 40% or more is in the interior of Gujarat and Sind. Rajkot shows a difference of 40%. In the evening most of Pakistan shows values lower by 30%. In the peninsula, the northwestern interior has over 30%, decreasing towards the south and the coast. In the interior of the south peninsula, morning humidities are higher by about 20–25%. Many hill stations all over the region show an increase in humidity in the afternoon as in winter, but some like Parachinar and Mahabaleswar have a 10% higher humidity in the morning.

Rainfall

Annual rainfall

Fig.10 shows the annual rainfall over the Indian subcontinent. Hills and mountain ranges cause striking variations in rainfall. On the southern slopes of the Khasi–Jaintia Hills, rainfall is over 1,000 cm at Cherrapunji, while to the north in the Brahmaputra valley it decreases to less than 200 cm. Cherrapunji's rainfall of 1,142 cm at an elevation of 1,313 m is obviously due to orographic lifting but its magnitude requires to be quantitatively explained. From the west coast along the slopes of the Western Ghats rainfall increases (Agumbe, 610 cm) and rapidly decreases on the eastern side (leeside in the southwest monsoon season). No definite information is available about the increase of rainfall with elevation and the height at which the rainfall attains the highest values. In the Himalayas with multiple ranges, there are complex variations in precipitation which have been only partially monitored by the precipitation network.

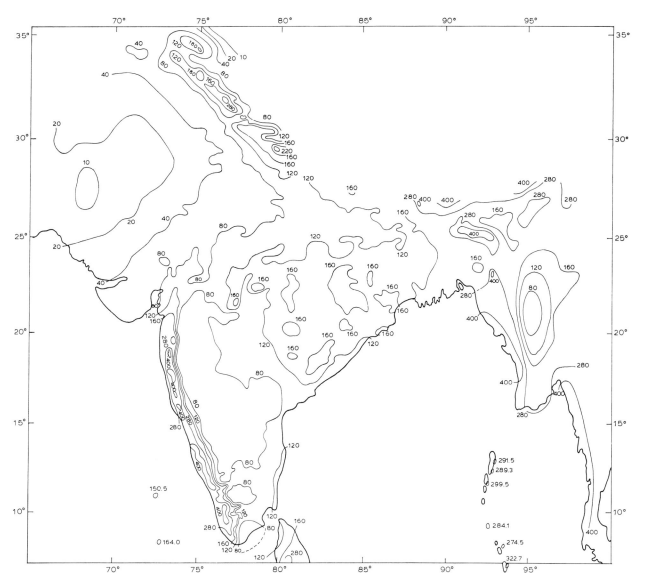

Fig.10. Mean annual rainfall (cm).

From the east coast rainfall decreases inland in the peninsula south of 17°N. This decrease continues up to the eastern side of the Western Ghats, with some increase over the Eastern Ghats. The low rainfall of about 80 cm at the extreme south of the peninsula and the Gulf of Mannar, between 8° and 10°N to the east of the Western Ghats, is interesting, as along the coast one degree to the north, rainfall is almost doubled. Between 17°N and 20°N, the Eastern Ghats are closer to the coast and higher than further south and rainfall increases from the coast to the hill ranges. Behind the first range there is a slight decrease and again an increase over the next range. From 80°E rainfall decreases continuously upto the east of the Western Ghats.

The region of lowest rainfall in the peninsula, with less than 50 cm, is found around 18°N in the upper basin of the river Bhima and its tributaries, just to the east of the Ghats.

From the coast of West Bengal and the hills of Orissa, rainfall decreases inland. Further to the west the Chota Nagpur Hills, the Maikala Range and the Mahadeo Hills cause a rainfall increase with smaller amounts occurring in the valleys in between. The effect of the Vindhyas and the rest of the Satpuras is not so marked, but the Gir Hills in Kathiawar have more rainfall than their surroundings. Mount Abu in the Aravallis has a rainfall of 160 cm while the surrounding plains have only 50–80 cm.

Across northern India a line of rainfall minima runs from 28.5°N 75°E to 25°N 88°E which is, paradoxically, close to the monsoon trough. The area to the south lies in the track of monsoon depressions which are responsible for much of the rainfall. Further to the north the influence of the Himalayas causes an increase of rainfall. Apart from this, there is a decrease of rainfall from east to west, from about 180 cm in Bangladesh to less than 10 cm in the Thar Desert in Sind.

The rainfall in the Himalayas will be dealt with in a separate section. In the eastern Himalayas there is more rainfall than in the western portions. In the east rainfall of 500 cm has been recorded, but only 280 cm in the west. In the western plateau of Pakistan rainfall is only 20–40 cm. Beyond the Great Himalaya, in the Karakoram Range, rainfall may be as low as 10 cm.

Over the Tripura, Manipur, Nagaland and Mizo hills, which have numerous hill ranges, rainfall is about 250 cm though higher values have probably been missed due to a poor network of raingauges. There is a rainshadow effect in the Assam Valley lying between the Himalayas to the north and other ranges to the south and east

The mountain ranges in the east more than double the annual rainfall in Bangladesh from west to east—157 cm at Pabna, 373 cm at Cox's Bazar, and 405 cm at Sylhet.

The Colombo–Jaffna belt on the west coast of Sri Lanka records about 250 cm of rain, slightly more than Nuwara Eliya at an elevation of 1,881 m. Rainfall decreases along the coast to the north and southeast to about 100 cm. Most of the east coast has a rainfall of about 170 cm.

Rainfall in the Andaman and Nicobar islands is about 300 cm, while in the Lakshadweep in the Arabian Sea it is only about 150 cm, though both island groups are in the same latitude belt. Kozhikode on the same latitude on the west coast of India, however, gets 300 cm which is comparable with that of the islands in the bay.

Number of rainy days

Fig.11 shows the distribution of the annual number of rainy days, when 2.5 mm of rain or more is recorded. The pattern of rainy days is similar to that of annual rainfall.

Cherrapunji has about 160 rainy days which appears to be about the maximum. Areas with less than 20 rainy days are confined to Kutch, West Rajasthan and adjoining parts of the Punjabs, Sind and Baluchistan and the Karakoram region. Some places in the Great Indian Desert have even less than 5 rainy days.

There seems to be a tendency for a maximum number of rainy days to occur between 9°N and 13°N. The maximum along the west coast is at Alleppay which has 137 days. Madras has a greater number of rainy days than to the south or the north on the east coast and so have Port Blair, and, Mergui and Victoria peak in Burma.

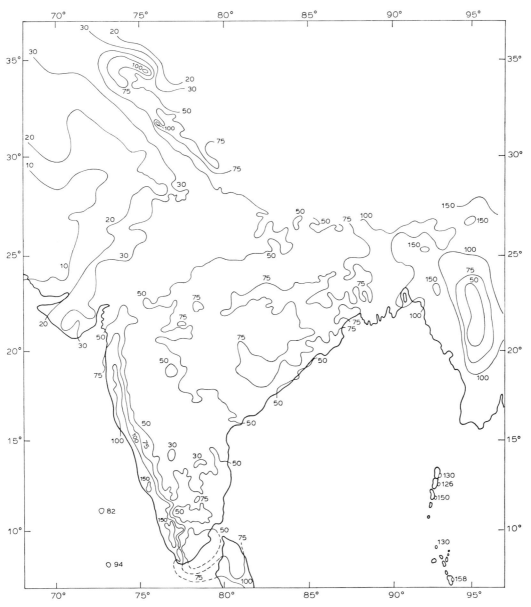

Fig.11. Mean annual numbers of rainy days.

Seasonal rainfall

The winter precipitation over the Indian subcontinent is a small percentage of the annual, except in Kashmir and surroundings. Yet this rainfall is very important for the winter crops of the northern parts of the country. Fig.12 shows the rainfall distribution. On the east coast of Sri Lanka, rainfall is more than 40 cm, decreasing to less than 20 cm along the west coast. Rainfall is more than 20 cm in Kashmir and in some places on the Nicobar Islands. In the northwest the 5 cm isohyet runs parallel to the mountain ranges. Other regions of over 5 cm of rain are the northern parts of Pakistan, northeast India east of 90°E, Bihar plateau and the adjoining parts of Madhya Pradesh and Orissa, the peninsula south of 10°N, the east coast strip south of 15°N, and the two island groups. In the hot weather period, the chief areas of rainfall are southwestern parts of Sri Lanka,

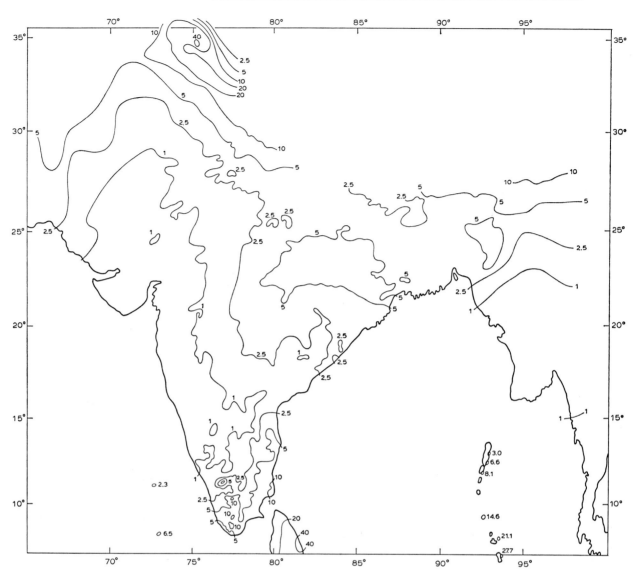

Fig.12. Seasonal rainfall (cm) for January and February.

Assam, Bangladesh, Jammu and Kashmir, Kerala and their surrounding areas. Fig.13 shows the rainfall during this season. Cherrapunji records over 200 cm already. Other rainy tracts are the northeastern parts of the subcontinent. Southwest Rajasthan and Sind are the driest areas. Southwestern parts of Sri Lanka get over 40 cm of rain.

The rainfall distribution in the principal rainy season of India, the southwest monsoon period, lasting from June to September, is shown in Fig.14. With the exception of Kashmir and surroundings, the extreme south peninsula (other than the west coast) and most of Sri Lanka, the annual rainfall is mainly accounted for by the falls in this season and hence the two are similar in distribution. Orographic influence is dominant in the distribution of rainfall in this season, as the prevailing winds blow almost at right angles to the Western Ghats and the Khasi–Jaintia Hills. Not only do the Himalayas get heavy rains but also the plains adjoining the foot hills right up to Peshawar receive more rain than the plains further south. Monsoon rains have been explained as due to moist air-

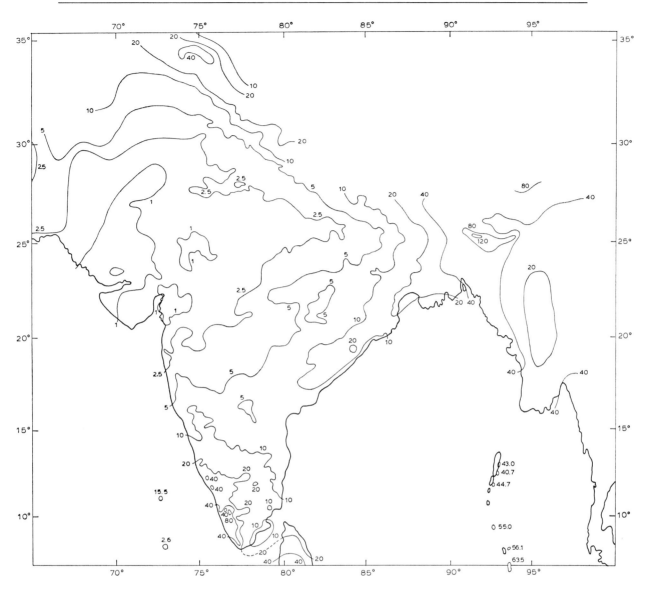

Fig.13. Seasonal rainfall (cm) for March–May.

mass entering a "box" with three sides, so that the air can escape only upwards after shedding great amounts of rain.

Rainfall decreases very rapidly southwards along the west coast from 9.5°N to Kanyakumari. The rainfall at Kanyakumari in this season is about the same as in the Great Indian Desert. To the east of the Western Ghats, between 8° and 10°N, rainfall decreases considerably and in the coastal strip rainfall is only 2 cm. This aridity covers the southeast tip of the peninsula and the northwestern parts of Sri Lanka. With all the significant amounts of rainfall occurring on the Ghats, a saving feature of economic interest is that all the important rivers of South India come out of the Western Ghats, flowing east through the plains where rainfall in this season is of the order of that in Rajasthan.

In the northwestern parts of the subcontinent rainfall progressively decreases westwards, from 40 cm in Rajasthan to 5 cm in Baluchistan. Southwest Sri Lanka and the hills get good rains at this time but not the other parts of the island.

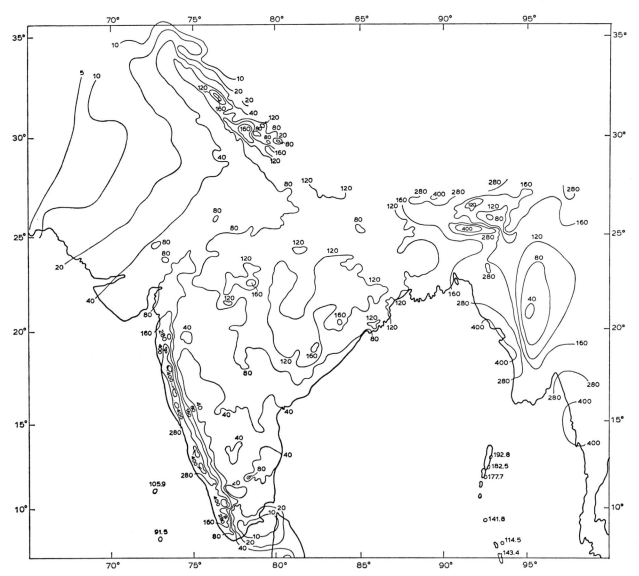

Fig.14. Seasonal rainfall (cm) for June–September.

In the post-monsoon period, Sri Lanka, the south peninsula, the east coast, Bangladesh, Assam, and parts of Kashmir are the chief areas of rainfall, as will be seen from Fig.15. From the east coast, rainfall decreases inland, markedly so in Tamilnadu and Andhra Pradesh. South of 15°N, rainfall again increases over and near the Western Ghats but decreases towards the west coast. The increase in rainfall near the Eastern Ghats is not marked. In Sri Lanka the east coast gets good rains and so does the southwest.

Though the rainfall has been discussed in terms of four periods of unequal duration, this classification is not quite appropriate in several border areas. Monsoon does not set in over the northwestern parts of the subcontinent in June. Also in these parts, December could rightly be regarded as winter.

In most parts of the tropics, one period or the other comes to be identified as the principal rainy season when a maximum in precipitation occurs. For most of the Indian subcontinent this is from June to September. But there are exceptions. The eastern coastal belt of the peninsula and the whole of Sri Lanka gets more rain from October to Decem-

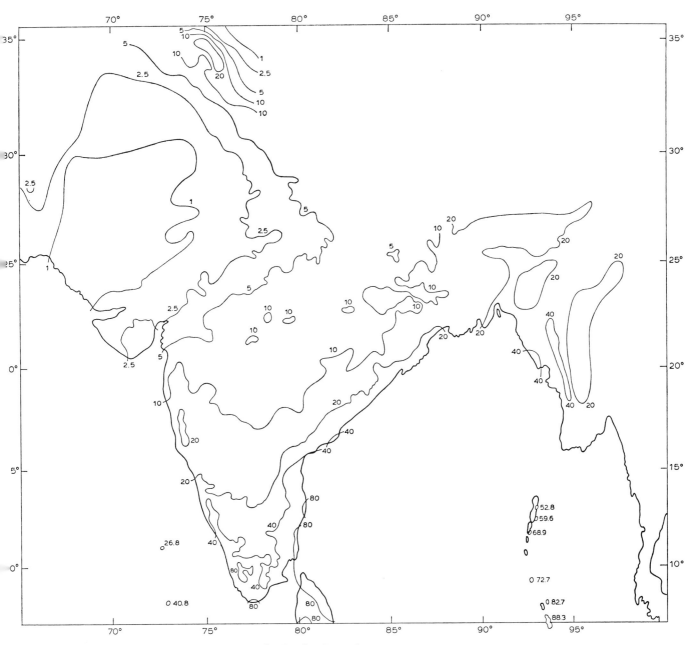

Fig.15. Seasonal rainfall (cm) for October–December.

ber, than in other periods. However, at the hill station of Nuwara Eliya, June–September is the main rainy season. The rains of October–December are popularly referred to as the "northeast monsoon" on account of the direction of the surface winds then blowing over most parts of the Bay of Bengal. In the northwestern hilly areas of the subcontinent interesting variations are noticed. To the east of 77°E, in the Himalayas and the Ladakh plateau, monsoon rainfall is more than to the west. At Leh the June–September rainfall is 5 cm, while in the remaining eight months it is 6 cm. To the west but north of 33°N, the hot weather period or spring (March–May) is the rainiest. Along the Sulaiman Range, intermittent incursions of the monsoon give more rain than in other periods. For the whole of Baluchistan, winter is the main rainy season.

Highest rainfall in 24 h

The highest rainfall recorded in a day is of considerable hydrological interest. Even desert areas may get heavy rains now and then, sometimes in excess of the mean annual rainfall. There is an element of chance in the magnitude of the highest rainfall recorded, even though observations may be available for over 50 years. But when the data of all the stations are examined together, some broad patterns emerge.

In most areas the highest probable rainfall in a day is between 20 and 30 cm. At Cherrapunji, this is about 100 cm. Dharampur in Gujarat had an equally high amount. The southwest monsoon period is the time for such heavy falls, but the period is different in areas where this is not the main rainy season.

Along the east coast, 50 cm may even be exceeded between 19°N and 11°N. Depressions and cyclones cause such falls, usually in September and October in the north and in November and December in the south. Even systems located far from the coast can cause such heavy rains. The heaviest fall of 40 cm at Cox's Bazar is also in October. The west coast from Surat to Bombay can experience 50 cm rain in a day. Further south up to 14°N it is over 30 cm in June and about 25 cm along the Kerala coast in May, at the time of the onset of the monsoon. In the Western Ghats the heaviest falls of 50–60 cm are in July.

On the plateau of the peninsula, south of about 20°N, the amount is between 20 and 30 cm in the eastern half, but lower (about 17 cm) in the west. Immediately on the leeside of the Ghats, it is only 11 cm near Poona.

In the track of monsoon depressions, the amount is generally about 25 cm with pockets of 30–40 cm to the east of 79°E. At Gujarat, the heaviest rainfall of between 30 and 40 cm in a day is generally recorded during the monsoon. The Assam Valley itself gets lower falls, between 15 and 25 cm. While the Indo–Gangetic plains of the north have falls between 20 and 30 cm, there are pockets of 30 and 40 cm in the northern mountains and the adjoining plains east of 75°E. Northwest Rajasthan gets in some parts only 10 cm.

In Kashmir, about 12 cm of rain can fall in a day in spring.

The heaviest falls in the plains of Punjab are between 15 and 20 cm in the north and 10 and 15 cm in south, all in the monsoon season. The coastal region of Sind can get 20–25 cm with monsoon depressions. Baluchistan can get only 3–8 cm in a day during winter.

In Sri Lanka, Jaffna has recorded 52 cm in November, Colombo 30 cm in May and Nuwara Eliya (1895 m) 24 cm in August. The manner in which monsoon precipitation occurs in the Indian subcontinent is interesting. Nearly half the amount is in heavy falls of 7 cm or more at Bombay and Madras in their monsoons. At a less rainy place like Delhi, this is still in amounts of 4 cm or more. However, at Srinagar 60% of precipitation, from December to May, is in amounts between 1 and 5 cm, and the rest in smaller falls.

Variability of rainfall

Maps of the coefficients of variation of the seasonal rainfall (standard deviation as a percentage of the mean rainfall) have been published for India by the India Meteorological Department. Usually the coefficient decreases with increase of the mean rainfall amount. The coefficient is 50–100% for the winter rainfall, the least near the northern hills and becoming 100% near the central parts. From March to May, the northeastern parts of the subcontinent, Kashmir and the highlands of the southern peninsula have a coefficient of variation of 30–40%, which increases to 100% towards the central parts and the coast. The reliability of southwest monsoon rains is of great importance. The areas of 75 cm or more rain have a coefficient of variation of 30% or less. The western desert in India has a variation of about 60% and the extreme southeastern tip of the peninsula 100%. The area of good post-monsoon rains in the southern peninsula has a coefficient of 40–50%. Northern parts of the subcontinent have then a coefficient of 60–80%, becoming 100% in the heart of the country.

Weather phenomena

The frequency and distribution of various weather phenomena are now considered Number of rainy days have been described in the section on rainfall.

Thunderstorms

Fig.16 shows the mean number of thunderstorms in a year. There is much subjectivity in recording thunderstorms which do not pass right over an observatory. The regions of high activity are the northeastern areas of the subcontinent, the Himalayas, east Madhya Pradesh and adjoining areas, south Kerala, northern parts of Pakistan, immediately east of the Aravalli and the west coast of Sri Lanka. The Andaman Islands and perhaps also the adjoining seas have a high susceptibility to thunderstorms. The smallest frequency is over northern Kashmir, Sind and southeastern Tamilnadu, the first two being also areas of very low rainfall. The narrow belt of rather few thunderstorms, about 100 km towards the plains from the foot of the Siwaliks, is very interesting. The west coast between 15° and 20°N has a relatively smaller number of thunderstorms, in spite of the heavy rainfall. Most of the monsoon in this region is in the form of showers and rain without thunder. Local orography controls the development of thunderstorms and small-scale features have not been included in the figure.

Table II shows the frequency of thunderstorms, squalls and number of rainy days for a few representative stations in the subcontinent. Northeastern parts of the subcontinent experience severe thundersqualls from March to May called the northwesters or kal baisakhis, noted for their destructiveness. One or two of them develop into tornadoes every year. At Dacca thunderstorms are frequent in April (9) and May (9) but decrease after the onset of the monsoon early in June, to 3 in July. There is again an increase to 7 in September, declining thereafter. The rainfall from March to May is mostly from thundershowers but not so the monsoon rainfall. Calcutta which is close to the monsoon trough, gets at least half its monsoon rain from thundershowers. September is the most thundery month and October half to that extent. Bhopal lying just to the south of the track of monsoon depressions has its maximum of thunderstorms in June when the monsoon sets in. About half the rain during the other monsoon months is from thundershowers. Kathmandu lying in a valley beyond the Mahabharat Range in the Himalayas, experiences thunderstorms from March to October with the maximum in May (11) and

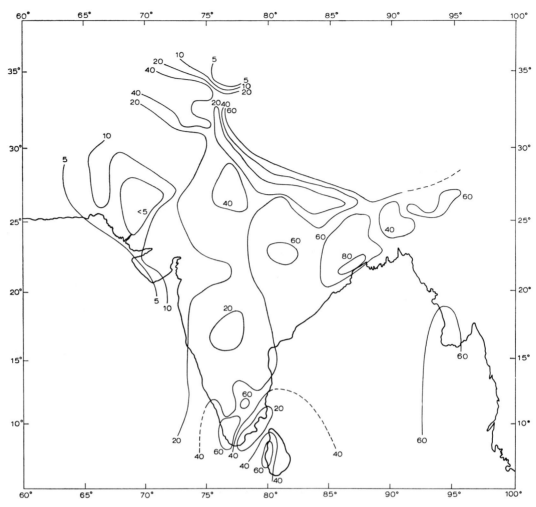

Fig.16. Mean annual number of thunderstorms.

8 to 9 in each of the monsoon months. The maximum thunderstorm activity in Delhi is in the mid-monsoon months of July and August. Peshawar has a high incidence of thunderstorms from March to October, often with squalls. Some of these do not give any rain. Quetta experiences thunderstorms mainly in March, April and July. Only 10% of the winter precipitation at this place is accompanied by thunder. Thunderstorms are frequent at Trivandrum from March to May and in October and November, but very few occur during the southwest monsoon. Bombay and Colombo are similar in the monsoon rains being generally without thunder. Colombo has thunderstorms throughout the year, except in the southwest monsoon, the maximum frequency being from March to May and in November. The rains of the post-monsoon period are more often accompanied by thunder, as for example at Madras, Colombo and Trivandrum. The other season of thunderstorms in Madras is during the southwest monsoon.

Duststorms

Both dust-raising winds and duststorms with convective clouds are classified under this category. A dry state of the ground and loose soil favour development of duststorms.

TABLE II

NUMBER OF THUNDERSTORMS, SQUALLS AND RAINY DAYS ($\geqslant 0.3$ mm)

Station		J.	F.	M.	A.	M.	J.	J.	A.	S.	O.	N.	D.	Year
Dacca (Narayan ganj)	Thunder-storms	0	1.7	4	9	9	7	3	3	7	2	0.7	0	46
	Squalls	0	0	0.3	0.5	0.8	0.1	0	0	0	0	0	0	2
	Rainy days	0.8	1.9	3.2	6.7	11.0	14.5	17.7	17.6	12.1	6.0	1.3	0.4	93
Calcutta	Thunder-storms	1	1	4	6	11	13	9	12	16	9	1	1	84
	Squalls	<1	<1	2	3	4	5	1	1	1	1	<1	<1	19
	Rainy days	1	3	4	5	10	17	23	24	20	12	2	1	122
Bhopal	Thunder-storms	1	1	2	2	2	9	7	6	5	1	0	<1	37
	Squalls	<1	<1	<1	<1	1	<1	1	<1	<1	0	0	0	3
	Rainy days	3	1	1	1	2	11	21	21	14	3	1	1	80
Kath-mandu	Thunder-storms	1	1	4	7	11	9	8	8	9	3	<1	<1	61
	Squalls	–	–	–	–	–	–	–	–	–	–	–	–	–
	Rainy days	0.8	5.5	1.7	3.2	10.5	15.8	22.7	20.8	13.3	3.2	0.2	0.8	98
New Delhi	Thunder-storms	1	2	3	3	5	5	7	8	4	1	<1	1	41
	Squalls	<1	<1	1	2	3	2	1	1	1	<1	<1	0	12
	Rainy days	3	3	3	1	3	5	14	14	7	2	1	1	57
Peshawar	Thunder-storms	0.2	1.5	5	6	7	6	9	8	5	4	0.8	0.3	53
	Squalls	0.5	0.8	1.9	3	6	4	4	3	1.6	1.7	0.4	0.3	27
	Rainy days	2.9	3.3	4.6	3.8	2.0	0.9	2.0	2.5	1.5	0.6	0.6	1.6	26
Quetta	Thunder-storms	0.6	0.7	4	4	1.1	1.7	5	0.9	0.1	0.1	0.3	0.6	19
	Squalls	0	0.3	0.5	1.0	0.1	0.3	0.3	0	0.2	0.4	0.2	0	3
	Rainy days	4.9	4.6	4.7	2.7	1.1	0.4	1.1	0.7	0.1	0.3	0.8	2.6	24
Trivan-drum	Thunder-storms	1	3	8	16	12	3	1	1	3	9	9	3	69
	Squalls	0	0	0	0	0	<1	<1	0	<1	0	0	0	<1
	Rainy days	3	3	5	10	15	24	21	17	14	16	15	6	149
Bombay	Thunder-storms	0	<1	<1	<1	2	5	1	1	3	3	1	<1	17
	Squalls	0	0	0	0	<1	4	5	4	1	1	0	0	15
	Rainy days	<1	<1	<1	1	2	20	29	27	21	5	2	<1	108
Madras	Thunder-storms	<1	<1	1	2	3	5	5	7	8	9	3	1	44
	Squalls	<1	0	<1	1	1	1	3	1	1	1	1	<1	11
	Rainy days	3	1	1	2	3	9	14	15	11	14	11	7	91
Colombo	Thunder-storms	3	5	11	19	9	2	1	2	2	8	10	8	80
	Squalls	–	–	–	–	–	–	–	–	–	–	–	–	–
	Rainy days	8	7	11	18	23	22	15	15	17	21	19	12	188

Mountain stations do not have duststorms, but plateau areas are affected by them. Quetta on the Pakistan plateau records on the average 6 duststorms per year.

The triangular area between Peshawar, Patna and Hyderabad (Pakistan) is most liable to duststorms. The average annual number is 13 at Peshawar, 6 at Patna and 3 at Hyderabad. The number increases to the west, up to 18 in Bikaner, 27 in Ganganager and 25 in Bannu. Parts of Baluchistan have an average number of about 30.

In the peninsula between about 70° and 75°E and 15° and 21°N, up to 5 duststorms occur per year.

The monthly frequencies of duststorms are given in Table III. New Delhi and Allahabad have a maximum frequency in April, May and June, after the onset of the monsoon duststorms disappear. At stations further to the west, duststorms continue during the monsoon months as well, ceasing only by November. No month can be said to be completely free from duststorms at Dalbandin.

TABLE III

NUMBER OF DUSTSTORMS

Station	Jan.	Feb.	Mar.	Apr.	May	June	July	Aug.	Sep.	Oct.	Nov.	Dec.
New Delhi	0	0	0.5	1	3	3	0.5	0	0.1	0.3	0	0
Peshawar	0	0.1	0.2	0.5	3	2	3	2	1.7	0.9	0.1	0
Bikaner	0.3	1.2	1.7	2	3	5	2	1.3	0.6	0.6	0	0.2
Bannu	0	0.3	1.5	1.2	5	4	6	4	3	0.5	0	0
Dalbandin	1	2	4	4	4	4	5	2	1.2	0.8	0.3	0.3
Allahabad	0	0.3	0.2	0.7	2	1.5	0.3	0	0	0.1	0	0
Quetta	0	0.1	0.4	0.3	0.7	1.0	0.3	0.7	1.1	0.9	0	0.1

Fog

In the plains, fog is a comparatively infrequent phenomenon. Over 90% of the area, the number of occasions may not be more than 5 per year and even this, only to the north of 24°N. Due to the short duration of fog in the mornings, all instances might not be recorded except by careful observers. Proximity of water bodies, such as lakes, rivers and backwaters enhance the frequencies. Hill stations may have over 100 days of fog in a year, mostly fog from clouds.

TABLE IV

NUMBER OF OCCASIONS OF FOG

Station	Jan.	Feb.	Mar.	Apr.	May	June	July	Aug.	Sep.	Oct.	Nov.	Dec.	Year
Gauhati	12	3	0.3	0.2	0	0	0	0	0.4	3	10	16	45
Calcutta	9	7	3	0.1	0	0	0	0	0.1	0.3	2	4	25
New Delhi	2	0.6	0	0	0	0	0	0	0	0	0.1	0.7	3
Simla	4	5	1.4	0.2	0.1	5	17	21	7	0.3	0	1.4	62
Quetta	1	0	0	0	0	0	0	0	0	0	0	0	1.0
Pasni	0	0.6	0.6	2	4	1	0.3	0.2	3	7	0.4	0.1	19
Bangalore	3	0.2	0	0.1	0	0.1	0.5	1.2	1.1	2	2	4	14
Kodaikanal	8	4	3	2	1.4	3	4	5	6	10	17	10	73
Chittagong	2	3	2	1.5	0.3	0.1	0	0	0.4	4	2	4	19
Darjeeling	5	6	5	1.9	11	19	21	22	16	3	3	2	115

In the northern parts of the subcontinent fog occurs in winter, most commonly in the rear of "western disturbances", the low pressure systems which move across the northern parts from west to east. For stations in the plains New Delhi is illustrative of this feature. The Assam Valley is liable to fog on more than one-third of the days from November to January (Gauhati). Calcutta and Chittagong have rather frequent occurrences of fog in the post-monsoon months and winter. This is a feature of the riverine and coastal areas of West Bengal and Bangladesh. Simla in the Himalayas has hill fog frequently in the monsoon months, with a lesser number of cases in winter. The western plateau of Pakistan (Quetta) has less than one case of fog per year. This is different from the hill stations in other areas of the subcontinent. In the southern peninsula, the Mysore plateau (Bangalore) has a higher frequency of fog from October to January. Kodaikanal has mostly hill fog from October to December during the season of the northeast monsoon rains. The Mekran coast (Pasni) has rather frequent fogs in April and May and September and October.

Hailstorms

In the plains of the Indian subcontinent, the frequency is about one per year, if not once in two years. Coastal areas south of 20°N and also the Gujarat coast are practically free of hail. Immediately to the east of the Western Ghats, hailstorms occur once in three or four years. The plains north of 20°N experience hail about once in a year or two, excepting Gujarat and in and around the desert areas where the frequency is still lower. The foot hills of the Himalayas get most hailstorms, particularly in the western Himalayas. The western plateau of Pakistan and the highlands of the south peninsula are liable to hail, more than the plains.

TABLE V

FREQUENCY OF HAILSTORMS

Station	Jan.	Feb.	Mar.	Apr.	May	June	July	Aug.	Sep.	Oct.	Nov.	Dec.
Gauhati	0.1	0.1	0.4	0.3	0	0	0	0	0	0	0	0
New Delhi	0.2	0.2	0.1	0.2	0.2	0	0	0	0	0	0	0
Quetta	0.4	0.7	1.9	0.6	0.4	0.1	0.1	0	0	0	0.1	0.4
Simla	2.0	3.0	2.0	2.0	1.8	0.4	0	0	0.2	1.7	0.9	1.7
Kodaikanal	0.1	0.1	0.1	0.6	1.4	0.1	0	0.2	0.2	0	0	0
Darjeeling	0.3	0.4	1.0	1.1	1.4	0.1	0	0	0	0	0	0
Nagpur	0	0	0	0	0	0	0	0	0	0	0.5	0.1

The pre-monsoon period with its high temperatures, is also the period of maximum frequency of hail. In winter, hail occurs to a lesser extent. In Simla hailstorms occur throughout the year except in the mid-monsoon. Kodaikanal experiences hail from January to June and in the late monsoon. At Quetta, hail can occur from January to July and in November and December.

Climate of the Himalayas

Influence of the Himalayas on the climate of India

The Himalayas function as a great climatic divide which exerts a dominating influence on the meteorological conditions of the Indian subcontinent to the south and the Central Asian areas of the north. In winter, the Great Himalayan Range serves as an effective barrier to the intensely cold continental air blowing southwards into the subcontinent. Fig.17 gives some values of the mean temperatures of January in the area between 25°N and about 50°N. New Delhi is warmer than inland stations in China in the same latitude, by about 9°C.

In the monsoon months, the Himalayas force the rain-bearing winds up the mountains to deposit most of their moisture on the Indian side. Not only do the Himalayas cause rain in the hills, but their influence on the rainfall in the plains immediately below the foot hills is found to be marked too. For instance monsoon rainfall substantially increases even in the plains as we approach the Siwaliks.

Perhaps the most important effect of the Himalayas on the summer monsoon of India is in the location of the "heat low" over Sind and adjoining areas and the monsoon "trough" across northern India. This termal low is known to be one of the main controlling factors of the monsoon circulation in the lower levels in the Indian subcontinent.

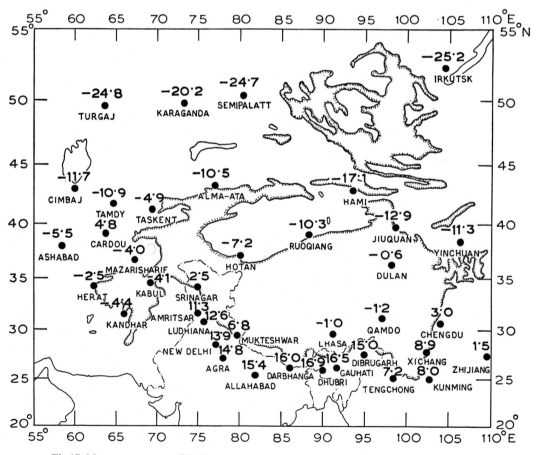

Fig.17. Mean temperatures (°C), January 1977.

If this low were to be located elsewhere, the distribution of the monsoon rainfall, could be quite different.

Already in 1930 BANERJEE argued that the position of the heat low over Sind and the monsoon trough over the northern plains was determined by the mountain and hill configuration around. In recent years, an interesting numerical simulation of the monsoon by HAHN and MANABE (1975) has been carried out in the United States at the Geophysical Fluid Dynamics Laboratory. Simulations were done with and without inclusion of the mountain topography of south Asia in the models. The mountain model has been found to be successful in simulating some of the important features of the south Asian monsoon circulation. In July, for instance, the computed south Asian low-pressure belt extends from Arabia across Asia into the west Pacific. In the model without mountains, the continental low occurs in the neighbourhood of 50°N and 125°E, i.e. in the extreme northeast China. Another important difference occurs between the two models in the distribution of precipitation. In the no-mountain model a desert-like climate is brought out in south Asia resulting from dry continental air from the northwest flowing towards the rain belt far to the south, whereas the mountain model indicates the extension of moist southwesterlies at the surface northwards towards the south Asia low-pressure belt, resulting in substantially more rainfall over the south Asian continent.

General climate

The Himalayas lie in the subtropical high-pressure belt where seasonal meridional migration of pressure and wind systems greatly alter the weather regimes between the different months. In winter, the middle latitude westerlies sweep over the ranges and precipitation comes from the "troughs of low pressure" in the westerly circulation. In the mid-troposphere (500 mbar) the subtropical ridge lies along 14°N. The ridge line gradually moves northward and by late June it is over Tibet. During the monsoon months the Himalayas are under the sway of easterlies, in contrast to winter.

The Tibetan Plateau with an elevation of about 5 km, strongly heated by insolation, is supposed to build up the strong "Tibetan high" in the upper troposphere, leading to easterlies all over the subcontinent at those levels. October usually brings westerly winds to the Himalayas, progressively from west to east.

The troughs in the westerlies are most marked in the winter and give more precipitation in the western Himalayas than in the eastern as the former are higher in latitude by four degrees, and also for other reasons. Simla's precipitation in winter is 11 cm, while Darjeeling gets only 5 cm. The same disturbances give more rains in the east in April and May as more moist air is coming from the Bay of Bengal. According to observations, winter precipitation reaches a maximum around 76°E in the Himalayas. This is interesting as this area is neither the highest section nor the northernmost part of the Himalayas. In the Assam Himalayas the monsoon starts a month earlier and ends a fortnight later, compared to the western part. This is reflected in Darjeeling's rainfall of 52 cm in June and 42 cm in September compared to 15 and 19 cm in the respective months at Simla. In the mid-monsoon months July and August, the western Himalayas are less rainy by about a third than the eastern parts. Darjeeling records 129 cm, Simla only 80 cm. The difference in monsoon rainfall between the eastern and western portions of the Himalayas is thus the reverse of the winter rainfall. The heaviest daily rainfall recorded is also much greater in the east.

In the "break monsoon" when the monsoon trough lies near the Himalayas, there is a temporary drought in the central parts of the country, but the Himalayas get very heavy rains.

Winter nights are colder at Simla than in Darjeeling, with greater probability of cold waves. May and June are substantially warmer at the western stations, with less frequent rains. Humidity is lower at Simla outside the monsoon period. The above comparison of the two stations shows the characteristic differences between the western and eastern Himalayas. The transition appears abrupt after 78°E. For example, the total annual rainfall is 276 cm at Darjeeling, 237 cm at Mussorie (78°E) but 148 cm at Simla (77°E).

Rainfall and orography

The Himalayan area offers very interesting opportunities to study the effect of orography on precipitation in relation to airflow and synoptic systems. While the problem is to understand the effect of each range running east–west, distinction between the influence of spurs and the main range becomes difficult. In some areas, rainfall may vary little, due to many local features. Dehra Dun (682 m) records an annual rainfall of 231 cm, and Mussoorie (2,042 m) within 10 km has only 6 cm more rain.

In Nepal, considerable rainfall data have been gathered for hydrological purposes which can be interpreted in relation to the three ranges: the Siwalik, the Mahabharat and the Great Himalaya. The Siwalik Range is not high and monsoon rainfall increases from 100 cm in the plains to about 140 cm near the Siwaliks. There is no rainshadow

TABLE VI

CLIMATIC AVERAGES OF DARJEELING (27°03′N 88°16′E, 2,217 m) AND SIMLA (31°06′N 77°10′E, 2,206 m)

	Jan.	Feb.	Mar.	Apr.	May	June	July	Aug.	Sep.	Oct.	Nov.	Dec.
Darjeeling												
Mean daily max. (°C)	9.3	11.1	14.8	18.0	18.6	19.3	19.3	19.8	19.9	18.6	15.3	11.9
Mean daily min. (°C)	3.0	4.3	7.7	10.8	12.9	14.7	15.4	15.4	14.6	11.5	7.4	4.4
Lowest min. (°C)	−7.2	−6.4	−4.6	0.2	3.2	6.6	3.9	8.3	8.0	3.2	4.4	4.6
Highest max. (°C)	18.9	18.3	23.3	26.7	25.7	26.7	25.7	28.5	27.5	25.3	23.4	20.0
Mean monthly precipitation (mm)	21.7	26.7	52.4	109.2	187.1	522.3	712.9	572.5	418.5	116.1	14.2	5.0
No. of days with precipitation ⩾0.3 mm	2	6	6	9	19	22	29	27	23	9	5	1
Heaviest fall in 24 h (mm)	38.1	42.9	72.9	135.1	232.9	454.1	200.9	237.5	492.8	334.5	219.7	32.2
Simla												
Mean daily max. (°C)	8.5	10.3	14.4	19.2	23.4	24.3	21.0	20.1	20.0	17.9	15.0	11.3
Mean daily min. (°C)	1.9	3.1	6.8	11.2	15.0	16.2	15.6	15.2	13.8	10.8	7.3	4.2
Lowest min. (°C)	−10.6	−8.5	−5.6	−1.1	4.4	7.8	9.6	10.6	5.0	0.2	−1.1	−6.1
Highest max. (°C)	18.9	20.6	24.5	28.3	30.0	30.6	28.9	27.8	25.0	23.9	21.4	20.4
Mean monthly precipitation (mm)	65.2	47.6	58.1	37.6	53.7	147.5	414.5	385.4	195.2	45.4	6.7	23.7
No. of days with precipitation ⩾ 0.3 mm	9	7	9	6	8	12	25	24	15	5	1	5
Heaviest fall in 24 h (mm)	78.7	63.5	63.0	39.6	97.8	122.2	167.2	227.1	135.9	113.0	68.8	76.5

effect as the Mahabharat Range is within 35 km. Rainfall up to 240 cm has been recorded on the southern side of the Mahabharat, decreasing by as much as 50% within 20 km on the northern side of the crest of the range. The Great Himalaya is about 100 km to the north with innumerable spurs and valleys. The Kathmandu Valley (1,324 m) records 111 cm. Again, rainfall increases on coming within 40 km of the crest of the Great Himalaya and values between 250 and 350 cm have been recorded. About 20–30 km from the crest, still to the south of the range, rainfall drops sharply to less than 100 cm and even down to about 60 cm. On the northern slopes of the Great Himalaya, rainfall is less than 70 cm and in some places as low as 12 cm. In Nepal, rainfall decreases from east to west in all areas. The interesting point about the airflow during the southwest monsoon is that it is broadly parallel to the ranges but orographic effects are very prominent. In the region of the Kumaon Himalayas there are again three ranges: the Siwaliks, the Lesser Himalaya and the Great Himalaya. The last two ranges are here closer to each other than in Nepal. Higher amounts of monsoon rain seem to occur between the Lesser Himalaya and the Siwaliks than between the Mahabharat and the Siwaliks. On the other hand, unlike Nepal, there is no evidence of precipitation of over 200 cm occurring between the Great Himalaya and the Lesser Himalaya. Unfortunately, the network of precipitation stations in the Kumaon and Punjab is unsatisfactory, being less dense than in Nepal. An interesting comparison is between Dehra Dun, Rajpur and Mussoorie, all lying within 15 km from each other, at elevations of 682, 1,219 and 2,115 m, respectively, which record monsoon precipitations of 185, 266 and 197 cm. Any conclusion about maximum precipitation occurring at about 1.2 km in relation to the slope of the Lesser Himalaya requires careful examination. On the whole, in the Kumaon region there does not seem to be any significant rainshadow effect up to the Great Himalaya.

In the Punjab, the area of the Himalayas between the Siwalik and the Pir Panjal ranges, monsoon precipitation as high as 250 cm occurs here and there to the east of 76°E but only 70–120 cm to the west. Between the Great Himalaya and the Pir Panjal, precipitation is over all less than 50 cm, in places as low as 15 cm. In the rain shadow of the Greater Himalaya, precipitation is less than 10 cm or even 5 cm.

The Siwalik Range does not present any obstruction to the monsoon flow extending to the north. In the area between this range and the Pir Panjal, Lesser Himalaya and the Mahabharat Ranges, precipitation upto 250 cm occurs to the east of 76°E. Unlike in Nepal, the effectiveness of the monsoon flow in causing precipitation between the Pir Panjal and the Lesser Himalaya on one side and the Great Himalaya on the other, decreases rapidly west of about 79°E. In the Assam area, apart from spurs, there is only the Great Himalaya running east–west. Rainfall reports from this region are few. However, monsoon precipitation of over 300 cm may be quite common.

It is interesting to note here, that Lhasa (91°E) beyond the Great Himalaya has its maximum rainfall during the monsoon period: July getting as much as 12 cm and the total rainfall during the monsoon months being 84% of the annual precipitation.

Coming to winter, the airflow is mainly westerly and precipitation is generally one order of magnitude lower. Precipitation occurs with southwest or south winds so that the windward and leeward faces of the mountain ranges may still be the same as in the monsoon. Sonemarg (2,667 m), to the south of the Great Himalaya, has a winter precipitation of 63 cm, while to the north of the range Dras (3,066 m) gets 27 cm and Kargil (2,676 m) further to the northeast only 10 cm. While near the Pir Panjal precipitation is 30–40 cm, it is less (20–30 cm) in the Kashmir Valley to the north. The increase in precipitation from the plains (10–12 cm) to the Pir Panjal is only gradual. While in the Kumaon area the plains and Siwaliks have a precipitation of about 10 cm, further north amounts of up to 20 cm have been recorded. Rain shadow effects do not appear to be marked. Dehra Dun, Rajpur and Mussoorie respectively record 15, 16 and 18 cm—almost the same amounts in contrast to the variations during the monsoon. Winter precipitation decreases to almost half from western Nepal to the eastern parts. There are, no doubt, marked increases near the peaks of the two northern ranges. In Arunachal Pradesh precipitation amounts up to 25 cm have been recorded.

In the hot weather period, rainfall increases from the Indo–Nepalese border eastwards as also in the western parts of Nepal, with lower values in between.

October and November are the driest months throughout the Himalayas and the precipitation during these two months expressed as a percentage of the annual normal precipitation is between 3 and 9%.

An interesting feature of the monthly distribution of rainfall in the western Himalayas is a maximum observed in March consistently at all stations with rainfall decreasing thereafter until June or July and again showing an increase in August. This August maximum is more marked in Srinagar and Leh than at other stations, indicating incursion of the monsoon current into the Kashmir Valley and the Ladakh region as well. In Simla, Dalhousie and Dharmasala a maximum is observed in March with a pronounced decrease in April, but the rainfall again shows an upward trend from April onwards. In the eastern Himalayas, however, no such decrease from March to April is observed. The rainfall continues to increase steadily from February onwards.

Variation of precipitation with elevation

DHAR and NARAYANAN (1965) have presented figures of precipitation at Chaurikharka (2,700 m) and Namche Bazar (3,300 m) in the upper reaches of the Dudh Kosi river,

TABLE VII

PRECIPITATION AT CHAURIKHARKA AND NAMCHE BAZAR

	Chaurikharka	Namche Bazar
Precipitation (cm):		
Dec.–Feb.	2.6	7.8
Mar.–May	20.2	10.3
June–Sep.	199.3	70.0
Oct.–Nov.	8.6	6.6
Annual	228.4	93.9
No. of rainy days:		
Dec.–Feb.	3.7	10.3
Mar.–May	19.9	18.2
June–Sep.	94.6	91.0
Oct.–Nov.	7.4	9.6
Heaviest precipitation in 24 h (cm):		
Dec.–Feb.	4.4	6.4
Mar.–May	5.9	6.6
June–Sep.	30.2	8.6
Oct.–Nov.	7.9	8.7

located respectively within 40 and 30 km from Mount Everest. Some interesting details are given in Table VII.

In winter precipitation is still increasing with height, upto 3,300 m, but in the monsoon it decreases very markedly between 2,700 m and 3,300 m. The zone of maximum precipitation appears to occur in the Alps at 2,100 m. (MILLER, 1959). PUGH (1962) estimated the annual precipitation in the Everest region (Khumpaghir 4,800 m) as 45 cm. INOUE (1976) of the Japanese Glaciological Expedition to Nepal has estimated that the annual precipitation in the Khumbu Himal region is only half that of the nearly 100 cm of precipitation at elevations between 3,500 m and 4,000 m. Most of the rainfall in the Khumbu region is during the monsoon.

Snowfall in the northern mountains

The name "Himalaya" of the northern mountain ranges of the subcontinent means "abode of snow". Besides the permanent snow in this region, snow accumulates in winter and melts in the subsequent months. However, observational data of snowfall are very meagre. As such, only qualitative description of the snowfall is possible.

Snowfall is apparently more to the west of 80°E or so. Annual snowfall in the Siwaliks may be of the order of 3 m. In the Pir Panjal and Lesser Himalayas, the total is of the order of 15 m. The Kashmir Valley has much less snow. In the Great Himalaya, snowfall should be much more. Beyond the Deosai and Zaskar mountains the amount of snowfall decreases markedly. In Pakistan, the ranges north of 34°N perhaps get about 15 m of snow. The amount decreases to about 1 m over north Baluchistan. The snowfall in the Himalayas begins in October and continues upto April and May reaching its maximum in January and February. Snowfall also occurs in the monsoon months but in very small amounts at high elevations. At places 2–3 km in elevation, 40% of the winter precipitation may still occur as rainfall.

Onset and withdrawal of the monsoon

The southwest monsoon rains mainly sustain the agriculture of the subcontinent and its population. The onset and withdrawal of the monsoon are phenomena of much interest to the subcontinent. Though change of wind to southwest, decrease in temperature, increase in rainfall etc. are associated with the onset of the monsoon, they are all not synchronous. Westerlies set in in the Arabian Sea in May but the rains only in the following month. On account of the preponderating importance of rains, meteorologists in this part of the world have fixed the dates of onset and withdrawal with reference to the rather sharp increase and decrease, respectively, seen in 5-day means of rainfall and changes in circulation. The monsoon rains are sometimes not easy to distinguish from premonsoon thundershowers.

The displacement of the monsoon airmass by continental airmass and development of anticyclonic flow determines the dates of withdrawal of monsoon over north and central India.

Fig.18 shows the isochrones of the onset of the monsoon. Starting from May 20 in the Andaman Islands and May 25 in Sri Lanka, the southern tip of the peninsula is reached

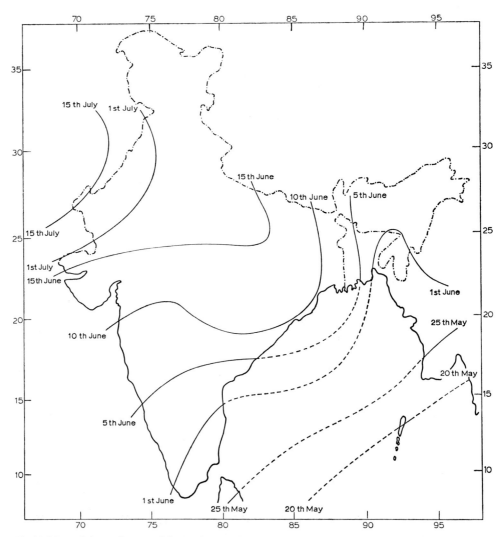

Fig.18. Normal dates of onset of the southwest monsoon.

by June 1. A month and a half later the monsoon reaches Pakistan. The withdrawal is shown in Fig.19. This commences by September 1, and by October 15 the monsoon has withdrawn even from the upper half of the peninsula. Thereafter, a transition to the northeast monsoon circulation takes place.

Synoptic systems

The important synoptic systems that influence the weather over the subcontinent are: (*1*) cyclonic storms; (*2*) low-pressure areas of the post-monsoon period; (*3*) western disturbances or troughs in the westerlies; (*4*) monsoon depressions; and (*5*) troughs off the west coast of the peninsula and mid-tropospheric cyclones.

Cyclonic storms

In the Bay of Bengal and the Arabian Sea a low pressure area with winds of 34 knots or more is called a cyclonic storm. Severe cyclonic storms (48 knots or more) occur in the

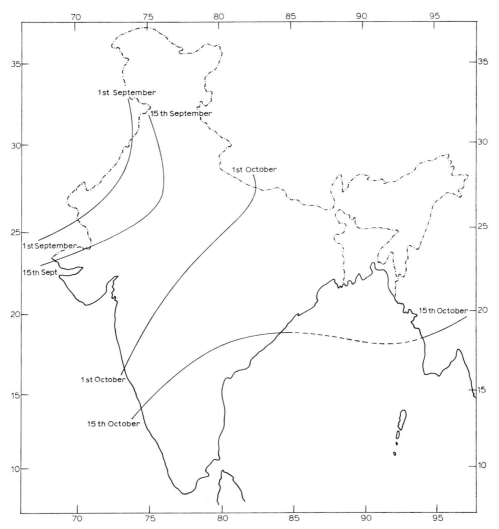

Fig.19. Normal dates of withdrawal of the southwest monsoon.

Bay of Bengal mainly in October and November and cause much devastation when they strike the coast due to high winds, tidal waves and heavy rains. River mouths are especially prone to high tidal waves, sometimes as high as 10 m. Some great killer cyclones were the Backergunge (Bangladesh) cyclone of November 1, 1876 with 9–12 m tidal waves and a loss of life of 100,000 due to drowning, the Calcutta cyclone of October 5, 1864 with a 12 m tidal wave and 50,000 dead, the Bangladesh cyclone of November 13 1971 with a peak surge of 7 m and a death toll of 300,000 and the Andhra Pradesh cyclone of November 17, 1977 with a storm surge of 5 m and 9,000 deaths.

Some statistics of cyclonic storms during the period 1891–1960 are given in Table VIII. The Bay of Bengal has four times as many cyclonic storms as the Arabian Sea. This is besides of the large number of monsoon depressions which form north of $17°-18°N$ in the Bay from June to September, while the Arabian Sea is conspicuous for the near absence of depressions during the monsoon season. Thus, the Bay of Bengal is much more cyclogenetic than the Arabian Sea.

The region of initial formation of cyclones is $5°-10°N$ in the Bay from January to March. It moves progressively northward, and by July and August depressions form only north

TABLE VIII

CHARACTERISTICS OF CYCLONIC STORMS OF THE BAY OF BENGAL AND THE ARABIAN SEA

	Jan.	Feb.	Mar.	Apr.	May	June	July	Aug.	Sep.	Oct.	Nov.	Dec.
Bay of Bengal												
Number of cyclonic storms in 70 years	4	1	4	18	28	34	38	25	27	53	56	26
Number of severe storms in 70 years	1	1	2	7	18	4	7	1	8	19	23	9
Latitude of formation	6°–11°N	5°N	5°–8°N	7°–14°N	9°–18°N	15°–22°N	North of 17°N	North of 17°N	North of 10°N	8°–19°N	7°–15°N	5°–15°N
Longitude of formation	85°–92°E	85°E	78°–94°E	82°–96°E	83°–96°E	85°–92°E	85°–92°E	85°–91°E	86°–94°E	82°–99°E	83°–98°E	82°–95°E
Place of striking coast	15°–19°N Indian coast	12°N	11°N	Burma coast south of 22°N; Indian coast south of 15°N	Indian coast north of 10°N; Bangladesh coast; Burma coast north of 15°N	Indian coast north of 12°N; Burma coast north of 20°N; Bangladesh coast.	Indian/Bangladesh coasts north of 18°N	Indian/Bangladesh coasts north of 18°N	Indian coast north of 16°N	Indian coast north of 10°N; Burma coast north of 16°N; Bangladesh coast.	whole of Indian/Bangladesh coasts and Burma coast north of 16°N	Bangladesh coast; Burma coast north of 16°N; Indian coast south of 13°N
Latitude of recurvature	–	–	–	North of 12°N	About 15°N	18°N	–	–	–	14°N	14°N	13°N
Arabian Sea												
Number of cyclonic storms in 70 years	2	0	0	5	13	13	3	1	4	17	21	3
Number of severe storms in 70 years	0	0	0	4	11	8	0	0	1	7	16	1
Latitude of formation				6°–10°N	8°–16°N	11°–21°N	North of 9°N		North of 12°N	7°–17°N	6°–13°N	6°–10°N
Longitude of formation				68°–75°E	56°–75°E	63°–73°E	Near-about 70°E		Along about 70°E	East of 58°E	East of 62°E	East of 63°E
Place of striking coast				Indian coast north of 21°N; Pakistan coast	Indian north of 11°N; Pakistan coast	Indian coast west of 73°E; Pakistan coast	Indian coast west of 73°E		Indian coast west of 71°E; Pakistan coast	Indian coast east of 68°E	Indian coast south of 23°N	Indian coast south of 23°N
Latitude of recurvature				17°N	17°N	19°N	–	–	22°N	16°N	14°N	–

of 17°–18°N. The region of formation shifts Equator–ward from September, and by December cyclones form only between 5° and 15°N. Similar changes take place in the Arabian Sea from 6° to 10°N in April to isolated cases north of 19°N in July and back to 6°–10°N in December.

From January to March, the Bay cyclones move between west and northwest and strike the coast of Tamilnadu. In April, after an initial northwest to north movement, most storms recurve northeastward and strike the coast of Burma. In May, some storms have a long enough west-northwest to northwest course to reach the east coast of India, but many recurve or move north-northeast from the beginning to strike the Bangladesh and Burma coasts. Most June storms move northwest. West-northwest is the course in July and August. Movement in September is west to northwest. The coast affected from June to September is mainly confined to Orissa and West Bengal.

In October and November, the course of cyclones ranges from west to northeast so that the whole coast from Arakan to Tamilnadu is vulnerable. Cyclonic storms that form at a lower latitude take a more west-northwesterly course and those originating at a higher latitude a northeasterly course. These are the months of the severest storms. December storms are equally distributed between storms having a west-northwest course and those recurving towards Burma. Illustrative tracks for October and November are given in Figs.20 and 21.

Quite a few of the Arabian Sea storms originate from Bay cyclones which have moved across the peninsula in a weakened form. Many storms in the Arabian Sea move north-northwest in May, October and November, striking the Arabian coast in the first two months. June storms move northwest to northeast and so do many in October and November.

The cyclones of the southwest monsoon season have a much longer life as depressions over land than initially over sea. October cyclones also continue over land for a few days after crossing the coast.

Low-pressure areas of the post-monsoon period

The cyclonic storms described in the previous section develop from low-pressure areas. However, there are cases when the low-pressure areas which form in the west-central or southwest Bay move westwards without intensification and affect weather over the southern peninsula and Sri Lanka. On an average, during October and November, there may be two or three such low-pressure areas in a month and in December one or two which do not develop further. In some cases, a trough extends northwards from the low. The cyclonic circulation of the lows, does not usually extend above the 850-mbar level. Rainfall due to some low-pressure areas is fairly widespread over the southern peninsula and sometimes even heavy.

Western disturbances

A "western disturbance" is a low or trough either at the surface or in the upper air in the westerly wind regime, north of the subtropical high pressure belt. Such systems, moving across Iran and Russian Turkistan, affect the subcontinent north of 30°N. Weak circulations, called induced lows, may simultaneously develop over the central parts of Pakistan and Rajasthan and move east-northeastwards. On an average about 7 or 8 western disturbances (including induced lows) move across Pakistan and India every month in the winter. In the hot weather period also western disturbances markedly control the weather of the northern parts, though the tracks of the disturbances are somewhat to the north.

In the southwest monsoon period, westerly troughs move mainly north of 40°N and their interaction with the monsoon circulation of the Indian subcontinent is varied.

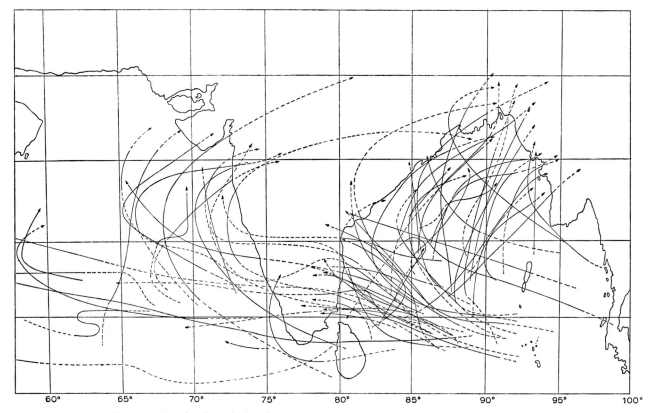

Fig.20. Tracks of cyclones in October.

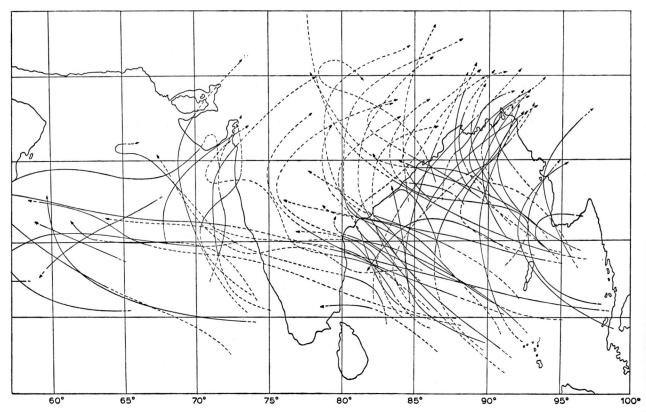

Fig.21. Tracks of cyclones in November.

TABLE IX

NUMBER OF CYCLONIC DISTURBANCES (1890–1970)

	June		July		August		September	
	D	S	D	S	D	S	D	S
Bay of Bengal	71	35	107	33	132	26	141	32
Land area	12	1	39	1	42	0	21	1

D = depression; S = cyclonic storm.

Monsoon depressions

Monsoon depressions are low-pressure areas with two or three closed isobars (at the 2-mbar interval) covering an area of about a 5° square, which form in the Bay of Bengal from June to September, north of 17°–18°N and move westnorthwest, at least upto the central parts of the subcontinent, before weakening or filling up. They give widespread rains in the southwest quadrant with many heavy falls. Some of them intensify into storms which has been mentioned in the earlier section. Quite a few depressions form in situ over land as well. The number of monsoon depressions and storms in the period 1890–1970 is given in Table IX.

Depressions and storms of July move mainly west to west-northwest over the Bay and across the country up to 25°N. In August their movement is mainly west-northwest. In higher latitudes, the movement becomes more northerly. In June, the movement is more spread out; besides, the other differences are change in movement to a northerly course at lower latitudes, and recurvature and movement towards the northeast while over the Bay. Early in June, with the advancing monsoon, depressions may take a north-northeast track and usher in the monsoon into the northeastern parts of the subcontinent. Even in July and August, some initial northerly tracks are occasionally seen before systems change to the usual west-northwestward movement. September is like June. While the tracks of the Bay depressions in July and August are within a narrow belt, they are very much spread out in June and September. Storms weaken as they cross the coast. Their duration as depression and storm is usually 3–4 days, though a few persist even upto 9 days.

Troughs off the west coast of the peninsula and mid-tropospheric cyclones

A weak trough develops frequently along and off the west coast of India, anywhere from north Kerala to south Gujarat, during the period of the southwest monsoon, and is responsible for the strengthening of the monsoon in terms of rainfall in the adjacent coastal belt. The troughs form more often near 13°–15°N and shift northward, though they may appear and disappear in situ over any area.

When such a trough is off north Konkan and Saurashtra, the system may have closed circulation in the upper air. Such systems have come to be known also as mid-tropospheric cyclones. Some workers claim that mid-tropospheric cyclones account for most of the monsoon rains of western India and the northeast Arabian Sea.

Radiation

The mean duration of sunshine as compiled by RAO and GANESAN (1971) is shown in Fig.22 for January, April, July and October. At Srinagar, the duration is as low as 2.5 h per day in January due to cloudiness from western disturbances, though 10 h is possible. Even at Peshawar it is only 5.6 h per day. A very large part of the Indian subcontinent experiences heavy clouding after the onset of the southwest monsoon and the hours of sunshine are about a third of the maximum possible. In August, Bombay has

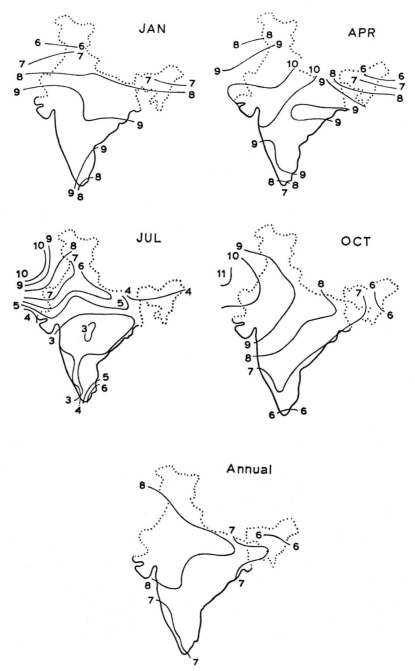

Fig.22. Mean sunshine hours.

only 17%, Nagpur 21% and stations in Kerala 20% of the possible sunshine. Karachi has also only 28% of the possible hours of bright sunshine in this month.

Fig.23 presents the total solar radiation received at the ground compiled by JAGANNA-THAN and GANESAN (1967). In some areas the solar radiation is less in June–September than in December–February, which is due to the heavy monsoon cloudiness. Thus, there is greater uniformity throughout the year in the radiation received except in the north-western parts.

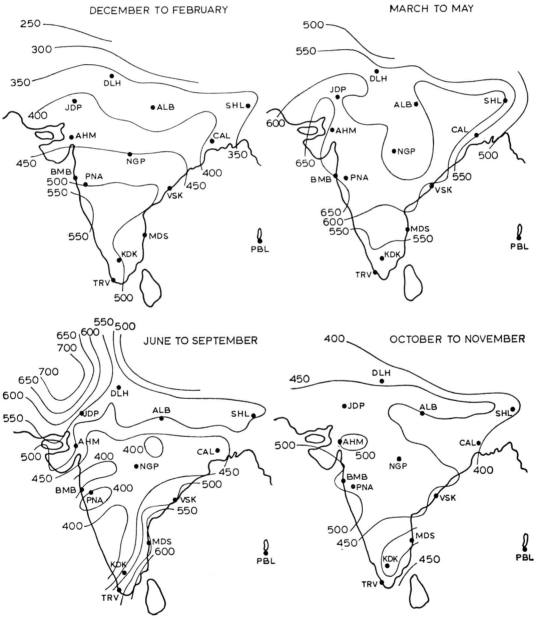

Fig.23. Mean solar radiation received at the surface (gcal.cm^{-2} day^{-1}).

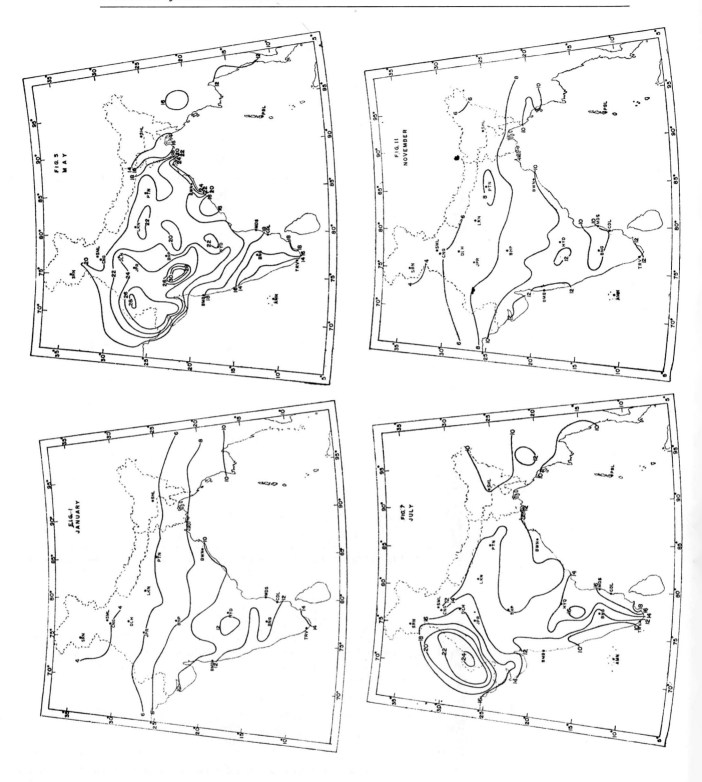

Evapotranspiration

RAO et al. (1971a) have calculated the potential evapotranspiration from climatological data of temperature, vapour pressure, cloudiness etc. using Penman's formula with some modifications. Fig.24 shows the potential evapotranspiration for January, May, July and November and for the year as a whole. May is the month of the maximum potential evapotranspiration in many parts and this situation continues in the northwest in June as well. The cloudiness of the southwest monsoon substantially reduces potential evapotranspiration at a time when the sun is very high. Thus the monsoon not only provides rains, but also conserves this water by restricting evapotranspiration. In July, the regions of very high potential evapotranspiration are Pakistan. Rajasthan and southern parts of Tamil Nadu, which do not have much cloudiness.

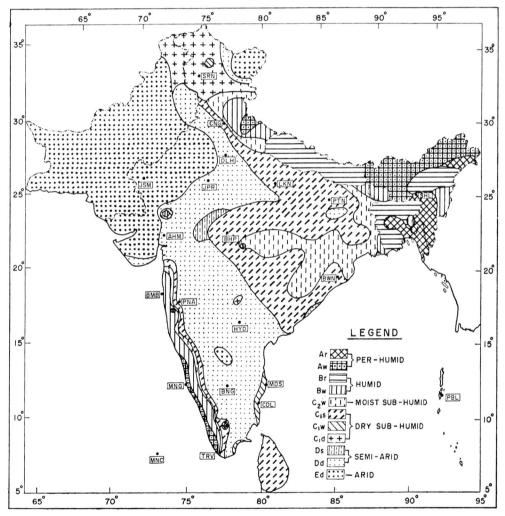

Fig.25. Climatic classification of the Indian subcontinent.

Climatic classification

Rao et al. (1971b) are the latest to make a climatic classification of India. They computed potential evapotranspiration as explained in the last section by Penman's method and used the values to classify the climate according to the revised procedure of Thornthwaite and Mather (1955). Fig.25 shows the moisture regime for the Indian subcontinent as prepared by R. P. Sarker (personal communication, 1979).

The west coast of the peninsula, the Western Ghats, the Himalayas, Assam and Meghalaya and small areas at high altitudes in the Aravallis, Vindhyas, etc., receiving abundant southwest monsoon rains have humid or per-humid climate. West Bengal, Orissa, Madhya Pradesh, Viharba and northern parts of Andhra Pradesh and the northern portions of the Punjab, Haryana and Uttar Pradesh have sub-humid (moist or dry)

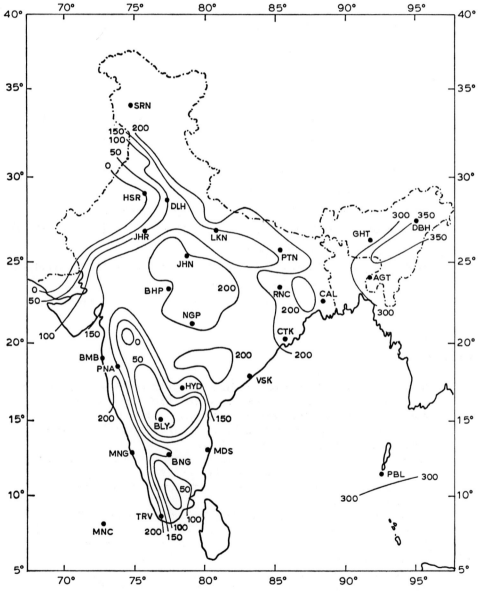

Fig.26. Water availability duration (days).

116

climate. A narrow coastal strip of Tamil Nadu also comes under this category. But more than half the subcontinent, particularly the interior of the peninsula and the northwestern parts of the subcontinent are arid or semi-arid. Practically the whole of the subcontinent comes under the mega thermal type of the thermal regime.

Areas classified as semi-arid support a fairly good density of population, which is dependent on agriculture. Irrigation from wells and small-sized reservoirs using natural slope of the ground no doubt provide water to the crops. RAMAN and SRINIVASAMURTHY (1971) have computed the duration of water availability periods in India in different regions. This is the period when the rainfall and water stored in the soil are able to meet the losses due to potential evapotranspiration. The water storage capacity of the soils has been broadly taken into consideration. As will be seen from Fig. 26, in some semi-arid areas water is available for more than 50 days. If the capacity of plants to withstand some moisture stress is considered, the period suitable for growing crops is even longer. The semi-arid areas are, however, liable to great variability of rainfall and in years of poor rains agriculture suffers badly.

Acknowledgements

The author is grateful to Mr. Muhammed Samiullah, Director-General of the Pakistan Meteorological Department, Mr. K. D. N. De Silva, Director of the Department of Meteorology, Sri Lanka, Wg. Cdr. M. S. Mawla, Director of the Bangladesh Meteorological Department and Dr. S. P. Adhikary, Chief Meteorologist of the Department of Irrigation, Hydrology and Meteorology, Nepal and the India Meteorological Department for the data provided by them relating to their countries.

The author thanks Dr. R. P. Sarkar of India for providing the climatic classification map for the whole of the subcontinent.

Discussions with Dr. A. S. Ramanathan and Mr. S. N. Tripathi were very useful in preparing this account and the author expresses his thanks to them.

The author is indebted to Mr. S. Jayaraman for typing the manuscript.

References

BANERJEE, S. K., 1930. The effect of the Indian mountain ranges on air motion. *Indian J. Phys.*, 5: 699–745.

BLANFORD, H. S., 1889. *Climates and Weather of India, Ceylon, Burma and the Storms of Indian Seas.*

Climatological Atlas for Airmen. 1943, India Meteorological Department.

Climatological Tables of Observatories in India. 1953, India Meteorological Department.

Climatological Tables of Observatories in India. 1931–1960, 1966, India Meteorological Department.

DHAR, O. N. and NARAYANAN, J., 1965. A study of precipitation distribution in the neighbourhood of Mount Everest. *Indian J. Meteorol. Geophys.*, 16: 229–240.

HAHN, D. G. and MANABE, S., 1975. The role of mountains in the South Asian monsoon circulation. *J. Atmos. Sci.*, 32: 1515–1541.

INOUE, J., 1976. Climate of Khumbu Himal. *J. Jpn. Soc. Snow Ice*, 38: 66–73.

JAGANNATHAN, P., 1968. Rainfall of India. *Forecasting manual rept., India Meteorol. Dept.*, IV–13: 1–12.

JAGANNATHAN, P. and GANESAN, H. R., 1967. Distribution of solar radiation in India. *Sci. Rept., India Meteorol. Dept.*, 13: 1–7.

MILLER, A. A., 1959. *Climatology.* Methuen, London, 39 pp.

Monthly and Annual Rainfall and Number of Rainy Days. 1901–1950. 1962, India Meteorol. Dept.

Monthly Meteorological Charts of the Indian Ocean. 1949, London Meteorol. Office.

PUGH, L. G. C. E., 1962. Himalayan scientific and mountaineering expedition 1960/61—The scientific programme. *Geogr. J.*, 128: 447–456.

Rainfall Atlas of India, 1972. India Meteorol. Dept.

RAMAN, C. R. V. and SRINIVASAMURTHY, B., 1971. Water availability periods for crop planning. *Sci. Rept., India Meteorol. Dept.*, 173: 1–9.

RAO, K. N. and GANESAN, H. R., 1971. Sunshine over India. *Sci. Rept., India Meteorol. Dept.*, 172: 1–9.

RAO, K. N., GEORGE, C. J. and RAMASASTRI, K. S., 1971a. Potential evapotranspiration over India. *Sci. Rept., India Meteorol. Dept.*, 136: 1–7.

RAO, K. N., GEORGE, C. J. and RAMASASTRI, K. S., 1971b. Climatic classification of India. *Sci. Rept., India Meteorol. Dept.*, 158: 1–9.

RAO, Y. P., 1976. *Southwest Monsoon*. India Meteorological Department, 367 pp.

RAO, Y. P. and RAMAMURTHI, K. S., 1968. Climate of India. *Forecasting Manual Rept., India Meteorol. Dept.*, I–2: 1–17.

SIMPSON, G., 1921. The southwest monsoon. *Q. J. R. Meteorol. Soc.*, 47: 151–172.

SUBRAHMANYAM, V. P., SUBBA RAO, B. and SUBRAMANIAM, A. R., 1965. Köppen and Thornthwaite system of climatic classification as applied to India. *Ann. Arid Zone*, 4: 1.

SRINIVASAN, V. and RAMAMURTHY, K., 1973. Northwest monsoon. *Forecasting Manual Rept., India Meteorol. Dept.*, IV–18.4: 1–60.

THORNTHWAITE, C. W. and MATHER, J. R., 1955. The water balance. In: *The Yearbook of Agriculture, 8.* U.S. Dept. of Agriculture, Washington D.C.

Tracks of Storms and Depressions in the Bay of Bengal and the Arabian Sea (1877–1960). 1964, India Meteorol. Dept.

Tracks of Storms and Depressions in the Bay of Bengal and the Arabian Sea (1961–1970). 1972, India Meteorol. Dept.

Appendix—Climatic tables

TABLE X

CLIMATIC TABLE FOR LEH

Latitude 34°09′N, longitude 77°34′E, elevation 3,514 m

Month	Mean sta. press. (mbar)	Temperature (°C)				Relative humidity (%)		Precipitation (mm)	
		daily max.	daily min.	extreme		03 h*¹	12 h	mean	max. in 24 h
				max.	min.				
Jan.	664.5	−2.8	−14.0	3.3	−28.3	61	51	11.8	24.4
Feb.	663.7	0.8	−11.8	12.8	−26.4	59	46	8.6	16.8
Mar.	664.3	6.4	−6.3	19.4	−19.4	55	43	11.9	16.0
Apr.	666.0	12.4	−1.2	23.9	−12.8	50	32	6.5	22.1
May	665.8	17.1	2.8	28.9	−4.4	39	27	6.5	22.3
June	664.1	21.1	6.7	33.9	−1.1	39	24	4.3	19.6
July	663.1	24.7	10.2	33.3	0.6	49	34	15.7	23.6
Aug.	663.4	24.2	9.6	32.2	2.8	54	36	19.5	51.3
Sep.	665.5	20.9	5.4	30.6	−4.4	47	32	12.2	25.9
Oct.	667.7	14.2	−0.9	25.6	−7.8	45	28	7.1	39.1
Nov.	667.3	7.8	−6.6	20.0	−13.9	45	34	2.9	16.2
Dec.	666.0	1.6	−11.1	12.8	−25.6	54	42	8.0	15.2
Annual	665.1	12.4	−1.4	33.9	−28.3	50	36	115.0	51.3

Month	Number of days with						Mean cloudi-ness (oktas)	Mean sun-shine hours	Wind		
	precip. ⩾ 0.3 mm	hail	thunder-storm	fog	dust-storm	squall			preval. direct.		mean speed (m/sec)
									03 h*¹	12 h	
Jan.	6	0	0	*	*	0	4.9		NE	SW	0.9
Feb.	4	0	0	0	0	*	4.7		NE	SW	1.1
Mar.	4	*	0	*	*	*	4.8		S	SW	1.5
Apr.	3	1	*	0	1	*	4.3		S	SW	1.9
May	3	1	1	0	*	*	4.1		S	SW	1.9
June	2	*	2	0	*	0	3.3		S	SW	1.8
July	4	*	1	0	*	0	3.7		S	SW	1.4
Aug.	5	*	1	0	0	0	4.1		S	SW	1.3
Sep.	3	*	*	0	0	0	2.9		S	SW	1.3
Oct.	2	0	0	0	0	0	2.3		S	SW	1.4
Nov.	1	0	0	0	0	0	2.9		NE	SW	1.4
Dec.	5	0	0	*	0	0	4.3		NE	SW	1.1
Annual	41	2	4	1	1	1	3.8		S	SW	1.4

*¹ Greenwich mean time. * = <1

Recording time climatic tables
Bangladesh, Pakistan, India: 1931–1960.
Sri Lanka: some elements are for 1931–1960, while others are for different periods.
Nepal: 9 to 4 years between 1968 and 1976.
The tables have been arranged according to latitude, from north to south.

TABLE XI

CLIMATIC TABLE FOR SRINAGAR

Latitude 34°05′N, longitude 74°50′E, elevation 1,586 m

Month	Mean sta. press. (mbar)	Temperature (°C)				Relative humidity (%)		Precipitation (mm)	
		daily max.	daily min.	extreme		03 h*¹	12 h	mean	max. in 24 h
				max.	min.				
Jan.	844.9	4.4	−2.3	17.2	−14.4	88	78	72.8	147.8
Feb.	843.2	7.9	−0.8	20.6	−13.8	87	69	72.3	66.3
Mar.	842.3	13.4	3.5	25.6	−5.6	84	60	104.1	70.1
Apr.	841.7	19.3	7.4	31.1	0.0	77	53	78.1	65.3
May	839.1	24.6	11.2	36.4	2.8	71	49	63.4	52.8
June	835.3	29.0	14.4	37.8	7.2	75	45	35.6	65.8
July	832.9	30.8	18.4	38.3	10.6	73	49	61.0	79.0
Aug.	834.2	29.9	17.9	36.7	10.0	79	53	62.8	67.3
Sep.	838.2	28.3	12.7	35.0	4.4	77	47	31.8	102.4
Oct.	843.1	22.6	5.7	33.9	−1.7	82	51	28.7	59.9
Nov.	845.0	15.5	−0.1	23.9	−7.8	85	54	17.5	64.3
Dec.	845.3	8.8	−1.8	18.3	−12.8	88	67	35.9	64.5
Annual	840.4	19.5	7.2	38.3	−20.0	81	56	664.0	147.8

Month	Number of days with						Mean cloudi-ness (oktas)	Mean sun-shine hours	Wind		
	precip. ⩾0.3 mm	hail	thunder-storm	fog	dust-storm	squall			preval. direct.		mean speed (m/sec)
									03 h*¹	12 h	
Jan.	11	0	*	1	0	*	6.6	2.5	SE	Var.	1.0
Feb.	8	0	*	1	0	0	5.9	4.0	SE	NW	1.2
Mar.	14	*	3	*	*	*	5.4	4.5	SE	NW	1.6
Apr.	12	*	7	0	*	*	4.6	6.1	SE	NW	1.5
May	10	*	10	0	*	*	4.0	7.7	SE	NW	1.2
June	7	0	7	0	1	*	3.1	8.2	SE	NW	1.1
July	9	0	7	0	*	*	3.8	8.2	SE	NW	1.1
Aug.	10	0	6	0	*	*	4.2	7.7	SE	NW	1.0
Sep.	6	0	4	0	*	*	2.7	8.0	SE	NW	1.0
Oct.	5	0	2	0	*	*	2.3	7.9	SE	NW	0.9
Nov.	4	0	0	*	0	0	2.9	6.7	SE	NW	0.8
Dec.	7	0	*	1	0	0	5.1	3.9	SE	NW	0.9
Annual	103	1	47	3	2	1	4.2	6.3	SE	NW	1.1

*¹ Greenwich mean time. * = <1.

TABLE XII

CLIMATIC TABLE FOR PESHAWAR

Latitude 34°01′N, longitude 71°35′E, elevation 359 m

Month	Mean sta. press. (mbar)	Temperature (°C)				Mean vap. press. (mbar)		Precipitation (mm)	
		daily max.	daily min.	extreme		03 h*¹	12 h	mean	max. in 24 h
				max.	min.				
Jan.	976.4	17.3	4.1	24	−3	6.9	8.2	38.6	84.1
Feb.	974.1	19.6	6.6	30	−1	8.1	8.4	41.1	61.2
Mar.	971.6	23.8	11.0	34	2	11.7	10.8	64.8	50.3
Apr.	968.2	29.8	16.0	42	7	14.4	13.0	41.9	54.4
May	962.6	36.4	21.8	48	12	14.8	12.5	14.5	24.6
June	957.2	40.2	25.6	48	13	17.5	14.3	6.6	29.7
July	956.4	38.3	26.9	46	21	26.5	25.1	33.8	76.2
Aug.	958.8	36.1	25.9	43	19	27.9	27.2	40.6	72.9
Sep.	963.4	35.2	22.6	41	14	21.3	20.8	14.2	44.5
Oct.	980.2	31.2	16.2	38	8	13.0	14.9	9.9	37.1
Nov.	974.8	25.5	9.3	33	1	7.8	10.5	9.9	50.5
Dec.	976.3	20.0	5.0	28	−2	6.6	8.8	15.2	41.4
Annual	967.5	29.4	16.0	48	−3	14.7	14.5	331.2	84.1

Month	Number of days with						Mean cloudi-ness (oktas)	Mean sun-shine hours	Wind		
	precip. ⩾0.3 mm	hail	thunder-storm	fog	dust-storm	squall			mean direct.		mean speed (m/sec)
									03 h*¹	12 h	
Jan.									S	E	1.1
Feb.											1.4
Mar.											1.5
Apr.									S	NE	1.6
May											2.3
June											2.5
July									N	NE	2.5
Aug.											2.2
Sep.											1.7
Oct.									SW	NE	0.9
Nov.											0.8
Dec.											0.8
Annual											1.6

*¹ Greenwich mean time.

TABLE XIII

CLIMATIC TABLE FOR RAWALPINDI

Latitude 33°35′N, longitude 73°03′E, elevation 511 m

Month	Mean sta. press. (mbar)	Temperature (°C)				Mean vap. press. (mbar)		Precipitation (mm)	
		daily max.	daily min.	extreme		03 h*[1]	12 h	mean	max. in 24 h
				max.	min.				
Jan.	958.2	16.0	2.2	24	−4	7.7	8.2	63.5	77.5
Feb.	956.1	19.4	5.6	31	−2	8.6	9.8	63.5	96.0
Mar.	953.7	24.0	10.2	36	1	11.0	10.6	81.3	93.2
Apr.	950.6	30.7	15.3	44	7	12.7	11.6	42.4	84.1
May	945.6	37.1	21.1	46	12	13.7	12.5	23.4	35.6
June	941.2	39.8	24.6	48	14	17.8	16.9	54.6	81.3
July	941.0	36.3	24.9	46	17	27.0	27.0	232.7	168.7
Aug.	943.0	34.0	23.8	42	14	28.4	28.8	258.1	150.6
Sep.	947.2	34.4	21.4	39	12	22.3	22.6	85.1	120.7
Oct.	953.0	31.6	14.6	38	7	13.4	14.1	21.1	68.6
Nov.	956.4	27.8	9.2	32	−1	8.4	10.1	11.9	53.3
Dec.	958.2	19.8	3.3	27	−3	7.3	8.6	22.6	51.1
Annual	950.3	29.2	14.6	48	−4	14.9	15.0	960.1	168.7

Month	Number of days with						Mean cloudiness (oktas)	Mean sunshine hours	Wind		
	precip. ⩾0.3 mm	hail	thunderstorm	fog	duststorm	squall			mean direct.		mean speed (m/sec)
									03 h*[1]	12 h	
Jan.									NW	W	0.4
Feb.											0.6
Mar.											0.6
Apr.									NW	W	0.6
May											0.7
June											0.6
July									E	S	0.5
Aug.											0.4
Sep.											0.5
Oct.									NE	S	0.4
Nov.											0.3
Dec.											0.4
Annual											0.5

*[1] Greenwich mean time.

TABLE XIV

CLIMATIC TABLE FOR D. I. KHAN

Latitude 31°49′N, longitude 70°55′E, elevation 174 m

Month	Mean sta. press. (mbar)	Temperature (°C)				Mean vap. press. (mbar)		Precipitation (mm)	
		daily max.	daily min.	extreme max.	extreme min.	03 h*1	12 h	mean	max. in 24 h
Jan.	997.2	19.6	4.6	27	−3	7.1	9.4	14.5	32.0
Feb.	994.4	22.8	7.4	33	−2	8.4	10.3	18.0	50.8
Mar.	991.3	27.4	13.0	38	0	12.4	13.0	27.2	50.0
Apr.	987.4	34.2	18.4	47	7	15.3	15.5	20.1	60.5
May	981.3	40.0	23.8	49	15	17.7	18.3	9.4	32.0
June	976.2	42.4	27.2	49	18	24.2	23.2	8.6	39.6
July	975.8	39.4	28.0	49	21	31.8	23.2	64.5	111.8
Aug.	978.2	38.0	27.0	48	18	31.5	31.7	36.1	101.6
Sep.	983.0	37.4	24.0	44	16	24.5	26.0	14.0	41.9
Oct.	990.0	34.2	17.0	41	8	15.7	18.6	2.3	20.3
Nov.	994.6	28.2	9.6	35	2	9.4	14.2	2.5	30.2
Dec.	997.0	22.2	5.4	29	−2	7.5	11.2	5.8	23.1
Annual	987.2	32.2	17.2	49	−3	17.1	18.5	223.0	111.8

Month	Number of days with						Mean cloudiness (oktas)	Mean sunshine hours	Wind mean direct. 03 h*1	12 h	mean speed (m/sec)
	precip. ≥0.3 mm	hail	thunderstorm	fog	duststorm	squall					
Jan.									NW	NW	0.4
Feb.											0.5
Mar.											0.6
Apr.									N	E	0.7
May											0.8
June											0.8
July									NE	SE	0.8
Aug.											0.7
Sep.											0.5
Oct.									NE	E	0.3
Nov.											0.3
Dec.											0.2
Annual											0.5

*1 Greenwich mean time.

123

TABLE XV

CLIMATIC TABLE FOR LAHORE

Latitude 31°33′N, longitude 74°20′E, elevation 214 m

Month	Mean sta. press. (mbar)	Temperature (°C)				Mean vap. press. (mbar)		Precipitation (mm)	
		daily max.	daily min.	extreme max.	extreme min.	03 h*1	12 h	mean	max. in 24 h
Jan.	992.0	19.3	5.1	28	−2	8.9	10.2	31.2	74.7
Feb.	989.4	22.4	8.0	33	0	10.5	10.7	23.1	52.6
Mar.	985.9	27.8	13.0	38	3	12.6	12.1	24.4	31.2
Apr.	982.5	34.8	18.4	46	10	15.5	12.6	15.7	41.7
May	976.9	40.2	23.6	48	15	17.2	15.6	8.1	18.8
June	972.6	41.0	26.8	47	18	22.6	20.9	38.9	67.3
July	972.6	37.0	27.2	46	21	29.8	30.5	121.7	135.9
Aug.	974.8	35.8	26.6	43	19	29.0	31.8	122.9	121.4
Sep.	979.3	35.9	24.1	42	17	25.5	26.7	80.0	228.1
Oct.	985.7	33.7	17.3	41	8	17.5	18.7	9.9	34.0
Nov.	989.8	27.8	9.6	35	2	11.5	13.9	3.6	57.9
Dec.	992.0	21.9	5.7	28	−1	9.7	11.9	10.7	24.9
Annual	982.8	31.5	17.1	48	−2	17.6	18.0	490.2	228.1

Month	Number of days with						Mean cloudiness (oktas)	Mean sunshine hours	Wind		
	precip. ⩾0.3 mm	hail	thunderstorm	fog	duststorm	squall			mean direct. 03 h*1	12 h	mean speed (m/sec)
Jan.									N	NW	0.7
Feb.											1.0
Mar.											1.3
Apr.									N	NW	1.3
May											1.4
June											1.4
July									SE	SE	1.3
Aug.											1.3
Sep.											0.9
Oct.									E	NW	0.5
Nov.											0.5
Dec.											0.3
Annual											0.9

*1 Greenwich mean time.

TABLE XVI

CLIMATIC TABLE FOR ZHOB (FORT SANDEMAN)

Latitude 31°21′N, longitude 69°27′E, elevation 1,406 m

Month	Mean sta. press. (mbar)	Temperature (°C)				Mean vap. press. (mbar)		Precipitation (mm)	
		daily max.	daily min.	extreme		03 h*1	12 h	mean	max. in 24 h
				max.	min.				
Jan.	861.0	12.4	−0.6	24	−9	4.7	4.5	24.1	22.9
Feb.	859.7	15.8	2.4	29	−7	5.0	5.3	30.2	26.7
Mar.	858.7	20.8	7.4	32	−4	7.0	5.7	41.4	57.1
Apr.	857.7	26.9	13.1	38	4	8.3	6.7	26.7	51.3
May	855.2	32.8	18.2	42	7	9.4	7.7	21.6	70.4
June	851.4	37.0	22.8	43	13	12.8	11.0	12.7	39.6
July	850.2	36.8	23.8	44	17	19.9	15.7	52.1	67.3
Aug.	851.6	35.9	22.9	41	17	18.9	14.1	47.5	48.3
Sep.	855.0	33.8	19.2	39	12	12.8	10.4	3.8	19.8
Oct.	860.2	28.2	12.0	35	2	7.2	5.6	1.3	8.4
Nov.	862.1	22.2	5.6	31	−4	4.1	2.9	2.8	20.8
Dec.	861.6	16.2	1.2	28	−9	4.4	4.4	14.5	27.9
Annual	857.0	26.6	12.4	44	−9	9.5	7.8	278.6	70.4

Month	Number of days with						Mean cloudi-ness (oktas)	Mean sun-shine hours	Wind		
	precip. ⩾0.3 mm	hail	thunder-storm	fog	dust-storm	squall			mean direct.		mean speed (m/sec)
									03 h*1	12 h	
Jan.									SE	NW	0.8
Feb.											1.1
Mar.											1.4
Apr.									SW	W	1.4
May											1.3
June											1.3
July									NE	NE	1.5
Aug.											1.1
Sep.											0.8
Oct.									SE	W	0.8
Nov.											0.6
Dec.											0.6
Annual											1.1

*1 Greenwich mean time.

TABLE XVII

CLIMATIC TABLE FOR LUDHIANA

Latitude 30°56′N, longitude 75°52′E, elevation 247 m

Month	Mean sta. press. (mbar)	Temperature (°C)				Relative humidity (%)		Precipitation (mm)	
		daily max.	daily min.	extreme		03 h*1	12 h	mean	max. in 24 h
				max.	min.				
Jan.	987.7	20.2	5.8	28.9	−2.2	83	53	34.6	119.6
Feb.	985.3	23.3	8.4	33.3	−1.1	78	44	33.7	71.9
Mar.	982.5	29.0	12.9	41.1	1.3	67	38	29.1	89.9
Apr.	978.7	36.0	18.5	46.1	8.9	47	27	11.0	55.9
May	973.5	41.2	24.2	48.3	11.7	37	22	9.5	31.2
June	969.5	41.1	27.1	47.9	18.0	49	32	54.4	108.7
July	969.7	36.0	26.7	47.3	17.4	74	60	187.5	163.1
Aug.	971.7	34.7	26.1	44.4	19.0	79	66	175.0	184.9
Sep.	975.9	35.3	23.9	41.7	15.2	74	53	118.5	206.5
Oct.	982.1	33.9	17.5	40.0	9.4	64	39	33.9	354.3
Nov.	986.1	28.8	10.1	35.8	−3.4	64	39	3.0	25.4
Dec.	987.9	22.9	6.2	29.4	−1.1	79	50	14.3	60.5
Annual	979.2	31.9	17.3	48.3	−3.4	66	44	704.5	354.3

Month	Number of days with						Mean cloudi-ness (oktas)	Mean sun-shine hours	Wind		
	precip. ≥0.3 mm	hail	thunder-storm	fog	dust-storm	squall			preval. direct.		mean speed (m/sec)
									03 h*1	12 h	
Jan.	4	0	1	*	0	0	2.9		NW	NW	0.7
Feb.	3	0	*	*	0	0	2.7		NW	NW	0.9
Mar.	4	*	1	0	0	0	2.5		NW	NW	1.1
Apr.	2	*	1	0	*	0	1.9		NW	NW	1.1
May	2	0	1	0	1	0	1.5		SE	NW	1.3
June	4	0	1	0	1	0	1.9		SE	NW	1.4
July	11	0	1	0	*	0	4.3		SE	SE	1.2
Aug.	10	0	1	0	0	0	4.3		SE	Var.	0.9
Sep.	6	0	*	0	*	0	2.6		SE	NW	0.8
Oct.	2	0	*	0	*	0	0.7		SE	NW	0.6
Nov.	1	0	*	0	0	0	0.9		WNW	NW	0.6
Dec.	2	0	0	*	0	0	2.2		NW	NW	0.5
Annual	50	*	8	1	3	0	2.4		SE	NW	0.9

*1 Greenwich mean time *= <1.

TABLE XVIII

CLIMATIC TABLE FOR AMBALA

Latitude 30°23′N, longitude 76°46′E, elevation 272 m

Month	Mean sta. press. (mbar)	Temperature (°C)				Relative humidity (%)		Precipitation (mm)	
		daily max.	daily min.	extreme		03 h*¹	12 h	mean	max. in 24 h
				max.	min.				
Jan.	984.5	20.3	6.8	28.9	−1.1	79	52	42.0	104.9
Feb.	981.9	23.8	8.5	33.9	−0.6	73	44	51.8	146.8
Mar.	978.9	29.6	14.1	41.7	3.9	61	33	28.2	103.9
Apr.	975.3	36.2	19.7	45.0	10.0	41	22	11.9	82.0
May	970.2	40.8	24.9	47.8	14.4	36	21	15.2	85.1
June	966.7	40.5	27.3	47.8	18.6	49	32	72.0	105.9
July	966.8	35.2	26.0	46.7	19.4	78	63	271.6	162.3
Aug.	968.7	33.8	25.4	43.9	20.0	82	68	249.5	228.9
Sep.	972.7	35.4	23.9	40.6	15.6	80	58	168.2	224.3
Oct.	978.7	33.2	16.4	39.4	8.3	68	41	27.3	138.4
Nov.	982.5	28.6	10.2	35.6	2.8	68	39	6.4	135.1
Dec.	984.3	23.2	7.1	29.4	−0.6	78	48	14.7	47.5
Annual	975.9	31.8	17.5	47.8	−1.1	66	43	958.8	228.9

Month	Number of days with						Mean cloudi-ness (oktas)	Mean sun-shine hours	Wind		
	precip. ⩾0.3 mm	hail	thunder-storm	fog	dust-storm	squall			preval. direct.		mean speed (m/sec)
									03 h*¹	12 h	
Jan.	4	0	*	*	0	0	2.6		NW	NW	1.5
Feb.	3	*	*	0	*	0	2.4		NW	NW	1.9
Mar.	3	*	1	0	*	0	2.1		NW	NW	2.0
Apr.	1	*	1	0	1	0	1.7		NW	NW	2.0
May	3	0	2	0	1	0	1.5		NW	NW	2.2
June	5	*	1	0	1	0	2.1		SE	NW	2.4
July	13	0	*	0	*	0	4.1		SE	SE	2.0
Aug.	13	0	*	0	0	0	4.2		SE	SE	1.5
Sep.	7	0	1	0	*	*	2.5		SE	NW	1.4
Oct.	2	0	*	0	0	0	0.7		SE	NW	1.2
Nov.	1	0	*	0	0	0	0.7		NW	NW	1.3
Dec.	2	*	*	0	0	0	1.7		NW	NW	1.4
Annual	57	1	7	*	3	*	2.2		NW	NW	1.7

*¹ Greenwich mean time * = <1.

TABLE XIX

CLIMATIC TABLE FOR QUETTA

Latitude 30°12′N, longitude 67°01′E, elevation 1,673 m

Month	Mean sta. press. (mbar)	Temperature (°C)				Mean vap. press. (mbar)		Precipitation (mm)	
		daily max.	daily min.	extreme		03 h*¹	12 h	mean	max. in 24 h
				max.	min.				
Jan.	833.0	9.6	−2.6	24	−16	5.2	5.6	52.3	49.3
Feb.	831.8	12.9	−0.3	27	−12	5.7	6.5	52.6	48.5
Mar.	831.6	17.6	3.6	28	−7	7.4	7.9	42.7	53.3
Apr.	831.3	23.6	7.4	34	−2	9.6	9.4	18.3	26.9
May	829.2	29.3	11.5	37	2	11.7	11.9	10.2	34.3
June	825.6	33.4	14.8	39	6	14.9	13.2	4.1	55.1
July	823.6	34.6	18.8	39	8	18.3	15.3	21.1	38.4
Aug.	825.2	33.8	16.8	39	7	16.6	14.3	8.9	25.4
Sep.	829.1	30.6	10.0	37	2	10.6	10.5	0.3	4.3
Oct.	833.5	24.6	4.0	33	−6	6.6	8.5	3.3	6.6
Nov.	835.1	18.2	−0.4	27	−13	5.1	5.6	4.8	20.1
Dec.	834.3	12.9	−2.3	23	−16	4.9	5.3	25.1	38.6
Annual	830.2	23.4	6.8	39	−16	9.7	9.5	243.6	55.1

Month	Number of days with						Mean cloudiness (oktas)	Mean sunshine hours	Wind		mean speed (m/sec)
	precip. ⩾0.3 mm	hail	thunderstorm	fog	duststorm	squall			preval. direct.		
									03 h*¹	12 h	
Jan.									SE	NW	1.4
Feb.											1.6
Mar.											1.5
Apr.									S	W	1.4
May											1.4
June											1.3
July									S	W	1.3
Aug.											1.3
Sep.											1.1
Oct.									SE	NW	1.1
Nov.											1.0
Dec.											1.1
Annual											1.3

*¹ Greenwich mean time.

TABLE XX

CLIMATIC TABLE FOR GANGANAGAR

Latitude 29°55′N, longitude 73°53′E, elevation 177 m

Month	Mean sta. press. (mbar)	Temperature (°C)				Relative humidity (%)		Precipitation (mm)	
		daily max.	daily min.	extreme		03 h*¹	12 h	mean	max. in 24 h
				max.	min.				
Jan.	996.1	20.5	4.7	36.1	−2.2	80	45	14.6	35.1
Feb.	993.4	24.1	7.5	35.0	−2.8	74	37	9.4	28.2
Mar.	990.1	29.6	11.1	41.1	0.6	63	31	13.6	47.0
Apr.	986.1	36.3	18.2	46.6	8.3	41	22	5.6	22.3
May	980.6	41.5	24.0	49.4	11.7	33	18	11.6	48.0
June	976.3	42.1	28.0	50.0	18.3	45	27	35.5	99.1
July	976.5	38.8	28.1	46.7	20.9	64	47	68.7	96.5
Aug.	978.3	37.3	26.9	42.8	21.7	68	56	76.8	251.7
Sep.	983.1	36.6	24.1	42.8	15.6	67	42	49.5	135.8
Oct.	988.9	35.0	17.0	41.1	6.7	57	30	2.8	10.6
Nov.	993.8	29.4	8.9	37.2	1.7	59	35	1.6	20.6
Dec.	996.1	23.3	5.5	31.1	−1.7	75	43	6.6	29.6
Annual	986.6	32.9	17.0	50.0	−2.8	61	36	296.3	251.7

Month	Number of days with						Mean cloudi-ness (oktas)	Mean sun-shine hours	Wind		
	precip. ⩾0.3 mm	hail	thunder-storm	fog	dust-storm	squall			preval. direct.		mean speed (m/sec)
									03 h*¹	12 h	
Jan.	3	0	1	4	0	0	2.5		Var.	NW	1.1
Feb.	2	0	*	1	*	0	2.2		SE	NW	1.4
Mar.	3	*	1	1	1	0	1.7		SE	NW	1.8
Apr.	1	0	1	*	2	0	1.3		SE	NW	1.9
May	2	*	1	1	4	0	0.9		SW	NW	2.2
June	2	0	1	*	4	0	0.9		SW	SW	3.0
July	6	0	2	0	3	0	2.3		SW	SW	2.6
Aug.	6	0	2	*	1	0	2.3		SW	SW	2.2
Sep.	3	0	1	*	1	0	1.3		SW	SW	1.7
Oct.	1	0	*	*	1	0	0.4		SE	NW	1.3
Nov.	*	0	*	1	0	0	0.7		SE	NW	0.9
Dec.	2	0	2	2	*	0	1.7		E	NW	0.9
Annual	31	*	13	10	17	0	1.5		SW	NW	1.7

*¹ Greenwich mean time. * = <1.

TABLE XXI

CLIMATIC TABLE FOR NOKKUNDI

Latitude 28°49′N, longitude 62°45′E, elevation 679 m

Month	Mean sta. press. (mbar)	Temperature (°C)				Mean vap. press. (mbar)		Precipitation (mm)	
		daily max.	daily min.	extreme		03 h*1	12 h	mean	max. in 24 h
				max.	min.				
Jan.	938.1	18.8	4.3	30	−10	5.7	7.0	15.0	53.6
Feb.	935.9	22.6	7.6	36	−9	6.5	8.1	7.6	20.1
Mar.	934.1	27.6	12.3	38	−2	9.1	9.8	6.3	24.1
Apr.	932.0	33.6	18.2	44	4	11.0	12.3	3.6	30.2
May	928.2	39.1	23.7	48	11	13.2	15.9	0.5	2.5
June	923.7	42.2	26.8	49	17	14.3	15.6	0.0	0.0
July	921.0	42.8	28.6	49	22	16.0	17.1	1.3	16.5
Aug.	923.2	41.4	27.0	48	19	14.0	15.3	0.0	0.0
Sep.	928.8	38.2	21.6	45	11	13.1	13.3	0.0	0.0
Oct.	934.9	33.4	15.6	42	5	7.7	11.1	0.0	0.0
Nov.	938.7	26.8	9.8	35	−6	5.5	7.4	0.8	13.2
Dec.	939.5	21.0	5.6	30	−7	5.1	7.2	4.3	23.4
Annual	931.6	32.2	16.8	49	−10	10.0	11.7	39.4	53.6

Month	Number of days with						Mean cloudi-ness (oktas)	Mean sun-shine hours	Wind		
	precip. ⩾0.3 mm	hail	thunder-storm	fog	dust-storm	squall			mean direct.		mean speed (m/sec)
									03 h*1	12 h	
Jan.									N	SE	1.8
Feb.											1.8
Mar.											2.0
Apr.									N	NW	2.0
May											2.3
June											3.6
July									N	NW	3.9
Aug.											4.0
Sep.											2.9
Oct.									N	NW	1.8
Nov.											1.3
Dec.											1.4
Annual											2.4

*1 Greenwich mean time.

130

TABLE XXII

CLIMATIC TABLE FOR NEW DELHI (Safdarjang)

Latitude 28°35′N, longitude 77°12′E, elevation 216 m

Month	Mean sta. press. (mbar)	Temperature (°C) daily max.	daily min.	extreme max.	min.	Relative humidity (%) 03 h*1	12 h	Precipitation (mm) mean	max. in 24 h	Mean evap. (mm)
Jan.	991.1	21.3	7.3	29.4	−0.6	72	41	24.9	116.8	3.4
Feb.	988.5	23.6	10.1	33.3	1.7	59	28	21.8	104.1	4.6
Mar.	985.4	30.2	15.1	40.6	4.4	47	21	16.5	62.2	7.0
Apr.	981.5	36.2	21.0	45.6	10.7	32	16	6.8	40.9	10.3
May	976.5	40.5	26.6	47.2	17.5	31	18	7.9	30.5	11.7
June	972.5	39.9	28.7	46.7	18.9	48	32	65.0	235.5	11.7
July	972.7	35.3	27.2	45.0	21.4	73	60	211.1	266.2	6.9
Aug.	974.8	33.7	26.1	40.0	21.2	77	65	172.9	181.6	5.0
Sep.	978.8	34.1	24.6	40.6	17.3	70	54	149.7	176.5	5.9
Oct.	985.1	33.1	18.7	39.4	9.4	54	35	31.2	172.7	6.1
Nov.	989.1	28.7	11.8	35.0	3.9	48	31	1.2	20.8	4.8
Dec.	991.1	23.4	8.0	28.9	1.1	63	38	5.2	53.3	3.5
Annual	982.3	31.7	18.8	47.2	−0.6	56	37	714.2	266.2	6.8

Month	Number of days with precip. ⩾0.3 mm	hail	thunder-storm	fog	dust-storm	squall	Mean cloudi-ness (oktas)	Mean sun-shine hours	Wind preval. direct. 03 h*1	12 h	mean speed (m/sec)
Jan.	3	*	1	3	0	*	2.9	7.7	W	var.	2.3
Feb.	3	*	2	1	0	*	2.5	8.9	W	NE	2.8
Mar.	3	*	3	*	1	1	2.5	8.3	W	NE	3.0
Apr.	1	*	3	0	1	2	1.9	9.4	W	NW	3.0
May	3	*	5	1	3	3	1.5	9.7	W	NW	3.6
June	5	0	5	*	3	2	2.7	7.5	W	NW	4.1
July	14	0	7	*	1	1	5.3	6.5	E	SE	2.9
Aug.	14	0	8	0	0	1	5.5	6.4	W	SE	2.5
Sep.	7	0	4	0	*	1	3.1	7.6	W	N	2.7
Oct.	2	0	1	*	*	*	0.9	9.1	W	N	1.7
Nov.	1	0	*	1	0	*	0.9	9.3	W	N	1.9
Dec.	1	0	1	2	0	0	2.2	8.3	W	N	2.1
Annual	57	1	41	7	8	12	2.7	8.2	W	NNE	2.7

*1 Greenwich mean time. * = <1.

TABLE XXIII

CLIMATIC TABLE FOR JACOBABAD

Latitude 28°18′N, longitude 68°28′E, elevation 56 m

Month	Mean sta. press. (mbar)	Temperature (°C)				Mean vap. press (mbar)		Precipitation (mm)	
		daily max.	daily min.	extreme max.	min.	03 h*¹	12 h	mean	max. in 24 h
Jan.	1010.4	22.4	7.1	31	−2	7.5	9.2	7.1	28.2
Feb.	1007.4	26.2	10.4	38	−1	10.9	10.5	8.6	24.9
Mar.	1004.0	31.9	16.1	42	3	12.6	13.3	7.6	38.7
Apr.	999.7	38.3	21.6	48	11	15.8	15.5	2.3	14.0
May	994.0	43.6	26.4	51	18	22.7	18.9	3.6	29.2
June	988.9	44.3	29.3	51	22	29.1	23.3	6.1	41.9
July	988.4	41.2	29.5	49	22	33.1	29.7	26.9	112.5
Aug.	990.8	38.9	28.3	44	22	31.6	30.8	21.8	58.7
Sep.	995.8	38.6	26.1	45	19	28.8	25.5	0.8	15.7
Oct.	1002.8	36.6	19.6	46	10	18.6	17.8	0.3	3.8
Nov.	1007.9	30.9	13.1	39	2	10.8	13.7	0.5	8.6
Dec.	1010.4	25.0	8.5	33	1	8.8	10.8	2.8	10.2
Annual	1000.0	34.8	19.6	51	−2	19.2	18.3	88.4	112.5

Month	Number of days with						Mean cloudi-ness (oktas)	Mean sun-shine hours	Wind		
	precip. ⩾0.3 mm	hail	thunder-storm	fog	dust-storm	squall			mean direct. 03 h*¹	12 h	mean speed (m/sec)
Jan.									N	NW	1.5
Feb.											1.8
Mar.											2.3
Apr.									E	SW	2.6
May											2.7
June											3.0
July									SE	SE	3.3
Aug.											3.2
Sep.											2.8
Oct.									E	S	1.7
Nov.											1.2
Dec.											1.2
Annual											2.2

*¹ Greenwich mean time.

TABLE XXIV

CLIMATIC TABLE FOR POKHARA AIRPORT

Latitude 28°13′N, longitude 84°0′E, elevation 827 m

Month	Mean sta. press. (mbar)	Temperature (°C)				Mean vap. press (mbar)	Precipitation (mm)	
		daily max.	daily min.	extreme			mean	max. in 24 h
				max.	min.			
Jan.	922.3	18.8	6.4	22.0	1.8	10.6	26	22
Feb.	920.8	21.2	6.0	28.2	3.0	10.6	25	40
Mar.	919.0	26.4	12.0	33.1	5.0	10.8	50	38
Apr.	915.6	29.8	15.4	37.4	6.0	14.4	87	55
May	914.2	29.6	18.0	35.0	8.0	19.4	292	135
June	911.0	29.6	20.0	33.4	12.0	24.0	569	158
July	910.7	29.1	20.9	32.4	13.0	25.7	809	173
Aug.	912.4	29.0	21.0	32.4	13.8	25.7	705	205
Sep.	916.3	27.8	20.2	31.0	15.9	24.0	581	148
Oct.	919.7	25.6	16.8	29.8	10.4	19.6	224	136
Nov.	922.0	22.8	10.8	27.0	4.0	14.7	19	35
Dec.	933.2	19.5	7.1	23.3	3.0	10.0	1	8
Annual	917.3	25.8	14.6	37.4	1.8	17.4	3,388	205

Month	Number of days with						Mean cloudi- ness (oktas)	Mean sun- shine hours	Wind		
	precip. ≥0.3 mm	hail	thunder- storm	fog	dust- storm	squall			mean direct.		mean speed (m/sec)
									03 h*1	12 h	
Jan.									NW	SE	0.4
Feb.									NNE	SE	0.7
Mar.									SSE	SE	0.9
Apr.									S	E	1.0
May									ESE	E	1.0
June									SE	E	0.8
July									E	E	0.9
Aug.									E	E	0.9
Sep.									SE	E	0.6
Oct.									S	SE	0.5
Nov.									S	SE	0.2
Dec.									S	SE	0.2
Annual											

*1 Greenwich mean time.

TABLE XXV

CLIMATIC TABLE FOR KATHMANDU AIRPORT

Latitude 27°42′N, longitude 85°22′E, elevation 1,336 m

Month	Mean sta. press. (mbar)	Temperature (°C)				Mean vap. press (mbar)	Precipitation (mm)	
		daily max.	daily min.	extreme			mean	max. in 24 h
				max.	min.			
Jan.	866.0	16.6	2.2	21.0	−2.3	9.4	18	16
Feb.	865.5	19.1	3.3	25.3	−1.9	8.7	17	23
Mar.	865.1	23.7	7.5	29.9	1.9	10.2	38	36
Apr.	863.5	26.8	11.6	31.7	4.4	12.4	48	46
May	861.3	27.7	15.5	32.4	8.7	16.3	90	64
June	858.4	27.4	19.0	33.0	14.2	21.3	248	84
July	858.0	27.0	19.8	30.6	18.0	23.5	386	103
Aug.	859.2	27.2	19.6	31.2	16.9	23.8	286	93
Sep.	862.5	25.9	18.1	30.0	13.6	22.4	179	62
Oct.	866.6	24.6	13.4	28.4	5.8	17.8	78	80
Nov.	867.2	21.3	6.9	25.5	1.3	13.2	6	14
Dec.	866.4	18.3	2.1	22.8	−1.6	9.8	1	5
Annual	863.3	23.8	11.6	33.0	−2.3	15.7	1,395	103

Month	Number of days with						Mean cloudi- ness (oktas)	Mean sun- shine hours	Wind		
	precip. ⩾0.3 mm	hail	thunder- storm	fog	dust- storm	squall			mean direct.		mean speed (m/sec)
									03 h[1]	12 h	
Jan.									SW	W	0.9
Feb.									S	W	1.4
Mar.									SW	S	2.4
Apr.									E	W	1.7
May									S	W	1.6
June									E	W	1.3
July									E	SW	1.2
Aug.									E	SW	1.1
Sep.									E	W	0.9
Oct.									SSE	W	0.7
Nov.									N	SW	0.5
Dec.									W	W	0.4
Annual											

[1] Greenwich mean time.

TABLE XXVI

CLIMATIC TABLE FOR BAHARAICH

Latitude 27°34′N, longitude 81°36′E, elevation 124 m

Month	Mean sta. press. (mbar)	Temperature (°C)				Relative humidity (%)		Precipitation (mm)	
		daily max.	daily min.	extreme		03 h*¹	12 h	mean	max. in 24 h
				max.	min.				
Jan.	1001.2	22.6	8.8	28.9	0.6	82	57	27.8	57.7
Feb.	998.5	25.6	10.9	34.4	0.6	74	47	17.0	49.5
Mar.	995.3	31.9	15.4	40.6	5.6	55	32	12.4	50.5
Apr.	991.3	37.4	20.9	44.4	11.1	43	24	7.3	61.0
May	986.9	39.8	25.6	45.6	15.6	50	31	23.0	88.9
June	983.5	37.6	27.0	47.6	18.3	68	51	141.7	269.5
July	983.9	33.0	26.3	44.4	18.7	81	73	361.3	186.2
Aug.	985.7	32.2	26.1	38.3	21.1	83	77	284.9	325.9
Sep.	989.4	32.7	25.1	38.3	18.3	80	72	209.6	236.2
Oct.	995.5	32.1	20.7	37.8	12.2	73	57	83.1	171.5
Nov.	999.3	28.6	13.4	33.9	5.0	72	51	1.5	78.0
Dec.	1001.3	24.3	9.4	31.7	1.7	79	56	7.4	81.0
Annual	992.6	31.5	19.1	47.6	0.6	70	52	1177.0	325.9

Month	Number of days with						Mean cloudi- ness (oktas)	Mean sun- shine hours	Wind		
	precip. ⩾0.3 mm	hail	thunder- storm	fog	dust- storm	squall			preval. direct.		mean speed (m/sec)
									03 h*¹	12 h	
Jan.	4	*	1	*	0	0	2.7		W	W	1.1
Feb.	2	*	1	0	0	0	2.1		W	W	1.3
Mar.	2	*	1	0	*	0	1.9		WNW	WNW	1.7
Apr.	1	0	1	0	*	0	1.3		E	W	2.0
May	2	0	3	0	1	0	1.2		E	W	2.3
June	8	0	4	0	1	0	3.3		E	E	2.3
July	15	0	4	0	0	0	5.9		E	E	2.2
Aug.	15	0	3	0	0	0	5.9		E	E	1.7
Sep.	10	0	3	0	0	0	4.3		ESE	E	1.5
Oct.	3	*	1	*	0	0	1.7		E	W	1.0
Nov.	*	0	0	*	0	0	0.7		NW	W	0.8
Dec.	1	0	*	1	*	0	1.4		NW	W	0.8
Annual	63	*	22	1	2	0	2.7		E	W	1.6

*¹ Greenwich mean time. * = <1.

TABLE XXVII

CLIMATIC TABLE FOR BHAIRAWA AIRPORT

Latitude 27°31′N, longitude 83°26′E, elevation 110 m

Month	Mean sta. press. (mbar)	Temperature (°C)				Mean vap. press (mbar)	Precipitation (mm)	
		daily max.	daily min.	extreme			mean	max. in 24 h
				max.	min.			
Jan.	1006.7	22.2	7.6	26.0	2.0	13.8	19	26
Feb.	1004.6	24.8	9.6	35.0	3.2	14.3	13	20
Mar.	1001.0	31.2	13.6	40.0	6.0	14.2	17	29
Apr.	995.2	36.2	19.6	41.6	10.0	15.6	8	11
May	992.8	36.6	23.4	43.0	16.0	21.6	45	23
June	988.2	34.6	24.2	42.4	13.0	27.5	244	112
July	989.0	32.6	25.0	37.8	20.0	31.4	507	171
Aug.	992.0	32.5	25.1	36.1	21.0	31.4	332	185
Sep.	996.2	31.6	23.8	35.0	19.9	30.5	237	148
Oct.	999.8	31.0	20.6	34.8	14.2	26.2	77	77
Nov.	1006.6	28.2	13.4	32.0	5.0	18.4	2	10
Dec.	1007.3	23.4	8.2	29.0	3.8	15.0	1	3
Annual	998.3	30.4	17.8	43.0	2.0	21.6	1,502	185

Month	Number of days with						Mean cloudi-ness (oktas)	Mean sun-shine hours	Wind		
	precip. ≥0.3 mm	hail	thunder-storm	fog	dust-storm	squall			preval. direct.		mean speed (m/sec)
									03 h*1	12 h	
Jan.									SE	W	0.4
Feb.									W	W	1.2
Mar.									SE	W	1.5
Apr.									SE	W	2.6
May									E	SE	2.7
June									SE	SE	2.2
July									E	SE	2.0
Aug.									E	E	1.8
Sep.									E	E	1.0
Oct.									E	SE	0.6
Nov.									S	W	0.1
Dec.									S	SW	0.1
Annual											

*1 Greenwich mean time.

TABLE XXVIII

CLIMATIC TABLE FOR AGRA

Latitude 27°10′N, longitude 78°02′E, elevation 169 m

Month	Mean sta. press. (mbar)	Temperature (°C)				Relative humidity (%)		Precipitation (mm)	
		daily max.	daily min.	extreme		03 h*1	12 h	mean	max. in 24 h
				max.	min.				
Jan.	996.9	22.2	7.4	31.1	−2.2	73	45	16.2	49.5
Feb.	994.2	25.7	10.3	35.6	−1.7	64	32	8.8	51.3
Mar.	991.2	31.9	15.7	42.8	5.6	47	24	10.9	41.3
Apr.	987.3	37.7	21.6	45.0	11.7	32	18	5.3	32.5
May	982.1	41.8	27.2	47.2	16.7	30	21	10.0	32.3
June	978.3	40.5	29.5	48.3	19.4	48	34	60.0	97.3
July	978.3	34.8	27.0	45.6	17.0	75	66	210.2	152.7
Aug.	980.5	32.8	25.8	42.2	20.8	81	75	263.2	149.9
Sep.	984.4	33.2	24.6	41.0	17.2	77	62	151.5	286.0
Oct.	990.8	33.3	19.1	41.1	9.4	59	40	23.5	169.7
Nov.	994.9	29.2	12.0	36.1	2.8	61	33	2.1	45.7
Dec.	997.1	24.1	8.2	30.0	−0.6	68	40	3.7	26.7
Annual	988.0	32.3	19.0	48.3	−2.2	60	41	765.4	286.0

Month	Number of days with						Mean cloudi-ness (oktas)	Mean sun-shine hours	Wind		
	precip. ≥0.3 mm	hail	thunder-storm	fog	dust-storm	squall			preval. direct.		mean speed (m/sec)
									03 h*1	12 h	
Jan.	2	*	*	1	0	0	2.2	8.1	W	NW	1.0
Feb.	2	0	1	*	0	*	1.8	9.2	W	NW	1.2
Mar.	2	*	2	*	1	*	1.7	9.1	W	NW	1.4
Apr.	1	0	1	0	1	1	1.5	9.8	W	NW	1.4
May	2	*	3	0	1	1	1.3	9.3	W	NW	1.6
June	4	*	3	*	2	1	2.9	7.4	W	W	1.9
July	15	*	2	1	*	*	5.3	5.5	E	E	1.6
Aug.	16	0	1	1	*	*	5.6	5.4	E	SW	1.4
Sep.	8	0	1	1	0	*	3.3	7.4	W	NW	1.3
Oct.	2	0	*	*	*	0	0.9	9.2	W	NW	0.9
Nov.	*	0	*	*	*	0	0.7	9.2	W	NW	0.7
Dec.	1	0	1	1	*	*	1.4	8.3	W	NW	0.8
Annual	55	1	16	5	5	4	2.4	8.1	W	NW	1.4

*1 Greenwich mean time. * = <1.

TABLE XXIX

CLIMATIC TABLE FOR SIMARA AIRPORT

Latitude 27°10′N, longitude 84°59′E, elevation 137 m

Month	Mean sta. press. (mbar)	Temperature (°C)				Mean vapour press. (mbar)	Precipitation (mm)	
		daily max.	daily min.	extreme			mean	max. in 24 h
				max.	min.			
Jan.	999.7	22.8	7.4	27.0	1.2	14.1	18	21
Feb.	996.0	25.0	8.8	32.2	1.6	13.7	20	28
Mar.	994.2	30.8	13.2	37.8	6.4	13.6	18	17
Apr.	989.8	35.6	19.2	40.2	11.8	16.5	21	22
May	987.5	35.0	23.4	40.2	15.8	24.3	99	64
June	984.0	33.8	25.4	42.0	21.4	30.0	298	90
July	983.2	32.0	25.6	36.0	21.9	32.0	533	240
Aug.	985.8	32.3	25.1	36.6	23.2	31.9	299	98
Sep.	989.6	31.6	24.0	35.8	19.6	30.3	278	108
Oct.	993.4	30.8	20.8	34.5	14.5	26.1	112	136
Nov.	997.4	28.3	13.5	32.0	8.6	19.5	3	6
Dec.	1000.0	24.2	7.8	28.6	4.0	16.0	1	2
Annual	991.7	30.2	17.8	42.0	1.2	22.3	1,700	240

Month	Number of days with						Mean cloudi-ness (oktas)	Mean sun-shine hours	Wind		
	precip. ≥0.3 mm	hail	thunder-storm	fog	dust-storm	squall			preval. direct.		mean speed (m/sec)
									03 h*[1]	12 h	
Jan.									N	S	0.6
Feb.									E	W	1.3
Mar.									E	W	1.7
Apr.									E	W	2.7
May									E	SE	3.3
June									E	E	2.3
July									E	E	2.3
Aug.									E	E	2.4
Sep.									E	E	1.9
Oct.									E	W	0.9
Nov.									E	W	0.4
Dec.									W	W	0.2
Annual											

*[1] Greenwich mean time.

TABLE XXX

CLIMATIC TABLE FOR LUCKNOW

Latitude 26°52′N, longitude 80°56′E, elevation 111 m

Month	Mean sta. press. (mbar)	Temperature (°C)				Relative humidity (%)		Precipitation (mm)		Mean evap. (mm)
		daily max.	daily min.	extreme		03 h*¹	12 h	mean	max. in 24 h	
				max.	min.					
Jan.	1002.0	23.3	8.9	30.6	1.1	82	55	23.7	24.4	2.3
Feb.	1000.5	26.4	11.5	35.0	1.7	70	43	17.2	16.8	3.7
Mar.	997.0	32.9	16.3	41.7	7.2	51	28	9.2	16.0	6.8
Apr.	993.0	38.3	21.3	45.6	11.4	39	23	6.4	22.1	9.7
May	988.3	41.2	26.5	47.2	17.8	44	27	11.7	22.3	11.0
June	984.7	39.3	28.0	48.3	19.4	61	45	93.9	19.6	8.9
July	984.9	33.6	26.6	45.6	21.1	82	76	299.0	23.6	5.5
Aug.	986.9	32.5	26.0	38.9	21.2	85	79	301.8	51.3	4.4
Sep.	991.0	33.0	25.1	39.4	17.6	82	74	181.7	25.9	4.2
Oct.	996.9	32.8	19.8	40.0	11.1	72	60	40.3	39.1	3.8
Nov.	1000.1	29.3	12.7	35.0	5.0	71	55	1.4	16.2	2.8
Dec.	1003.1	24.8	9.1	33.3	1.7	81	58	6.1	15.2	2.3
Annual	994.3	32.3	19.4	48.3	1.1	68	52	992.4	51.3	5.4

Month	Number of days with						Mean cloudi-ness (oktas)	Mean sun-shine hours	Wind		
	precip. ⩾0.3 mm	hail	thunder-storm	fog	dust-storm	squall			preval. direct.		mean speed (m/sec)
									03 h*¹	12 h	
Jan.	3	*	1	3	*	0	2.0		W	W	0.6
Feb.	3	*	1	1	0	0	1.9		W	W	0.8
Mar.	2	0	3	*	*	*	1.6		W	W	1.0
Apr.	1	0	1	*	1	0	1.1		W	W	1.1
May	2	0	1	0	1	0	1.1		E	W	1.2
June	7	0	3	*	1	0	3.4		E	E	1.4
July	18	0	3	0	0	0	5.9		E	E	1.1
Aug.	19	1	3	0	0	*	5.7		E	E	0.9
Sep.	11	*	1	0	0	0	3.9		E	E	0.8
Oct.	3	*	*	*	*	0	1.2		W	W	0.5
Nov.	*	0	0	*	0	0	0.6		W	Calm	0.4
Dec.	1	0	*	2	0	0	1.1		W	Calm	0.5
Annual	70	1	18	6	2	*	2.5		W	W	0.9

*¹ Greenwich mean time. * = <1.

139

TABLE XXXI

CLIMATIC TABLE FOR JAIPUR (Samgamer)

Latitude 26°49′N, longitude 75°48′E, elevation 390 m

Month	Mean sta. press. (mbar)	Temperature (°C)				Relative humidity (%)		Precipitation (mm)	
		daily max.	daily min.	extreme		03 h[*1]	12 h	mean	max. in 24 h
				max.	min.				
Jan.	971.0	22.0	8.3	29.6	−2.0	60	35	14.0	45.2
Feb.	968.8	25.4	10.7	36.7	2.2	49	28	8.4	37.3
Mar.	966.4	30.9	15.5	42.8	7.8	38	22	8.6	33.8
Apr.	963.4	36.5	21.0	44.9	11.4	29	18	4.2	20.6
May	958.9	40.6	25.8	47.8	16.6	33	19	10.0	30.2
June	955.3	39.2	27.3	47.2	19.7	47	33	54.0	87.1
July	954.8	34.1	25.6	46.7	20.2	75	62	193.2	165.9
Aug.	956.8	31.9	24.3	40.2	18.9	82	71	239.0	188.4
Sep.	960.7	33.2	23.0	38.0	16.8	74	57	89.8	187.5
Oct.	966.5	33.2	18.3	37.9	11.1	51	32	19.3	114.3
Nov.	969.9	29.0	12.0	34.6	5.8	48	30	3.4	32.3
Dec.	971.3	24.4	9.1	31.3	0.0	56	35	4.2	41.4
Annual	963.6	31.7	18.4	47.8	−2.0	53	37	648.1	188.4

Month	Number of days with						Mean cloudi-ness (oktas)	Mean sun-shine hours	Wind		
	precip. ≥0.3 mm	hail	thunder-storm	fog	dust-storm	squall			preval. direct.		mean speed (m/sec)
									03 h[*1]	12 h	
Jan.	2	*	2	1	0	*	2.7	8.6	E	NW	3.1
Feb.	2	*	1	*	0	0	1.9	9.3	E	NW	3.2
Mar.	1	*	2	0	*	*	1.9	8.9	E	NW	3.3
Apr.	1	*	3	*	1	1	1.7	9.3	NW	WNW	3.6
May	2	0	5	0	2	*	1.2	9.1	W	W	4.6
June	7	1	9	*	3	1	2.7	7.1	W	W	4.9
July	15	1	9	0	1	1	5.5	4.6	W	W	4.5
Aug.	16	0	9	1	*	1	5.9	4.6	W	W	3.4
Sep.	9	*	6	0	0	1	3.5	7.3	NW	W	3.4
Oct.	2	0	2	*	0	*	1.1	9.6	NW	NW	2.8
Nov.	1	0	1	0	*	*	0.9	9.7	E	NW	2.1
Dec.	1	*	1	0	0	*	1.9	8.9	E	NW	2.3
Annual	59	2	49	2	6	4	2.5	8.1	NW	NW	3.4

[*1] Greenwich mean time. * = <1.

TABLE XXXII

CLIMATIC TABLE FOR BIRATNAGAR AIRPORT

Latitude 26°29′N, longitude 87°16′E, elevation 72 m

Month	Mean sta. press. (mbar)	Temperature (°C)				Mean vapour press. (mbar)	Precipitation (mm)	
		daily max.	daily min.	extreme			mean	max. in 24 h
				max.	min.			
Jan.	1005.8	23.4	8.8	26.3	3.2	14.2	16	31
Feb.	1003.4	25.8	10.8	32.8	4.0	14.4	11	20
Mar.	1000.5	31.4	14.8	37.9	8.0	15.2	15	22
Apr.	996.3	33.6	20.6	41.3	13.0	19.4	52	43
May	994.8	32.4	23.4	38.8	20.0	25.4	132	80
June	991.1	32.0	24.8	38.0	20.3	30.5	483	159
July	991.2	31.6	25.2	36.2	19.0	32.3	503	202
Aug.	992.7	31.6	25.2	36.6	21.2	31.6	358	145
Sep.	996.7	30.8	24.2	34.8	21.5	30.1	272	142
Oct.	998.0	30.8	21.2	33.9	15.0	26.8	105	84
Nov.	1003.7	28.6	14.4	31.8	7.8	19.0	15	31
Dec.	1006.5	24.8	9.0	30.2	5.9	14.8	0	0
Annual	998.4	29.8	18.6	41.3	3.2	22.8	1889	202

Month	Number of days with						Mean cloudi-ness (oktas)	Mean sun-shine hours	Wind		
	precip. ≥0.3 mm	hail	thunder-storm	fog	dust-storm	squall			mean direct.		mean speed (m/sec)
									03 h*1	12 h	
Jan.									E	WSW	0.8
Feb.									W	WSW	1.8
Mar.									E	W	2.4
Apr.									E	W	3.9
May									E	E	3.8
June									E	E	2.7
July									E	E	2.3
Aug.									E	E	2.9
Sep.									E	E	2.3
Oct.									E	W	1.4
Nov.									E	W	0.6
Dec.									E	W	0.6
Annual											

*1 Greenwich mean time.

TABLE XXXIII

CLIMATIC TABLE FOR GAUHATI

Latitude 26°11′N, longitude 91°45′E, elevation 55 m

Month	Mean sta. press. (mbar)	Temperature (°C) daily max.	daily min.	extreme max.	min.	Relative humidity (%) 03 h*1	12 h	Precipitation (mm) mean	max. in 24 h	Mean evap. (mm)
Jan.	1009.4	24.0	11.0	28.4	5.1	88	67	11.4	41.1	2.0
Feb.	1006.7	26.3	12.8	32.2	5.3	78	57	18.3	53.3	3.2
Mar.	1003.7	30.2	16.5	37.4	9.5	71	50	53.4	56.1	4.8
Apr.	1001.1	31.6	20.3	39.5	10.3	72	57	125.9	75.7	5.3
May	998.6	31.0	22.7	40.3	16.4	81	72	273.6	185.4	4.7
June	994.1	31.5	24.7	36.6	21.1	85	79	293.4	194.3	4.0
July	994.0	32.1	25.8	36.7	21.9	85	79	301.5	232.9	4.0
Aug.	995.3	32.2	25.8	36.3	22.3	84	79	263.0	187.7	4.0
Sep.	998.7	32.1	25.2	35.9	21.2	84	82	190.1	133.9	3.7
Oct.	1003.9	30.5	22.0	34.4	15.2	84	77	90.1	94.3	3.1
Nov.	1007.3	27.7	16.9	32.5	10.0	87	74	11.5	36.8	2.4
Dec.	1009.1	24.9	12.5	30.9	4.9	89	72	5.0	21.3	1.9
Annual	1001.8	29.5	19.7	40.3	4.9	82	70	1637.2	232.9	3.6

Month	Number of days with precip. ≥0.3 mm	hail	thunder- storm	fog	dust- storm	squall	Mean cloudi- ness (oktas)	Mean sun- shine hours	Wind preval. direct. 03 h*1	12 h	mean speed (m/sec)
Jan.	2	0	*	9	0	0	2.7	7.9	NE	NE	0.7
Feb.	4	*	1	2	*	0	2.7	8.3	NE	NE	0.9
Mar.	6	*	3	*	1	0	2.6	7.8	NE	NE	1.1
Apr.	11	1	8	0	1	0	3.5	7.8	NE	NE	1.3
May	19	*	9	0	0	0	5.1	7.1	NE	NE	1.1
June	17	0	5	0	0	0	5.6	4.9	NE	NE	1.0
July	17	0	5	0	0	0	6.4	4.8	NE	SW	0.8
Aug.	17	0	6	0	*	0	6.1	5.1	NE	SW	0.9
Sep.	14	0	4	*	0	0	5.7	5.7	NE	SW	0.8
Oct.	8	0	1	2	*	0	3.9	7.1	NE	NE	0.8
Nov.	2	0	*	9	0	0	2.7	8.4	NE	NE	0.7
Dec.	1	0	*	15	0	0	2.5	7.9	NE	NE	0.7
Annual	118	1	43	37	3	0	4.1	6.9	NE	NE	0.9

*1 Greenwich mean time. * = <1.

TABLE XXXIV

CLIMATIC TABLE FOR BARMER

Latitude 25°45′N, longitude 71°23′E, elevation 194 m

Month	Mean sta. press. (mbar)	Temperature (°C)				Relative humidity (%)		Precipitation (mm)	
		daily max.	daily min.	extreme max.	extreme min.	03 h[*1]	12 h	mean	max. in 24 h
Jan.	993.0	24.7	10.2	33.5	−1.7	53	31	4.7	59.9
Feb.	990.7	28.6	13.3	39.4	4.0	50	28	3.9	23.6
Mar.	987.8	33.9	18.7	43.3	8.9	47	26	3.4	24.6
Apr.	984.5	38.9	24.1	48.3	12.2	46	25	2.8	23.1
May	980.3	41.9	26.8	48.9	16.7	57	25	9.2	36.6
June	976.5	40.3	27.3	48.2	16.2	69	35	21.5	55.1
July	975.6	36.1	26.3	44.4	19.4	77	52	87.4	124.5
Aug.	977.9	33.8	25.1	43.3	20.0	81	59	139.1	255.5
Sep.	982.1	35.3	24.5	42.8	16.7	75	47	33.1	55.4
Oct.	987.5	36.0	21.7	42.8	13.9	54	30	2.0	18.0
Nov.	991.3	31.5	15.9	38.5	6.7	46	29	1.6	27.6
Dec.	993.1	27.0	12.1	35.0	2.3	51	31	1.6	14.0
Annual	985.0	34.0	22.2	48.9	−1.7	59	35	310.6	255.5

Month	Number of days with						Mean cloudiness (oktas)	Mean sunshine hours	Wind		
	precip. ≥0.3 mm	hail	thunderstorm	fog	duststorm	squall			preval. direct. 03 h[*1]	12 h	mean speed (m/sec)
Jan.	1	*	*	*	0	0	1.8		NW	NE	2.1
Feb.	1	0	*	0	*	*	1.7		NW	NE	2.1
Mar.	1	*	1	*	*	0	1.9		NW	W	2.5
Apr.	1	*	1	0	*	0	1.5		WNW	SW	2.9
May	1	0	1	0	*	0	0.9		SW	SW	3.6
June	2	0	2	0	2	0	2.2		SW	SW	3.9
July	9	0	3	0	0	0	4.9		SW	SW	3.4
Aug.	8	0	3	0	0	0	5.1		SW	SW	2.9
Sep.	4	0	3	0	0	0	2.8		SW	SW	2.7
Oct.	*	0	1	0	0	0	1.0		NW	SW	2.0
Nov.	1	0	*	0	0	0	1.0		NW	NE	1.5
Dec.	*	*	*	*	0	0	1.7		NW	NE	1.8
Annual	30	*	16	*	3	*	2.2		NW	SW	2.6

*1 Greenwich mean time. * = <1.

143

TABLE XXXV

CLIMATIC TABLE FOR PATNA

Latitude 25°37′N, longitude 85°10′E, elevation 53 m

Month	Mean sta. press. (mbar)	Temperature (°C)				Relative humidity (%)		Precipitation (mm)	
		daily max.	daily min.	extreme		03 h*1	12 h	mean	max. in 24 h
				max.	min.				
Jan.	1009.9	23.6	11.0	30.3	2.8	71	53	21.1	57.4
Feb.	1006.9	26.3	13.4	35.1	2.2	62	43	20.2	40.6
Mar.	1003.3	32.9	18.6	40.6	7.8	45	27	6.7	53.1
Apr.	999.5	37.6	23.3	43.5	14.4	41	24	8.2	46.7
May	995.3	38.9	26.0	45.6	17.2	56	36	28.3	109.2
June	991.7	36.7	27.1	46.1	20.0	71	57	139.0	350.5
July	992.1	32.9	26.7	41.7	21.1	81	75	265.8	177.8
Aug.	993.8	32.1	26.6	37.1	21.6	83	78	307.1	165.3
Sep.	997.5	32.3	26.3	37.8	19.0	80	75	242.5	366.0
Oct.	1003.7	31.9	23.0	36.1	14.2	70	62	62.8	158.2
Nov.	1007.7	28.9	16.1	34.3	8.1	62	52	5.7	63.7
Dec.	1009.9	24.9	11.7	30.6	4.1	67	53	2.4	73.7
Annual	1009.9	31.6	20.8	46.1	2.2	66	53	1109.3	366.0

Month	Number of days with						Mean cloudi-ness (oktas)	Mean sun-shine hours	Wind		
	precip. ⩾0.3 mm	hail	thunder-storm	fog	dust-storm	squall			preval. direct.		mean speed (m/sec)
									03 h*1	12 h	
Jan.	3	0	1	3	0	0	2.0	9.0	SW	W	1.4
Feb.	3	0	1	1	*	0	1.8	9.5	SW	W	1.7
Mar.	1	*	1	*	1	*	1.9	8.9	SW	WNW	2.1
Apr.	1	0	1	*	1	0	1.5	9.8	NE	NW	2.4
May	3	0	3	0	2	*	1.9	10.4	NE	NE	2.7
June	10	*	6	0	1	0	4.8	7.7	NE	NE	2.5
July	18	0	5	0	*	0	5.5	5.8	E	E	2.4
Aug.	18	*	5	0	0	0	6.5	6.5	E	E	2.1
Sep.	13	0	5	0	0	0	5.4	7.0	E	E	1.9
Oct.	5	0	1	*	0	0	2.3	8.6	SW	NE	1.3
Nov.	1	0	0	*	0	0	1.1	9.5	SW	W	1.0
Dec.	1	0	*	1	0	0	1.1	8.7	SW	W	1.1
Annual	77	1	29	6	4	*	2.9	8.4	SW	W	1.9

*1 Greenwich mean time. * = <1.

TABLE XXXVI

CLIMATIC TABLE FOR SHILLONG

Latitude 25°34′N, longitude 91°53′E, elevation 1,500 m

Month	Mean sta. press. (mbar)	Temperature (°C)				Relative humidity (%)		Precipitation (mm)		Mean evap. (mm)
		daily max.	daily min.	extreme max.	extreme min.	03 h*1	12 h	mean	max. in 24 h	
Jan.	851.3	15.5	3.6	22.0	−4.5	65	83	15.2	52.3	1.6
Feb.	850.0	17.1	6.4	26.0	−3.9	56	71	28.5	41.1	2.4
Mar.	849.5	21.5	10.5	28.9	−0.6	44	57	59.4	189.5	3.5
Apr.	848.4	23.8	14.1	30.2	6.6	51	62	136.4	117.9	3.7
May	846.2	23.7	15.5	30.7	5.6	69	77	325.4	169.7	3.3
June	843.5	23.7	17.4	29.1	11.7	81	84	544.6	415.3	2.1
July	843.5	24.1	18.1	29.4	14.5	81	83	394.9	205.7	2.1
Aug.	844.7	24.1	17.8	29.4	12.3	81	84	334.6	118.1	2.2
Sep.	847.3	23.6	16.6	30.2	11.7	79	89	314.9	256.2	2.0
Oct.	850.9	21.8	12.9	27.2	5.2	71	89	220.2	296.2	1.9
Nov.	851.7	18.9	7.7	25.6	−1.6	63	86	34.9	96.0	1.8
Dec.	851.7	16.4	4.5	22.8	−3.1	64	85	6.3	41.1	1.6
Annual	848.2	21.2	12.1	30.7	−4.5	67	79	2415.3	415.3	2.3

Month	Number of days with						Mean cloudi- ness (oktas)	Mean sun- shine hours	Wind		
	precip. ≥0.3 mm	hail	thunder- storm	fog	dust- storm	squall			preval. direct.		mean speed (m/sec)
									03 h*1	12 h	
Jan.	3	0	*	1	0	0	3.7		SW	N	0.7
Feb.	5	*	1	*	0	0	3.7		SW	SW	1.1
Mar.	7	1	4	*	0	0	3.1		SW	SW	1.7
Apr.	13	1	10	*	0	0	3.9		SW	SW	2.2
May	22	*	14	1	0	0	5.6		SW	SW	1.9
June	25	0	10	1	0	0	7.1		SW	SW	1.3
July	27	0	5	1	0	0	7.5		SW	SW	1.1
Aug.	23	*	7	1	0	*	7.0		var.	SW	0.9
Sep.	22	0	10	2	0	0	6.5		E	SW	0.7
Oct.	17	*	6	1	0	0	5.1		E	SW	0.6
Nov.	5	0	1	1	0	0	4.1		E	N	0.6
Dec.	2	*	*	1	0	0	3.7		SE	N	0.6
Annual	171	2	68	11	0	*	5.1		SW	SW	0.1

*1 Greenwich mean time. * = <1.

TABLE XXXVII

CLIMATIC TABLE FOR VARANASI

Latitude 25°18′N, longitude 83°01′E, elevation 76 m

Month	Mean sta. press. (mbar)	Temperature (°C)				Relative humidity (%)		Precipitation (mm)	
		daily max.	daily min.	extreme		03 h*¹	12 h	mean	max. in 24 h
				max.	min.				
Jan.	1007.1	23.4	9.5	31.1	2.5	80	51	25.0	69.6
Feb.	1004.3	26.6	12.0	36.1	1.7	68	37	21.9	67.1
Mar.	1001.9	33.4	17.2	41.1	6.7	47	24	11.9	37.1
Apr.	997.0	38.6	22.4	44.4	11.1	36	20	3.9	34.5
May	991.8	41.5	27.0	47.2	17.3	44	24	12.3	51.6
June	988.3	39.1	28.3	47.2	20.6	59	45	91.9	159.5
July	988.7	33.5	26.5	45.0	20.0	81	73	306.2	288.3
Aug.	990.7	32.2	26.0	40.0	22.2	85	79	342.2	321.6
Sep.	994.5	32.7	25.4	38.3	17.8	82	74	225.7	349.5
Oct.	1000.9	32.5	20.7	39.4	11.7	73	55	59.5	138.9
Nov.	1005.1	28.6	13.4	35.6	6.7	67	47	9.6	74.9
Dec.	1007.3	24.4	9.7	32.8	2.2	76	51	3.3	53.1
Annual	998.1	32.2	19.8	47.2	1.7	67	48	1113.4	349.5

Month	Number of days with						Mean cloudi-ness (oktas)	Mean sun-shine hours	Wind		
	precip. ⩾0.3 mm	hail	thunder-storm	fog	dust-storm	squall			preval. direct.		mean speed (m/sec)
									03 h*¹	12 h	
Jan.	3	0	*	2	0	0	2.1		SW	W	1.8
Feb.	2	*	1	1	*	0	1.8		SW	W	2.2
Mar.	2	*	1	*	*	0	1.6		SW	W	2.5
Apr.	1	0	1	*	*	*	1.4		SW	NW	2.6
May	2	0	1	*	*	0	1.5		E	NW	2.8
June	7	0	4	1	*	0	4.5		E	NW	2.7
July	18	0	7	0	0	0	6.4		E	NE	2.5
Aug.	18	*	7	0	0	0	6.3		E	W	2.2
Sep.	13	0	6	0	0	0	4.9		SW	var.	1.8
Oct.	4	0	1	0	0	0	2.3		SW	NW	1.5
Nov.	1	0	0	*	0	0	0.9		SW	WNW	1.4
Dec.	1	0	0	1	0	0	1.3		SW	W	1.6
Annual	71	*	31	5	1	*	2.9		SW	W	2.1

*¹ Greenwich mean time. * = <1.

TABLE XXXVIII

CLIMATIC TABLE FOR CHERRAPUNJI

Latitude 25°15′N, longitude 91°44′E, elevation 1,313 m

Month	Mean sta. press. (mbar)	Temperature (°C)				Relative humidity (%)		Precipitation (mm)	
		daily max.	daily min.	extreme		03 h[*1]	12 h	mean	max. in 24 h
				max.	min.				
Jan.	870.5	15.8	7.6	26.7	0.0	63	80	19.8	85.3
Feb.	869.1	16.9	10.5	28.9	0.6	61	73	37.3	91.9
Mar.	868.6	20.5	12.9	30.6	0.6	61	67	178.9	306.1
Apr.	867.5	22.0	15.1	28.3	3.9	75	78	605.2	462.3
May	865.3	22.1	16.3	30.2	3.3	84	88	1705.1	812.0
June	862.5	22.9	17.3	29.2	9.2	92	92	2921.5	973.8
July	862.3	22.2	18.4	28.6	10.0	93	94	2456.7	838.2
Aug.	863.5	22.5	18.4	29.2	12.8	91	93	1827.5	682.7
Sep.	866.1	22.9	18.1	28.9	12.8	87	91	1167.7	632.2
Oct.	869.4	22.4	15.9	29.9	10.0	75	87	447.4	590.5
Nov.	870.4	19.7	11.9	26.7	6.7	65	83	46.7	332.2
Dec.	870.7	17.0	8.8	24.0	3.3	64	82	4.9	189.7
Annual	867.1	20.6	14.3	30.6	0.0	76	84	11418.7	973.8

Month	Number of days with						Mean cloudi-ness (oktas)	Mean sun-shine hours	Wind		
	precip. ⩾0.3 mm	hail	thunder-storm	fog	dust-storm	squall			preval. direct.		mean speed (m/sec)
									03 h[*1]	12 h	
Jan.	2	*	*	2	0	*	2.5		NE	SW	1.9
Feb.	4	*	1	2	0	*	2.7		SW	SW	2.6
Mar.	8	*	4	3	0	*	3.0		SW	SW	3.3
Apr.	15	1	8	6	*	1	4.5		SW	SW	3.3
May	25	*	9	9	0	0	5.1		SW	SW	3.3
June	26	*	5	15	0	*	6.5		SW	SW	3.3
July	30	*	3	23	0	0	6.8		SW	SW	3.2
Aug.	28	*	4	19	0	0	6.3		NE	SW	2.4
Sep.	21	*	5	12	0	0	5.6		NE	SW	1.9
Oct.	13	0	2	5	*	0	3.9		NE	SW	1.7
Nov.	3	0	*	3	*	*	3.7		NE	SW	1.5
Dec.	9	0	*	2	0	*	2.6		NE	SW	1.5
Annual	186	2	41	101	*	2	4.4		SW	SW	2.5

[*1] Greenwich mean time. * = <1.

TABLE XXXIX

CLIMATIC TABLE FOR JIWANI

Latitude 25°04′N, longitude 61°48′E, elevation 56 m

Month	Mean sta. press. (mbar)	Temperature (°C)				Mean vap. press (mbar)		Precipitation (mm)	
		daily max.	daily min.	extreme		03 h*¹	12 h	mean	max. in 24 h
				max.	min.				
Jan.	1009.8	23.7	14.2	28	2	14.3	15.6	67.6	110.5
Feb.	1008.2	25.2	15.6	31	7	15.9	16.5	18.0	62.0
Mar.	1005.6	28.0	18.6	36	11	19.9	19.6	10.4	42.9
Apr.	1002.6	31.4	22.1	42	13	24.9	23.8	8.6	42.4
May	997.8	33.9	25.4	43	18	30.2	30.8	0.8	7.6
June	992.6	34.2	27.6	44	22	33.9	34.2	0.0	0.0
July	991.2	32.2	27.2	41	23	32.6	32.4	8.6	41.4
Aug.	993.2	31.0	26.1	37	23	30.6	31.1	1.8	21.3
Sep.	998.6	31.1	24.7	41	19	28.7	28.6	0.3	1.5
Oct.	1004.8	31.8	22.1	39	16	26.0	25.8	0.3	1.8
Nov.	1008.2	29.4	18.4	38	6	19.0	19.5	9.9	40.6
Dec.	1009.9	25.8	14.8	32	6	13.6	14.7	22.6	90.9
Annual	1001.9	29.8	21.4	44	2	24.1	24.4	148.8	110.5

Month	Number of days with						Mean cloudi-ness (oktas)	Mean sun-shine hours	Wind		
	precip. ⩾0.3 mm	hail	thunder-storm	fog	dust-storm	squall			mean direct.		mean speed (m/sec)
									03 h*¹	12 h	
Jan.									NE	SW	2.8
Feb.											3.0
Mar.											3.6
Apr.									W	SW	3.7
May											4.1
June											4.1
July									S	SE	4.0
Aug.											3.8
Sep.											3.4
Oct.									NW	SW	2.7
Nov.											2.6
Dec.											2.8
Annual											3.4

*¹ Greenwich mean time. * = <1.

TABLE XL

CLIMATIC TABLE FOR KARACHI AIRPORT

Latitude 24°54′N, longitude 67°09′E, elevation 22 m

Month	Mean sta. press. (mbar)	Temperature (°C)				Relative humidity (%)		Precipitation (mm)	
		daily max.	daily min.	extreme		03 h*¹	12 h	mean	max. in 24 h
				max.	min.				
Jan.	1013.6	25.0	10.4	32	0	9.1	10.5	7.6	26.7
Feb.	1011.4	27.4	12.8	35	3	12.0	12.5	12.7	57.1
Mar.	1008.7	31.4	17.6	39	9	17.0	16.1	4.6	35.1
Apr.	1005.2	34.7	21.7	44	13	22.8	21.0	2.3	25.9
May	1000.9	35.3	25.7	48	18	28.9	28.1	1.3	25.4
June	996.2	34.8	28.0	47	22	31.8	32.4	8.9	50.0
July	995.4	35.7	24.3	42	22	31.7	31.6	101.1	119.9
Aug.	997.6	31.4	26.2	39	23	29.8	29.6	47.5	152.4
Sep.	1002.0	32.1	25.1	43	18	28.3	28.2	23.4	111.8
Oct.	1007.4	34.4	20.8	43	10	21.6	20.7	8.3	55.6
Nov.	1011.5	32.1	15.7	38	6	13.4	14.0	3.0	55.6
Dec.	1013.4	27.2	11.7	33	2	11.3	12.4	5.6	38.1
Annual	1005.2	31.6	20.2	48	0	21.5	21.5	221.2	152.4

Month	Number of days with						Mean cloudi-ness (oktas)	Mean sun-shine hours	Wind		
	precip. ≥0.3 mm	hail	thunder-storm	fog	dust-storm	squall			mean direct.		mean speed (m/sec)
									03 h*¹	12 h	
Jan.									N	SW	2.5
Feb.											2.7
Mar.											3.2
Apr.									W	SW	4.2
May											5.4
June											5.6
July									W	W	5.8
Aug.											5.3
Sep.											4.6
Oct.									W	SW	2.9
Nov.											2.1
Dec.											2.0
Annual											3.8

*¹ Greenwich mean time.

TABLE XLI

CLIMATIC TABLE FOR PABNA

Latitude 24°01′N, longitude 89°14′E, elevation 15 m

Month	Mean sta. press. (mbar)	Temperature (°C)				Mean vap. press (mbar)		Precipitation (mm)	
		daily max.	daily min.	extreme		03 h*1	12 h	mean	max. in 24 h
				max.	min.				
Jan.	1014.2	25.7	11.5	30.6	3.9	14.9	16.9	10.9	76.2
Feb.	1010.5	28.3	13.7	36.1	6.1	16.3	17.2	20.6	47.2
Mar.	1008.0	33.4	18.2	41.1	8.9	20.4	17.8	35.3	59.4
Apr.	1004.2	36.3	22.7	42.8	15.6	27.6	21.2	55.1	68.8
May	1000.0	35.1	24.6	43.9	17.8	32.1	28.7	180.8	215.1
June	997.8	33.2	25.6	40.0	21.1	34.3	34.7	292.6	243.8
July	997.0	31.8	25.9	36.1	21.7	34.0	34.4	267.2	211.1
Aug.	998.5	31.8	26.1	35.6	21.7	33.7	34.5	288.8	204.5
Sep.	1001.6	32.3	25.9	36.1	21.7	33.0	33.1	234.7	207.5
Oct.	1007.8	31.9	23.3	36.1	16.1	29.3	30.1	168.7	231.6
Nov.	1012.0	29.3	17.4	34.4	10.0	21.1	22.6	18.8	62.7
Dec.	1013.5	26.6	12.9	32.2	6.1	16.3	18.2	1.5	22.9
Annual	1005.4	31.3	20.7	43.9	3.9	26.0	25.8	1575.0	243.8

Month	Number of days with						Mean cloudi-ness (oktas)	Mean sun-shine hours	Wind		
	precip. ⩾0.3 mm	hail	thunder-storm	fog	dust-storm	squall			preval. direct.		mean speed (m/sec)
									03 h*1	12 h	
Jan.											1.0
Feb.											
Mar.											
Apr.											1.8
May											
June											
July											2.1
Aug.											
Sep.											
Oct.											1.7
Nov.											
Dec.											
Annual											

*1 Greenwich mean time.

TABLE XLII

CLIMATIC TABLE FOR DACCA

Latitude 23°46'N, longitude 90°23'E, elevation 8 m

Month	Mean sta. press. (mbar)	Temperature (°C)				Mean vap. press (mbar)		Precipitation (mm)	
		daily max.	daily min.	extreme		03 h*¹	12 h	mean	max. in 24 h
				max.	min.				
Jan.	1014.3	25.5	11.7	30.6	6.7	15.4	14.9	17.8	40.9
Feb.	1011.0	28.1	13.4	34.4	8.3	15.7	13.8	31.2	47.5
Mar.	1009.3	32.5	16.1	39.4	12.8	20.6	17.6	58.2	51.0
Apr.	1005.6	35.1	23.5	42.2	17.8	27.7	24.2	102.6	48.8
May	1001.6	33.7	25.4	42.2	18.9	31.9	31.4	194.3	119.6
June	999.0	31.7	25.9	36.1	21.7	33.9	32.7	321.8	114.8
July	999.2	30.7	26.0	33.9	23.9	32.9	32.4	436.9	325.6
Aug.	1000.2	31.1	26.2	36.1	23.3	33.4	32.7	304.8	164.8
Sep.	1003.2	31.2	25.8	35.0	22.8	32.9	32.3	235.7	124.5
Oct.	1008.4	30.9	23.7	33.9	17.2	29.2	28.8	168.7	106.9
Nov.	1012.0	28.7	17.6	31.1	11.7	21.9	21.1	27.7	43.9
Dec.	1014.1	26.3	12.7	29.4	7.2	17.7	16.7	2.3	14.5
Annual	1006.5	30.4	20.9	42.2	6.7	26.1	24.9	1902.0	325.6

Month	Number of days with						Mean cloudi-ness (oktas)	Mean sun-shine hours	Wind	
	precip. ⩾0.3 mm	hail	thunder-storm	fog	dust-storm	squall			preval. direct. 03 h*¹ 12 h	mean speed (m/sec)
Jan.										1.1
Feb.										
Mar.										
Apr.										2.9
May										
June										
July										2.7
Aug.										
Sep.										
Oct.										1.2
Nov.										
Dec.										
Annual										

*¹ Greenwich mean time.

TABLE XLIII

CLIMATIC TABLE FOR RANCHI

Latitude 23°23′N, longitude 85°20′E, elevation 655 m

Month	Mean sta. press. (mbar)	Temperature (°C) daily max.	daily min.	extreme max.	min.	Relative humidity (%) 03 h[1]	12 h	Precipitation (mm) mean	max. in 24 h
Jan.	941.1	23.6	9.9	31.4	3.9	62	51	24.1	52.1
Feb.	939.3	25.9	12.2	35.0	2.8	55	41	38.7	78.5
Mar.	937.1	31.2	16.7	39.4	7.8	41	31	28.3	82.3
Apr.	934.5	35.9	21.1	41.7	9.5	38	28	22.6	58.7
May	930.8	37.9	24.0	43.3	15.6	46	35	49.9	104.7
June	926.9	34.2	23.9	43.1	14.8	67	62	202.3	165.6
July	926.6	29.1	22.6	38.3	13.2	86	84	360.3	215.9
Aug.	929.2	28.9	22.4	35.5	13.2	86	85	353.8	147.3
Sep.	931.8	29.2	21.9	37.9	14.9	83	83	256.2	152.9
Oct.	937.2	28.5	18.5	36.8	7.1	71	68	105.2	231.1
Nov.	940.2	25.7	13.0	33.7	5.6	57	55	17.4	79.5
Dec.	941.4	23.7	9.9	30.1	3.6	60	51	3.9	32.3
Annual	934.7	29.5	18.0	43.3	2.8	63	56	1462.7	231.1

Month	Number of days with precip. ≥0.3 mm	hail	thunder-storm	fog	dust-storm	squall	Mean cloudi-ness (oktas)	Mean sun-shine hours	Wind preval. direct. 03 h[1]	12 h	mean speed (m/sec)
Jan.	3	0	1	*	0	0	1.9		NNW	NW	1.0
Feb.	4	*	1	*	*	0	1.7		NW	NW	1.3
Mar.	3	*	2	0	*	0	1.7		NW	NW	1.6
Apr.	3	*	3	0	*	0	1.9		SSW	NW	1.6
May	4	0	4	0	*	0	2.7		S	NW	1.7
June	13	*	4	0	0	0	6.2		S	S	1.9
July	20	0	3	*	0	0	7.1		S	S	1.8
Aug.	20	*	2	0	0	0	6.9		S	S	1.6
Sep.	15	0	3	0	0	0	5.7		S	SE	1.4
Oct.	7	0	1	0	0	0	3.1		N	N	0.9
Nov.	1	0	*	0	0	0	1.5		N	N	0.9
Dec.	1	0	*	*	0	0	1.1		N	N	0.9
Annual	94	1	24	1	1	0	3.5		S	NW	1.4

[1] Greenwich mean time. * = <1.

TABLE XLIV

CLIMATIC TABLE FOR BHOPAL (BAIRAGARH)

Latitude 23°17′N, longitude 77°21′E, elevation 523 m

Month	Mean sta. press. (mbar)	Temperature (°C)				Relative humidity (%)		Precipitation (mm)	
		daily max.	daily min.	extreme max.	extreme min.	03 h*1	12 h	mean	max. in 24 h
Jan.	955.8	25.7	10.4	32.2	0.6	60	35	16.8	34.3
Feb.	953.9	28.5	12.5	36.1	1.7	47	23	4.5	15.5
Mar.	951.7	33.6	17.1	40.4	7.8	31	17	9.8	35.1
Apr.	949.3	37.8	21.2	44.2	12.2	25	14	3.3	13.5
May	945.6	40.7	26.4	45.6	17.2	32	16	11.1	72.6
June	942.4	36.9	25.4	45.1	19.5	63	41	136.6	120.9
July	941.9	29.9	23.2	41.2	19.0	86	72	428.5	218.2
Aug.	943.6	28.6	22.5	35.5	19.4	88	76	307.7	188.5
Sep.	946.7	30.1	21.9	36.7	13.8	83	66	232.0	233.2
Oct.	952.1	31.3	18.0	37.8	11.7	62	41	36.9	123.7
Nov.	955.3	28.5	13.3	34.0	6.1	53	33	14.7	68.3
Dec.	956.2	26.1	10.6	32.8	3.1	59	39	7.0	31.7
Annual	949.5	31.5	18.5	45.6	0.6	57	39	1208.9	233.2

Month	Number of days with						Mean cloudiness (oktas)	Mean sunshine hours	Wind		
	precip. ≥0.3 mm	hail	thunderstorm	fog	duststorm	squall			preval. direct. 03 h*1	12 h	mean speed (m/sec)
Jan.	3	*	1	2	2	*	2.1		NE	NE	1.6
Feb.	1	0	1	1	0	*	1.6		NE	NW	1.8
Mar.	1	*	2	*	*	*	1.7		NE	NW	2.0
Apr.	1	0	2	0	*	*	1.7		W	W	2.4
May	2	*	2	*	*	1	1.9		W	W	3.3
June	11	0	9	*	*	*	4.7		W	W	3.6
July	21	0	7	0	0	1	7.0		W	W	3.7
Aug.	21	0	6	*	0	*	6.9		W	W	3.1
Sep.	14	0	5	*	0	*	5.2		W	WNW	2.5
Oct.	3	0	1	*	0	0	2.3		NE	NE	1.4
Nov.	1	0	0	1	0	0	1.4		NE	NE	1.2
Dec.	1	0	*	1	0	0	1.7		NE	NE	1.2
Annual	81	*	37	6	3	3	3.1		W	W	2.3

*1 Greenwich mean time. * = <1.

153

TABLE XLV

CLIMATIC TABLE FOR BHUJ

Latitude 23°15′N, longitude 69°48′E, elevation 80 m

Month	Mean sta. press. (mbar)	Temperature (°C)				Relative humidity (%)		Precipitation (mm)	
		daily max.	daily min.	extreme		03 h*1	12 h	mean	max. in 24 h
				max.	min.				
Jan.	1005.9	26.1	10.1	34.1	0.4	45	24	2.9	14.0
Feb.	1003.9	29.1	12.9	38.9	1.0	52	17	4.7	61.5
Mar.	1001.5	34.0	18.3	41.3	6.4	59	26	1.3	30.0
Apr.	998.7	37.6	22.7	45.6	13.9	57	22	0.8	67.1
May	995.1	38.7	25.6	47.8	18.1	66	33	8.3	186.9
June	991.1	36.7	27.4	46.1	21.2	72	50	22.1	129.8
July	990.1	33.0	26.3	40.6	20.6	77	64	164.0	467.9
Aug.	992.3	31.7	25.3	38.3	21.2	80	62	96.2	199.4
Sep.	996.3	33.2	24.1	42.8	17.9	77	54	38.4	170.2
Oct.	1000.9	35.6	21.5	42.2	12.2	64	26	6.7	118.6
Nov.	1004.3	32.3	15.6	37.8	6.6	55	27	1.8	55.4
Dec.	1006.1	28.0	11.2	34.9	4.0	56	25	1.5	36.8
Annual	998.9	33.0	20.1	47.8	0.4	63	36	348.7	467.9

Month	Number of days with						Mean cloudi-ness (oktas)	Mean sun-shine hours	Wind		
	precip. ⩾0.3 mm	hail	thunder-storm	fog	dust-storm	squall			preval. direct.		mean speed (m/sec)
									03 h*1	12 h	
Jan.	1	0	*	0	0	0	1.3		var.	N	2.2
Feb.	*	0	*	*	0	0	1.1		SW	NE	2.3
Mar.	*	0	*	*	0	0	1.1		SW	W	2.4
Apr.	*	0	*	0	*	0	0.9		W	W	3.1
May	*	0	*	0	0	0	0.8		SW	W	4.9
June	2	0	1	0	*	0	3.5		SW	SW	6.1
July	10	0	2	0	0	0	5.9		SW	SW	5.6
Aug.	7	0	1	0	0	0	5.9		SW	SW	4.7
Sep.	4	0	1	0	0	0	3.5		WSW	SW	3.6
Oct.	1	0	1	0	0	0	1.2		W	W	2.4
Nov.	1	0	*	*	0	0	0.9		Calm	NE	1.9
Dec.	*	0	*	*	0	0	1.1		NE	NE	1.6
Annual	26	0	8	1	*	0	2.3		SW	WSW	3 4

*1 Greenwich mean time. * = <1.

TABLE XLVI

CLIMATIC TABLE FOR JESSORE

Latitude 23°10′N, longitude 89°13′E, elevation 8 m

Month	Mean sta. press. (mbar)	Temperature (°C)				Mean vap. press (mbar)		Precipitation (mm)	
		daily max.	daily min.	extreme		03 h*1	12 h	mean	max. in 24 h
				max.	min.				
Jan.	1014.6	25.6	10.3	32.2	2.2	15.6	16.4	13.7	42.2
Feb.	1012.0	28.4	13.0	36.7	3.3	17.3	17.9	21.6	53.8
Mar.	1008.8	33.4	18.4	41.1	8.3	22.9	21.1	34.5	62.2
Apr.	1005.8	35.9	23.1	42.8	13.9	29.5	26.5	88.1	120.1
May	1002.2	35.0	24.8	42.2	17.8	33.3	31.7	189.0	140.7
June	999.6	32.9	25.6	40.0	21.1	35.0	33.9	274.6	189.7
July	998.4	31.4	25.6	35.0	21.1	33.6	33.1	314.4	143.8
Aug.	999.9	31.6	25.7	36.1	22.2	34.0	34.3	307.1	180.3
Sep.	1003.2	32.1	25.3	36.7	22.2	34.2	33.8	188.0	124.2
Oct.	1008.6	31.6	22.8	37.2	14.4	30.2	30.2	135.9	184.9
Nov.	1012.7	29.1	16.3	35.0	6.7	21.6	22.5	22.4	79.0
Dec.	1014.4	26.5	11.3	30.6	4.4	16.7	17.8	1.5	13.0
Annual	1006.6	31.1	20.2	42.8	2.2	27.0	26.5	1590.8	189.7

Month	Number of days with						Mean cloudi-ness (oktas)	Mean sun-shine hours	Wind		
	precip. ⩾0.3 mm	hail	thunder-storm	fog	dust-storm	squall			preval. direct.		mean speed (m/sec)
									03 h*1	12 h	
Jan.											0.9
Feb.											
Mar.											
Apr.											2.4
May											
June											
July											1.9
Aug.											
Sep.											
Oct.											1.2
Nov.											
Dec.											
Annual											

*1 Greenwich mean time.

TABLE XLVII

CLIMATIC TABLE FOR AHMEDABAD

Latitude 23°04′N, longitude 72°38′E, elevation 55 m

Month	Mean sta. press. (mbar)	Temperature (°C)				Relative humidity (%)		Precipitation (mm)		Mean evap. (mm)
		daily max.	daily min.	extreme		03 h*1	12 h	mean	max. in 24 h	
				max.	min.					
Jan.	1008.3	28.7	11.9	36.1	3.3	55	28	3.9	30.7	5.0
Feb.	1006.4	31.0	14.5	40.6	2.2	52	24	0.3	26.4	6.6
Mar.	1004.0	35.7	18.6	43.9	9.4	47	20	0.9	12.2	9.0
Apr.	1001.3	39.7	23.0	46.2	12.8	49	18	1.9	21.6	11.7
May	997.9	40.7	26.3	47.8	19.1	68	21	4.5	46.2	13.1
June	994.1	38.0	27.4	47.2	19.4	77	41	100.0	130.8	10.3
July	993.7	33.2	25.7	42.2	21.1	86	68	316.3	414.8	6.2
Aug.	995.7	31.8	24.6	38.9	21.4	84	69	213.3	150.6	4.6
Sep.	999.1	33.1	24.2	41.7	17.2	80	60	162.8	257.8	5.8
Oct.	1003.7	35.6	21.2	42.8	13.5	64	35	13.1	52.8	6.6
Nov.	1006.9	33.0	16.1	38.9	8.3	52	29	5.4	53.3	6.0
Dec.	1008.4	29.6	12.6	35.6	5.4	56	29	0.7	14.0	5.1
Annual	1001.7	34.2	20.5	47.8	2.2	64	37	823.1	414.8	7.5

Month	Number of days with						Mean cloudi-ness (oktas)	Mean sun-shine hours	Wind		mean speed (m/sec)
	precip. ≥0.3 mm	hail	thunder-storm	fog	dust-storm	squall			preval. direct.		
									03 h*1	12 h	
Jan.	1	0	*	0	0	0	1.4	9.6	NE	NW	1.5
Feb.	*	0	*	0	0	0	1.3	10.2	NE	NW	1.4
Mar.	*	0	*	0	0	0	1.3	9.3	NW	NW	2.0
Apr.	1	0	1	0	*	*	1.4	10.0	NW	W	2.2
May	1	0	1	*	*	*	1.5	10.6	WSW	W	2.6
June	7	0	4	0	1	1	4.3	8.8	SW	SW	3.0
July	17	0	3	0	0	0	6.8	4.6	SW	SW	3.0
Aug.	17	0	2	0	0	*	6.7	4.3	SW	SW	2.3
Sep.	9	0	3	0	0	*	4.5	6.7	W	W	1.9
Oct.	1	0	1	0	0	*	1.7	9.5	NE	NE	1.3
Nov.	1	0	*	0	0	*	1.3	9.7	NE	NE	1.1
Dec.	*	0	*	*	0	0	1.3	9.5	NE	NE	0.7
Annual	54	0	16	*	1	1	2.8	8.0	WSW	W	1.9

*1 Greenwich mean time. * = <1.

156

TABLE XLVIII

CLIMATIC TABLE FOR CALCUTTA

Latitude 22°32′N, longitude 88°20′E, elevation 6 m

Month	Mean sta. press. (mbar)	Temperature (°C)				Relative humidity (%)		Precipitation (mm)	
		daily max.	daily min.	extreme		03 h*¹	12 h	mean	max. in 24 h
				max.	min.				
Jan.	1014.9	26.8	13.6	32.8	6.7	78	56	13.8	46.7
Feb.	1012.2	29.5	16.5	36.7	7.2	75	48	24.2	80.8
Mar.	1009.0	34.3	21.5	41.1	10.0	71	46	26.5	69.9
Apr.	1005.8	36.3	25.0	43.3	16.1	71	56	42.7	107.4
May	1001.7	35.8	26.5	43.7	18.3	74	67	120.6	156.2
June	998.2	34.1	26.7	43.9	20.4	80	77	259.1	303.5
July	998.0	32.0	26.3	39.9	22.8	84	82	300.6	183.6
Aug.	999.6	32.0	26.3	36.1	22.6	85	83	306.3	253.0
Sep.	1003.1	32.3	26.1	36.8	22.2	84	82	289.7	369.1
Oct.	1008.9	31.8	23.9	35.6	17.2	80	77	160.2	206.0
Nov.	1012.5	29.5	18.4	33.9	10.6	74	67	34.9	85.1
Dec.	1014.7	27.0	14.2	32.5	7.2	76	63	3.2	53.1
Annual	1006.5	31.8	22.1	43.9	6.7	78	67	1581.8	369.1

Month	Number of days with						Mean cloudiness (oktas)	Mean sunshine hours	Wind		
	precip. ≥0.3 mm	hail	thunderstorm	fog	duststorm	squall			preval. direct.		mean speed (m/sec)
									03 h*¹	12 h	
Jan.	1	0	1	8	0	*	1.5	7.9	N	NW	0.8
Feb.	3	0	1	7	0	*	1.9	8.6	var.	NW	1.0
Mar.	4	*	4	4	*	2	2.2	8.2	SW	SW	1.4
Apr.	5	*	6	1	*	3	2.9	8.6	SW	S	2.1
May	10	*	11	0	0	4	4.0	8.2	SW	S	2.4
June	17	0	13	0	0	5	6.2	4.4	SW	S	1.9
July	23	0	9	0	0	1	6.9	3.8	SW	S	1.7
Aug.	24	0	12	0	0	1	6.8	3.9	SW	S	1.5
Sep.	20	0	16	*	0	1	6.1	4.5	S	S	1.2
Oct.	12	0	9	1	0	1	3.9	6.3	NE	var.	0.9
Nov.	2	0	1	2	0	*	1.9	8.3	N	N	0.7
Dec.	1	0	1	6	0	*	1.4	8.2	N	NW	0.7
Annual	122	*	83	28	*	19	3.9	6.7	SW	S	1.4

*¹ Greenwich mean time. * = <1.

157

TABLE XLIX

CLIMATIC TABLE FOR CHITTAGONG

Latitude 22°30′N, longitude 92°12′E, elevation 63 m

Month	Mean sta. press. (mbar)	Temperature (°C)				Mean vap. press. (mbar)		Precipitation (mm)	
		daily max.	daily min.	extreme		03 h*1	12 h	mean	max. in 24 h
				max.	min.				
Jan.	1014.6	25.9	13.8	30.6	9.4	15.9	17.4	10.4	32.5
Feb.	1012.6	27.5	15.8	32.8	11.1	17.8	19.9	7.6	26.7
Mar.	1010.2	30.4	20.1	35.6	12.2	23.2	25.4	88.9	152.7
Apr.	1008.2	32.0	23.4	37.8	14.4	29.8	30.0	67.8	87.9
May	1004.5	32.1	24.9	39.4	20.0	31.5	30.1	283.7	150.1
June	1001.2	31.1	25.2	35.0	21.1	32.4	32.3	569.2	230.6
July	1001.2	30.5	24.8	35.0	21.7	31.7	32.4	624.1	417.1
Aug.	1002.2	30.8	25.3	35.6	22.2	32.1	32.3	564.6	256.3
Sep.	1005.0	30.5	25.0	35.0	21.7	31.0	33.0	305.8	375.4
Oct.	1009.3	30.8	23.7	35.0	18.3	30.3	30.2	290.8	389.4
Nov.	1012.2	29.2	18.8	33.3	13.3	22.4	23.6	50.0	77.0
Dec.	1014.2	26.3	15.3	31.7	11.1	18.0	18.6	10.4	6.4
Annual	1008.0	29.7	21.4	39.4	9.4	26.3	27.1	2873.3	417.1

Month	Number of days with						Mean cloudiness (oktas)	Mean sunshine hours	Wind	
	precip. ≥0.3 mm	hail	thunderstorm	fog	duststorm	squall			preval. direct. 03 h*1 12 h	mean speed (m/sec)
Jan.										1.4
Feb.										
Mar.										
Apr.										3.3
May										
June										
July										3.6
Aug.										
Sep.										
Oct.										1.4
Nov.										
Dec.										
Annual										

*1 Greenwich mean time.

TABLE L

CLIMATIC TABLE FOR RAJKOT

Latitude 22°18′N, longitude 70°47′E, elevation 138 m

Month	Mean sta. press. (mbar)	Temperature (°C)				Relative humidity (%)		Precipitation (mm)	
		daily max.	daily min.	extreme		03 h*¹	12 h	mean	max. in 24 h
				max.	min.				
Jan.	998.5	28.1	10.7	36.1	−0.6	51	22	1.1	13.2
Feb.	997.0	30.7	13.1	40.0	1.1	57	19	0.4	21.3
Mar.	995.1	35.3	17.2	43.9	6.1	63	17	1.0	21.6
Apr.	992.5	38.8	21.3	44.4	10.0	66	18	2.5	29.2
May	989.5	40.5	24.7	47.8	16.1	72	27	6.6	117.3
June	985.4	37.8	26.2	45.0	20.0	76	48	99.3	218.9
July	984.6	32.6	24.9	40.6	19.4	85	68	292.8	375.2
Aug.	986.6	31.6	24.0	37.8	20.6	86	65	143.0	233.2
Sep.	990.1	32.9	22.9	42.8	16.7	85	57	93.3	184.7
Oct.	994.2	35.4	20.9	41.7	12.2	70	31	25.4	105.6
Nov.	997.1	33.2	16.5	38.3	7.2	50	24	4.8	99.8
Dec.	998.9	29.6	12.3	36.1	2.8	49	25	3.6	22.9
Annual	992.5	33.9	19.6	47.8	−0.6	67	35	673.8	375.2

Month	Number of days with						Mean cloudi-ness (oktas)	Mean sun-shine hours	Wind		
	precip. ⩾0.3 mm	hail	thunder-storm	fog	dust-storm	squall			preval. direct.		mean speed (m/sec)
									03 h*¹	12 h	
Jan.	1	0	*	2	0	0	1.3		NE	NE	3.6
Feb.	*	0	0	2	*	0	1.3		NE	NW	3.8
Mar.	*	0	*	2	0	0	1.5		NW	NW	4.6
Apr.	1	*	1	1	0	0	1.5		W	NW	5.2
May	1	0	1	0	*	0	1.9		WSW	W	7.2
June	7	0	5	0	*	0	4.7		SW	SW	7.8
July	17	0	2	0	*	*	6.5		SW	SW	7.8
Aug.	13	0	1	0	0	0	6.5		SW	SW	6.4
Sep.	8	*	2	*	0	0	4.7		SW	W	4.8
Oct.	2	0	1	1	0	0	2.1		NW	NE	3.4
Nov.	1	0	*	1	0	0	1.3		NE	NE	2.9
Dec.	*	0	0	2	0	0	1.3		NE	NE	3.1
Annual	50	*	13	11	1	*	2.9		SW	W	5.1

*¹ Greenwich mean time. * = <1.

TABLE LI

CLIMATIC TABLE FOR COX'S BAZAR
Latitude 21°26′N, longitude 91°58′E, elevation 12 m

Month	Mean sta. press. (mbar)	Temperature (°C)				Mean vap. press. (mbar)		Precipitation (mm)	
		daily max.	daily min.	extreme		03 h*1	12 h	mean	max. in 24 h
				max.	min.				
Jan.	1013.6	26.6	13.3	31.7	7.8	16.7	17.8	10.7	56.1
Feb.	1011.9	27.8	15.4	33.9	9.4	18.5	19.8	12.2	55.9
Mar.	1009.8	30.4	19.6	37.2	11.7	24.5	26.0	32.3	162.8
Apr.	1007.8	32.0	23.2	37.2	16.1	29.4	29.4	80.0	206.0
May	1004.3	32.2	24.8	36.7	17.2	31.7	31.8	292.6	194.8
June	1000.8	30.4	24.9	40.6	20.6	31.1	32.4	770.6	279.4
July	1001.0	29.7	24.7	33.9	21.7	32.4	32.2	933.4	317.5
Aug.	1002.0	29.7	24.7	33.9	21.1	32.4	32.1	780.0	289.8
Sep.	1004.6	30.4	24.6	33.9	21.7	32.1	32.3	443.2	238.0
Oct.	1008.6	31.0	23.4	35.0	17.2	30.1	29.5	275.1	398.3
Nov.	1011.3	29.4	19.3	33.9	13.3	23.9	22.8	63.2	329.4
Dec.	1013.2	27.0	15.2	31.7	8.9	19.0	19.1	32.8	385.1
Annual	1007.4	29.7	21.1	40.6	7.8	26.9	27.0	3726.1	398.3

Month	Number of days with						Mean cloudiness (oktas)	Mean sunshine hours	Wind		
	precip. ⩾0.3 mm	hail	thunderstorm	fog	duststorm	squall			preval. direct.		mean speed (m/sec)
									03 h*1	12 h	
Jan.											0.8
Feb.											
Mar.											
Apr.											1.8
May											
June											
July											2.3
Aug.											
Sep.											
Oct.											1.2
Nov.											
Dec.											
Annual											

*1 Greenwich mean time.

TABLE LII

CLIMATIC TABLE FOR RAIPUR

Latitude 21°14′N, longitude 81°29′E, elevation 298 m

Month	Mean sta. press. (mbar)	Temperature (°C)				Relative humidity (%)		Precipitation (mm)	
		daily max.	daily min.	extreme		03 h*1	12 h	mean	max. in 24 h
				max.	min.				
Jan.	980.9	27.7	13.5	35.0	5.0	62	39	12.2	55.4
Feb.	978.5	30.3	16.2	37.8	5.0	55	31	20.4	57.4
Mar.	976.3	34.7	20.5	43.3	8.3	41	24	23.3	55.9
Apr.	972.9	39.2	25.1	46.1	15.0	36	21	15.1	38.3
May	969.5	42.3	28.8	47.7	14.4	35	22	16.8	80.3
June	965.9	37.5	26.8	47.2	16.1	61	49	193.6	197.6
July	966.2	30.3	24.1	38.9	20.0	85	78	391.8	213.1
Aug.	967.5	30.1	24.1	37.5	20.0	86	78	393.6	370.3
Sep.	970.5	31.0	24.1	37.2	18.3	82	74	249.4	148.6
Oct.	975.9	31.2	21.5	37.8	13.9	73	60	62.4	148.6
Nov.	979.6	29.1	16.0	35.6	8.3	63	46	7.8	70.4
Dec.	981.3	27.3	13.2	32.3	3.9	62	43	1.8	52.1
Annual	973.7	32.6	21.2	47.7	3.9	62	47	1388.2	370.3

Month	Number of days with						Mean cloudi- ness (oktas)	Mean sun- shine hours	Wind		
	precip. ⩾0.3 mm	hail	thunder- storm	fog	dust- storm	squall			preval. direct.		mean speed (m/sec)
									03 h*1	12 h	
Jan.	1	0	*	*	0	0	1.5		NE	NNE	1.4
Feb.	3	*	1	*	0	0	1.5		NNE	N	1.7
Mar.	3	*	3	0	*	0	1.7		NE	WNW	1.9
Apr.	3	*	3	0	*	0	2.2		SW	W	2.3
May	3	*	4	0	1	0	2.9		SW	NW	3.0
June	13	0	8	0	1	0	5.8		SW	SW	3.4
July	21	0	6	0	0	0	7.1		SW	WSW	3.3
Aug.	20	0	5	0	0	0	6.9		W	W	2.9
Sep.	16	0	7	0	0	*	5.8		SW	W	2.1
Oct.	5	0	2	*	0	0	3.1		NE	NE	1.7
Nov.	1	0	*	0	0	0	1.9		NE	NE	1.1
Dec.	*	0	0	0	0	0	1.3		NE	NE	1.2
Annual	90	1	39	1	2	*	3.5		SW	W	2.2

*1 Greenwich mean time. * = <1.

TABLE LIII

CLIMATIC TABLE FOR CUTTACK

Latitude 20°48′N, longitude 85°56′E, elevation 27 m

Month	Mean sta. press. (mbar)	Temperature (°C)				Relative humidity (%)		Precipitation (mm)	
		daily max.	daily min.	extreme		03 h*1	12 h	mean	max. in 24 h
				max.	min.				
Jan.	1012.2	28.9	15.7	39.0	7.8	80	48	10.4	61.0
Feb.	1009.7	31.5	18.2	39.1	10.6	76	43	28.5	98.0
Mar.	1006.9	35.9	22.1	42.8	14.4	73	41	19.5	99.1
Apr.	1003.8	38.3	25.3	45.0	14.9	71	50	27.0	94.5
May	999.5	38.8	26.9	47.7	18.9	71	58	71.8	142.7
June	996.2	35.8	26.5	47.2	20.0	76	69	214.6	205.7
July	996.1	31.6	25.6	40.0	20.5	83	81	355.1	210.8
Aug.	997.7	31.6	25.6	37.2	21.0	83	81	364.5	320.8
Sep.	1000.7	32.2	25.5	37.6	21.7	83	80	252.1	249.2
Oct.	1006.3	32.0	23.7	40.0	16.7	79	72	167.6	292.6
Nov.	1010.2	30.1	18.8	35.5	10.6	74	59	41.4	195.6
Dec.	1012.3	28.4	15.5	33.6	8.9	77	52	4.7	54.9
Annual	1004.3	32.9	22.5	47.7	7.8	77	61	1557.2	320.8

Month	Number of days with						Mean cloudi-ness (oktas)	Mean sun-shine hours	Wind		
	precip. ≥0.3 mm	hail	thunder-storm	fog	dust-storm	squall			preval. direct.		mean speed (m/sec)
									03 h*1	12 h	
Jan.	1	0	*	3	0	0	1.7	8.9	var.	ESE	0.8
Feb.	2	0	1	3	*	0	2.1	9.0	W	S	1.1
Mar.	2	0	2	1	*	0	2.5	8.5	SW	S	1.6
Apr.	3	*	4	3	3	0	3.1	8.4	SW	S	2.1
May	6	*	7	0	1	0	4.3	8.7	SW	SW	2.5
June	14	0	9	0	*	0	5.7	5.0	SW	SW	2.0
July	21	0	5	0	0	0	6.7	3.5	SW	SW	1.9
Aug.	21	0	7	0	0	0	6.6	4.5	W	SW	1.7
Sep.	17	0	7	0	0	0	5.8	5.1	W	SW	1.3
Oct.	10	0	5	0	0	0	4.3	6.9	NE	NE	1.2
Nov.	2	0	1	3	0	0	2.5	9.2	var.	NE	0.9
Dec.	1	0	*	1	0	0	1.7	9.1	W	E	0.7
Annual	100	*	48	13	5	0	3.9	7.2	SW	SW	1.5

*1 Greenwich mean time. * = <1.

TABLE LIV

CLIMATIC TABLE FOR AKOLA

Latitude 20°42′N, longitude 77°02′E, elevation 282 m

Month	Mean sta. press. (mbar)	Temperature (°C)				Relative humidity (%)		Precipitation (mm)	
		daily max.	daily min.	extreme		03 h*¹	12 h	mean	max. in 24 h
				max.	min.				
Jan.	981.5	30.2	13.7	36.2	3.9	57	29	9.1	49.0
Feb.	979.6	32.8	15.2	40.0	2.2	46	22	7.7	42.2
Mar.	977.3	37.1	19.4	44.4	5.6	34	18	7.5	38.6
Apr.	974.7	40.5	24.6	46.1	11.1	30	16	6.6	58.7
May	971.3	42.4	28.1	47.8	11.9	39	19	10.9	44.7
June	968.9	37.4	26.2	47.2	20.0	66	43	146.0	188.0
July	969.1	31.4	23.9	40.6	20.6	81	68	260.7	188.5
Aug.	970.5	30.7	23.5	37.8	18.3	82	67	170.1	224.5
Sep.	972.7	31.5	23.1	40.0	15.4	81	64	177.9	365.4
Oct.	977.5	33.3	20.0	40.0	10.0	66	41	46.3	110.5
Nov.	980.9	31.0	15.0	36.5	5.6	59	35	27.6	112.3
Dec.	982.1	29.6	12.6	36.7	3.9	60	32	6.4	65.0
Annual	975.5	34.0	20.4	47.8	2.2	58	38	876.8	365.4

Month	Number of days with						Mean cloudi-ness (oktas)	Mean sun-shine hours	Wind		
	precip. ≥0.3 mm	hail	thunder-storm	fog	dust-storm	squall			preval. direct.		mean speed (m/sec)
									03 h*¹	12 h	
Jan.	2	0	*	0	0	0	1.5	9.8	E	NE	1.5
Feb.	1	0	1	0	0	0	1.3	10.3	E	NW	1.6
Mar.	1	*	1	0	*	0	1.4	9.4	NW	NW	1.8
Apr.	1	*	1	0	*	0	2.1	10.4	NW	NW	2.3
May	2	0	3	0	0	0	2.5	9.5	NW	NW	3.7
June	11	0	6	0	1	0	4.5	7.0	NW	NW	4.1
July	18	0	3	0	0	0	6.3	3.3	W	W	3.6
Aug.	14	0	2	0	0	0	5.9	3.7	W	NW	3.3
Sep.	12	0	4	0	0	0	5.1	5.6	WNW	NW	2.7
Oct.	4	0	1	0	0	0	2.9	8.8	W	NE	1.4
Nov.	2	0	1	*	0	0	2.0	9.4	E	NE	1.3
Dec.	1	0	*	*	0	0	1.7	9.5	E	NE	1.2
Annual	68	1	24	*	1	0	3.1	8.0	W	NW	2.4

*¹ Greenwich mean time. * = <1.

163

TABLE LV

CLIMATIC TABLE FOR JAGDALPUR

Latitude 19°05′N, longitude 82°02′E, elevation 553 m

Month	Mean stat. press. (mbar)	Temperature (°C)				Relative humidity (%)		Precipitation (mm)	
		daily max.	daily min.	extreme		03 h*¹	12 h	mean	max. in 24 h
				max.	min.				
Jan.	952.1	28.5	11.0	33.1	2.8	73	41	5.3	40.6
Feb.	950.3	31.0	14.3	36.7	5.0	66	36	14.5	120.4
Mar.	948.5	34.7	18.4	40.6	9.4	54	28	17.1	45.7
Apr.	946.2	36.9	22.2	43.3	13.9	54	31	51.1	54.4
May	942.5	38.3	24.6	46.1	17.2	53	37	65.6	64.3
June	940.0	33.5	23.9	44.4	17.2	71	61	211.8	133.1
July	940.0	28.1	22.2	38.9	18.3	86	82	397.9	180.9
Aug.	941.2	28.4	22.2	33.9	16.7	86	81	381.2	203.2
Sep.	943.5	29.5	22.3	35.0	17.8	85	80	245.5	163.8
Oct.	947.9	29.5	19.9	34.5	11.1	80	67	115.8	136.9
Nov.	950.8	28.1	14.7	33.3	5.6	76	56	24.4	102.9
Dec.	952.1	27.4	11.3	32.2	3.9	76	50	3.9	38.3
Annual	946.3	31.2	18.9	46.1	2.8	72	54	1534.1	203.2

Month	Number of days with						Mean cloudiness (oktas)	Mean sunshine hours	Wind		
	precip. ⩾0.3 mm	hail	thunderstorm	fog	duststorm	squall			preval. direct.		mean speed (m/sec)
									03 h*¹	12 h	
Jan.	1	0	*	1	0	0	1.9		Calm	NE	0.8
Feb.	2	0	1	*	0	0	1.8		NE	SW	1.1
Mar.	3	*	3	0	0	0	2.1		SW	SW	1.3
Apr.	7	1	8	0	0	*	3.1		SW	SW	1.6
May	8	*	9	0	*	1	4.0		SW	SW	2.1
June	17	*	9	0	0	*	6.0		SW	SW	2.0
July	25	*	4	*	0	0	7.2		SW	SW	2.5
Aug.	25	0	6	*	0	0	6.9		SW	W	2.2
Sep.	20	0	8	1	0	0	6.3		SW	SW	1.6
Oct.	10	0	4	1	0	0	4.5		NE	NE	1.0
Nov.	2	0	1	4	0	0	2.9		NE	NE	0.8
Dec.	1	0	0	6	0	0	2.1		Calm	NE	0.7
Annual	120	1	53	14	*	1	4.1		SW	SW	1.4

*¹ Greenwich mean time. * = <1.

TABLE LVI

CLIMATIC TABLE FOR BOMBAY

Latitude 18°54′N, longitude 72°49′E, elevation 11 m

Month	Mean stat. press. (mbar)	Temperature (°C) daily max.	daily min.	extreme max.	min.	Relative humidity (%) 03 h*1	12 h	Precipitation (mm) mean	max. in 24 h	Mean evap. (mm)
Jan.	1011.3	29.1	19.4	35.6	11.7	71	63	2.0	49.3	3.4
Feb.	1010.8	29.5	20.3	38.3	11.7	72	62	1.1	41.7	4.1
Mar.	1009.1	31.0	22.7	39.7	16.7	72	63	0.4	34.3	4.9
Apr.	1007.5	32.3	25.1	40.6	20.0	73	66	2.8	37.3	5.5
May	1005.5	33.3	26.9	39.7	22.8	73	68	16.0	126.2	5.9
June	1002.5	31.9	26.3	37.2	21.1	80	78	520.3	408.9	4.6
July	1002.3	29.8	25.1	35.6	21.7	85	85	709.5	304.8	3.1
Aug.	1003.8	29.5	24.8	33.0	20.0	85	84	439.3	287.0	3.1
Sep.	1006.0	30.1	24.7	35.0	20.0	85	80	297.0	548.1	3.4
Oct.	1008.3	31.9	24.6	37.0	20.6	80	74	88.0	148.6	3.9
Nov.	1010.3	32.3	22.8	36.9	17.8	73	67	20.6	122.7	3.8
Dec.	1011.7	30.9	20.8	35.7	12.3	70	64	2.2	24.4	3.2
Annual	1007.5	31.0	23.6	40.6	11.7	77	71	2099.2	548.1	4.1

Month	Number of days with precip. ≥0.3 mm	hail	thunder- storm	fog	dust- storm	squall	Mean cloudi- ness (oktas)	Mean sun- shine hours	Wind preval. direct. 03 h*1	12 h	mean speed (m/sec)
Jan.	*	0	0	0	0	0	1.2	9.2	NE	NW	2.5
Feb.	*	0	*	0	0	0	0.9	9.6	NE	NW	2.6
Mar.	*	0	*	0	0	0	1.1	9.3	N	NW	2.9
Apr.	1	0	*	0	0	0	1.7	9.4	N	NW	2.9
May	2	0	2	0	0	*	2.9	9.3	WNW	W	2.8
June	20	0	5	0	0	4	5.6	5.6	W	W	3.6
July	29	0	1	0	0	5	7.3	2.3	W	W	4.1
Aug.	27	0	1	0	0	4	6.9	2.7	W	W	3.7
Sep.	21	0	3	0	0	1	5.5	4.9	W	W	2.8
Oct.	5	0	3	0	0	1	3.0	8.2	E	NW	2.4
Nov.	2	0	1	0	0	0	2.1	9.1	ENE	NW	2.3
Dec.	*	0	*	0	0	0	1.4	9.1	ENE	NW	2.4
Annual	108	0	17	0	0	15	3.3	7.4	NE	NW	2.9

*1 Greenwich mean time. * = <1.

TABLE LVII

CLIMATIC TABLE FOR POONA

Latitude 18°32′N, longitude 73°51′E, elevation 559 m

Month	Mean stat. press. (mbar)	Temperature (°C)				Relative humidity (%)		Precipitation (mm)		Mean evap. (mm)
		daily max.	daily min.	extreme max.	min.	03 h*1	12 h	mean	max. in 24 h	
Jan.	950.3	30.7	12.0	35.0	1.7	74	30	1.9	22.3	
Feb.	949.1	32.9	13.3	38.9	3.9	64	23	0.3	16.3	
Mar.	947.7	36.1	16.8	42.8	7.2	52	20	3.1	35.1	
Apr.	946.1	37.9	20.6	43.3	10.6	50	26	17.6	51.1	
May	944.1	37.2	22.6	43.3	13.8	58	36	34.7	82.5	
June	941.9	31.9	23.0	41.7	17.2	74	63	102.8	97.0	
July	941.7	27.8	22.0	36.0	18.9	83	78	186.8	130.4	
Aug.	942.9	27.7	21.5	35.0	17.2	85	77	106.4	108.7	
Sep.	944.9	29.2	20.8	36.1	15.1	82	71	127.3	132.3	
Oct.	947.5	31.8	19.3	37.8	9.4	79	52	91.9	149.1	
Nov.	950.0	30.8	15.0	36.1	4.6	73	40	37.0	96.8	
Dec.	950.7	30.1	12.0	35.0	3.3	75	35	4.9	42.4	
Annual	946.4	32.0	18.2	43.3	1.7	71	46	714.7	149.1	

Month	Number of days with						Mean cloudi-ness (oktas)	Mean sun-shine hours	Wind		mean speed (m/sec)
	precip. ⩾0.3 mm	hail	thunder-storm	fog	dust-storm	squall			preval. direct. 03 h*1	12 h	
Jan.	1	0	*	*	0	0	1.7	9.7	Calm	W	0.8
Feb.	*	0	*	0	0	0	1.1	10.3	SW	W	1.0
Mar.	1	*	1	0	0	*	1.3	9.8	SW	W	1.2
Apr.	3	*	4	0	*	1	2.1	10.0	SW	W	1.5
May	4	1	5	0	0	1	2.5	9.5	W	W	2.0
June	13	0	3	0	0	1	5.7	6.2	W	W	2.5
July	24	0	*	0	0	*	7.1	3.0	W	W	2.6
Aug.	23	0	1	0	0	*	6.9	3.7	W	W	2.1
Sep.	14	0	4	*	0	1	5.8	5.5	W	W	1.6
Oct.	8	0	5	1	*	*	3.8	8.2	SW	W	1.0
Nov.	3	0	2	1	0	0	2.7	9.1	Calm	E	0.8
Dec.	1	0	*	*	0	0	1.9	9.3	Calm	E	0.7
Annual	94	1	26	3	*	4	3.5	7.8	W	W	1.5

*1 Greenwich mean time. * = <1.

TABLE LVIII

CLIMATIC TABLE FOR VISAKHAPATNAM

Latitude 17°43′N, longitude 83°14′E, elevation 3 m

Month	Mean stat. press. (mbar)	Temperature (°C)				Relative humidity (%)		Precipitation (mm)		Mean evap. (mm)
		daily max.	daily min.	extreme		03 h*1	12 h	mean	max. in 24 h	
				max.	min.					
Jan.	1014.4	27.7	17.5	33.1	10.5	77	78	7.2	132.1	5.5
Feb.	1012.3	29.2	19.3	38.0	13.3	77	73	14.9	64.5	6.5
Mar.	1010.3	31.2	22.6	39.2	14.4	74	71	8.7	64.5	7.9
Apr.	1007.5	32.8	25.9	40.5	18.3	73	80	12.7	73.9	8.8
May	1003.4	34.0	27.8	44.3	20.0	75	83	53.5	145.3	9.0
June	1000.0	33.7	27.4	45.3	21.1	80	83	87.8	166.1	8.0
July	1000.3	31.7	26.0	39.4	21.3	84	82	121.9	145.0	6.2
Aug.	1001.5	32.0	26.0	38.3	21.1	82	83	132.2	121.4	5.9
Sep.	1003.9	31.6	25.6	37.8	21.5	81	84	167.3	148.6	5.2
Oct.	1008.5	30.9	24.5	37.2	17.8	78	79	259.3	293.3	5.0
Nov.	1012.1	29.3	21.2	33.9	12.9	68	73	90.6	270.5	5.8
Dec.	1014.1	27.7	18.3	32.8	11.3	70	74	17.5	191.3	5.7
Annual	1007.3	31.0	23.5	45.3	10.5	77	79	973.6	293.3	6.6

Month	Number of days with						Mean cloudiness (oktas)	Mean sunshine hours	Wind		
	precip. ⩾0.3 mm	hail	thunderstorm	fog	duststorm	squall			preval. direct.		mean speed (m/sec.)
									03 h*1	12 h	
Jan.	1	0	*	*	0	*	2.0		NW	E	1.7
Feb.	1	0	*	*	0	*	1.9		W	S	1.8
Mar.	1	0	1	0	0	*	2.1		W	S	2.7
Apr.	2	0	4	0	0	1	3.5		SW	SW	4.1
May	5	0	7	0	0	2	4.5		SW	SW	4.4
June	9	0	7	0	*	2	6.3		SW	SW	3.8
July	14	0	5	0	0	1	6.9		SW	SW	4.6
Aug.	14	0	6	0	0	1	6.5		W	SW	3.7
Sep.	15	0	11	0	0	1	6.1		W	SW	2.6
Oct.	13	0	7	0	0	1	4.9		W	E	2.3
Nov.	4	0	1	0	0	*	3.6		N	E	2.3
Dec.	1	0	*	0	0	*	2.6		NNW	E	2.1
Annual	80	0	49	*	*	11	4.3		W	SW	3.0

*1 Greenwich mean time. * = <1.

TABLE LIX

CLIMATIC TABLE FOR HYDERABAD (Begumpet)
Latitude 17°27′N, longitude 78°28′E, elevation 545 m

Month	Mean stat. press. (mbar)	Temperature (°C)				Relative humidity (%)		Precipitation (mm)		Mean evap. (mm)
		daily max.	daily min.	extreme max.	min.	03 h*1	12 h	mean	max. in 24 h	
Jan.	952.5	28.6	14.6	35.0	6.1	79	36	1.7	93.2	6.0
Feb.	950.8	31.2	16.7	37.2	8.9	64	35	11.4	42.9	7.6
Mar.	949.1	34.8	20.0	42.2	13.2	54	30	13.4	103.1	9.7
Apr.	946.9	36.9	23.7	43.3	16.1	51	31	24.1	60.7	10.4
May	943.9	38.7	26.2	44.4	18.0	50	33	30.0	65.0	11.8
June	942.3	34.1	24.1	43.9	17.8	71	54	107.4	122.7	8.8
July	942.3	29.8	22.3	37.2	18.6	83	69	165.0	109.2	6.8
Aug.	943.5	29.5	22.1	36.1	19.2	82	70	146.9	190.5	6.1
Sep.	945.2	29.7	21.6	36.1	17.8	82	71	163.3	153.2	5.9
Oct.	948.6	30.3	19.8	36.7	11.7	73	58	70.8	117.1	5.8
Nov.	951.3	28.7	16.0	33.9	7.4	68	48	24.9	95.5	5.7
Dec.	952.7	27.8	13.4	33.3	7.1	71	42	5.5	44.5	5.4
Annual	947.4	31.7	20.0	44.4	6.1	69	48	764.4	190.5	7.5

Month	Number of days with						Mean cloudi-ness (oktas)	Mean sun-shine hours	Wind		mean speed (m/sec)
	precip. ≥0.3 mm	hail	thunder-storm	fog	dust-storm	squall			preval. direct. 03 h*1	12 h	
Jan.	*	0	*	1	*	0	2.3		SE	E	2.3
Feb.	1	0	1	*	*	0	2.1		SE	ESE	2.5
Mar.	2	*	2	0	*	1	1.9		SE	SE	2.7
Apr.	4	0	5	0	0	2	3.1		SE	SE	3.0
May	4	0	6	0	*	2	3.9		W	NW	3.4
June	11	0	5	0	0	3	6.0		W	W	6.6
July	19	0	1	0	0	2	7.0		W	W	6.1
Aug.	17	0	2	0	0	1	6.7		W	W	5.1
Sep.	13	0	4	0	0	1	6.1		W	W	3.5
Oct.	6	0	3	*	0	*	4.5		E	NE	2.5
Nov.	3	0	*	*	*	0	3.6		ENE	NE	2.2
Dec.	1	0	0	1	*	0	2.5		SE	E	2.1
Annual	81	*	29	2	1	11	4.1		W	E	3.5

*1 Greenwich mean time. * = <1.

TABLE LX

CLIMATIC TABLE FOR KURNOOL

Latitude 15°50′N, longitude 78°04′E, elevation 281 m

Month	Mean stat. press. (mbar)	Temperature (°C)				Relative humidity (%)		Precipitation (mm)		Mean evap. (mm)
		daily max.	daily min.	extreme		03 h*¹	12 h	mean	max. in 24 h	
				max.	min.					
Jan.	981.3	31.3	17.0	36.5	8.3	70	32	0.2	41.1	
Feb.	979.4	34.3	19.3	39.8	11.1	57	24	5.0	66.8	
Mar.	977.3	37.5	22.5	41.9	12.8	48	21	9.7	50.0	
Apr.	975.3	39.3	26.0	44.6	15.5	49	24	21.6	42.2	
May	972.1	40.0	27.2	45.6	19.4	54	27	44.4	165.6	
June	971.8	35.6	25.0	44.4	19.6	69	46	90.5	81.3	
July	972.2	32.5	23.8	38.5	19.2	75	57	129.6	132.1	
Aug.	972.9	32.1	23.5	37.8	19.9	75	56	121.6	121.9	
Sep.	974.5	31.9	23.3	37.8	17.0	76	57	147.1	200.1	
Oct.	977.3	32.4	22.4	38.3	13.0	74	53	79.2	146.3	
Nov.	979.8	31.0	19.2	38.8	9.7	72	44	21.8	80.5	
Dec.	981.5	30.3	16.6	34.4	6.7	72	37	3.0	71.1	
Annual	976.3	34.0	22.1	45.6	6.7	66	40	673.7	200.1	

Month	Number of days with						Mean cloudi-ness (oktas)	Mean sun-shine hours	Wind		
	precip. ≥0.3 mm	hail	thunder-storm	fog	dust-storm	squall			preval. direct.		mean speed (m/sec)
									03 h*¹	12 h	
Jan.	0	0	0	0	0	0	2.0		E	E	1.7
Feb.	1	0	*	0	*	*	1.9		ESE	E	1.9
Mar.	1	0	1	0	0	*	1.7		SE	E	2.1
Apr.	3	0	3	0	1	1	2.9		W	E	2.4
May	4	0	5	0	1	1	4.1		W	W	3.7
June	9	0	3	0	*	5	6.2		W	W	5.9
July	14	0	1	0	0	4	7.0		W	W	5.9
Aug.	14	0	2	0	*	3	6.7		W	W	5.1
Sep.	12	0	4	0	0	1	6.2		W	W	3.5
Oct.	7	0	4	0	0	*	4.9		W	E	1.7
Nov.	3	0	1	0	0	0	3.9		E	E	1.4
Dec.	1	0	*	0	0	0	2.5		E	E	1.4
Annual	69	0	24	0	2	15	4.7		W	E	2.8

*¹ Greenwich mean time. * = <1.

TABLE LXI

CLIMATIC TABLE FOR MARMAGAO

Latitude 15°25′N, longitude 73°47′E, elevation 62 m

Month	Mean stat. press. (mbar)	Temperature (°C)				Relative humidity (%)		Precipitation (mm)		Mean evap. (mm)
		daily max.	daily min.	extreme		03 h[*1]	12 h	mean	max. in 24 h	
				max.	min.					
Jan.	1005.1	29.7	21.4	34.9	16.7	66	62	1.8	31.7	4.6
Feb.	1004.3	29.0	21.9	37.2	17.2	72	66	0.	0	5.6
Mar.	1003.1	30.0	23.9	37.2	19.4	73	69	0.4	4.6	5.6
Apr.	1001.9	30.9	26.1	35.4	20.6	72	69	20.3	103.4	6.3
May	1000.1	31.3	26.9	34.7	21.0	73	72	81.3	307.1	6.2
June	998.7	29.4	24.7	35.0	17.0	83	83	777.8	196.3	4.2
July	999.1	28.0	24.0	31.7	20.2	86	88	905.1	259.1	3.6
Aug.	1000.0	27.8	23.9	32.4	19.3	87	86	412.9	153.9	3.7
Sep.	1001.4	28.1	23.8	32.6	21.1	87	84	225.9	175.5	4.0
Oct.	1002.3	29.8	23.9	36.3	18.3	82	78	138.7	159.8	4.2
Nov.	1003.7	31.0	22.8	36.5	12.2	69	65	42.6	72.1	4.6
Dec.	1004.9	30.5	21.5	36.6	17.2	62	60	4.9	26.7	4.4
Annual	1002.1	29.5	23.7	37.2	16.7	76	73	2611.7	307.1	4.7

Month	Number of days with						Mean cloudi- ness (oktas)	Mean sun- shine hours	Wind		
	precip. ⩾0.3 mm	hail	thunder- storm	fog	dust- storm	squall			preval. direct.		mean speed (m/sec)
									03 h[*1]	12 h	
Jan.	*	0	*	0	*	*	1.7		E	W	2.9
Feb.	0	0	0	*	0	*	1.8		E	W	3.2
Mar.	*	0	*	*	0	*	2.2		NE	W	3.5
Apr.	1	0	3	0	0	*	3.3		N	W	3.7
May	4	0	4	0	0	*	4.2		NW	W	4.1
June	24	0	4	0	*	1	6.0		SW	W	5.3
July	29	0	*	0	0	1	6.5		W	W	6.4
Aug.	26	0	1	0	0	*	5.9		W	W	5.2
Sep.	18	*	2	0	0	*	5.4		W	W	3.1
Oct.	9	0	9	*	0	*	4.5		E	W	2.7
Nov.	3	0	3	*	0	*	3.2		E	W	2.6
Dec.	1	0	0	*	0	0	2.3		E	W	2.7
Annual	115	*	27	1	*	4	3.9		E	W	3.8

[*1] Greenwich mean time. * = <1.

TABLE LXII

CLIMATIC TABLE FOR NELLORE

Latitude 14°27′N, longitude 79°59′E, elevation 20 m

Month	Mean stat. press. (mbar)	Temperature (°C)				Relative humidity (%)		Precipitation (mm)		Mean evap. (mm)
		daily max.	daily min.	extreme		03 h*¹	12 h	mean	max. in 24 h	
				max.	min.					
Jan.	1011.8	29.8	20.0	35.6	15.0	84	64	8.3	94.2	
Feb.	1010.0	32.0	21.1	39.4	16.1	79	59	4.6	116.8	
Mar.	1007.9	34.5	23.1	43.9	17.2	74	59	6.9	59.2	
Apr.	1005.3	37.1	25.7	45.6	20.6	69	62	14.9	73.7	
May	1001.4	39.6	27.8	46.7	20.2	60	54	54.0	185.9	
June	1000.3	38.2	28.2	46.7	21.1	57	46	40.6	83.8	
July	1001.0	35.6	26.7	42.2	22.2	65	51	73.2	95.3	
Aug.	1001.7	35.2	26.5	40.6	21.7	66	54	82.5	75.2	
Sep.	1003.1	34.7	26.0	41.7	21.5	69	60	98.0	133.6	
Oct.	1006.3	32.5	24.7	39.4	18.9	79	72	266.5	444.0	
Nov.	1009.4	29.6	22.3	36.7	16.7	84	75	292.7	356.9	
Dec.	1011.4	28.7	20.4	35.0	14.4	84	70	89.8	189.2	
Annual	1005.8	34.0	24.4	46.7	14.4	73	61	1032.0	444.0	

Month	Number of days with						Mean cloudi-ness (oktas)	Mean sun-shine hours	Wind		mean speed (m/sec)
	precip. ≥0.3 mm	hail	thunder-storm	fog	dust-storm	squall			preval. direct.		
									03 h*¹	12 h	
Jan.	1	0	0	*	0	0	3.0		NW	E	1.5
Feb.	1	0	*	0	0	0	2.5		S	E	1.8
Mar.	1	0	1	0	0	0	2.1		S	SE	2.4
Apr.	1	0	1	0	0	0	3.1		S	SE	2.8
May	3	0	2	0	*	0	4.2		S	SE	2.9
June	6	0	2	0	0	0	5.9		W	W	2.7
July	12	0	1	0	0	*	6.7		W	W	2.7
Aug.	10	0	3	0	0	*	6.1		W	W	2.6
Sep.	8	0	3	0	0	0	5.5		W	W	2.1
Oct.	11	0	4	0	0	0	5.1		NW	E	1.6
Nov.	9	0	1	*	0	0	4.9		NW	NE	1.7
Dec.	4	0	*	0	0	0	4.0		NW	NE	1.7
Annual	67	0	18	*	*	*	4.4		W	E	2.2

*¹ Greenwich mean time. * = <1.

TABLE LXIII

CLIMATIC TABLE FOR MADRAS (Minambakkam)

Latitude 13°00'N, longitude 80°11'E, elevation 16 m

Month	Mean stat. press. (mbar)	Temperature (°C)				Relative humidity (%)		Precipitation (mm)		Mean evap. (mm)
		daily max.	daily min.	extreme max.	extreme min.	03 h*1	12 h	mean	max. in 24 h	
Jan.	1012.1	28.8	20.3	32.8	13.9	83	67	23.8	212.9	
Feb.	1010.6	30.6	21.1	36.7	15.0	80	63	6.8	123.2	
Mar.	1008.9	32.7	23.1	40.6	16.7	77	64	15.1	88.1	
Apr.	1006.3	34.9	26.0	42.8	20.0	72	68	24.7	96.3	
May	1002.7	37.6	27.8	45.0	21.1	63	66	51.7	214.9	
June	1001.7	37.3	27.6	43.3	20.6	58	59	52.6	59.2	
July	1002.5	35.2	26.3	41.1	21.7	65	61	83.5	116.3	
Aug.	1003.0	34.5	25.8	40.0	20.6	69	64	124.3	91.7	
Sep.	1004.3	33.9	25.4	38.9	20.6	73	69	118.0	100.3	
Oct.	1006.9	31.8	24.4	39.4	16.7	81	76	267.0	233.7	
Nov.	1009.5	29.2	22.5	34.4	15.0	83	76	308.7	236.2	
Dec.	1011.5	28.2	21.0	32.8	13.9	84	71	139.1	261.6	
Annual	1006.6	32.9	24.3	45.0	13.9	74	67	1215.3	261.6	

Month	Number of days with						Mean cloudi-ness (oktas)	Mean sun-shine hours	Wind		mean speed (m/sec)
	precip. ⩾0.3 mm	hail	thunder-storm	fog	dust-storm	squall			preval. direct. 03 h*1	12 h	
Jan.	3	0	*	1	0	*	3.5	8.5	NW	NE	2.5
Feb.	1	0	*	1	0	0	2.7	9.7	NW	E	2.6
Mar.	1	0	1	*	0	*	2.3	9.7	SW	SE	2.8
Apr.	2	0	2	*	0	1	3.7	9.5	S	SE	2.9
May	3	0	3	0	*	1	4.5	8.7	S	SE	3.6
June	9	0	5	0	*	1	6.1	6.6	W	SE	4.6
July	14	0	5	0	0	3	6.9	4.8	W	SE	4.1
Aug.	15	0	7	0	0	1	6.1	5.6	W	SE	3.8
Sep.	11	0	8	0	0	1	5.7	6.3	W	SE	3.1
Oct.	14	0	9	*	0	1	5.5	6.5	W	ESE	2.6
Nov.	11	0	3	*	0	1	5.7	6.7	NW	NE	3.3
Dec.	7	0	1	*	0	*	4.7	7.4	NW	NE	3.5
Annual	91	0	44	3	*	11	4.7	7.5	W	SE	3.3

*1 Greenwich mean time. * = <1.

TABLE LXIV

CLIMATIC TABLE FOR BANGALORE

Latitude 12°58′N, longitude 77°35′E, elevation 921 m

Month	Mean stat. press. (mbar)	Temperature (°C)				Relative humidity (%)		Precipitation (mm)		Mean evap. (mm)
		daily max.	daily min.	extreme		03 h*¹	12 h	mean	max. in 24 h	
				max.	min.					
Jan.	911.5	26.9	15.0	32.2	7.8	77	40	3.3	65.8	4.6
Feb.	910.7	29.7	16.5	34.5	9.4	67	29	10.2	67.3	6.2
Mar.	909.5	32.3	19.0	37.2	11.1	63	24	6.1	50.8	7.8
Apr.	908.3	33.4	21.2	38.3	14.4	70	34	45.7	90.7	7.3
May	906.5	32.7	21.1	38.9	16.7	75	46	116.5	153.9	5.9
June	905.8	28.9	19.7	37.8	16.7	82	62	80.1	101.6	4.7
July	906.0	27.2	19.2	33.3	16.1	86	68	116.6	105.4	3.9
Aug.	906.3	27.3	19.2	33.3	14.4	86	66	147.1	162.1	3.6
Sep.	907.3	27.6	18.9	33.3	15.0	85	62	142.7	124.7	3.7
Oct.	908.7	27.5	18.9	32.2	13.2	83	64	184.9	116.3	3.5
Nov.	910.1	26.3	17.2	31.1	14.1	78	59	54.3	114.5	3.6
Dec.	911.3	25.7	15.3	31.1	8.9	78	51	16.2	67.3	3.6
Annual	908.5	28.8	18.4	38.9	7.8	77	50	923.7	162.1	4.8

Month	Number of days with						Mean cloudi-ness (oktas)	Mean sun-shine hours	Wind		
	precip. ⩾0.3 mm	hail	thunder-storm	fog	dust-storm	squall			preval. direct.		mean speed (m/sec)
									03 h*¹	12 h	
Jan.	1	0	0	3	0	0	3.3		E	E	2.9
Feb.	1	0	1	*	0	0	2.7		E	E	2.7
Mar.	1	0	1	*	0	*	2.5		SE	E	2.6
Apr.	5	*	7	*	1	*	4.3		SW	E	2.5
May	11	*	12	*	*	1	5.3		W	W	3.3
June	13	0	4	*	0	1	6.9		W	W	4.7
July	18	0	2	*	0	1	7.5		W	W	4.9
Aug.	18	0	4	1	0	1	7.3		W	W	4.2
Sep.	15	0	4	1	0	*	6.9		W	W	3.4
Oct.	15	0	7	1	0	*	6.2		W	NE	2.3
Nov.	7	0	1	2	1	0	5.1		E	E	2.4
Dec.	3	0	*	3	*	0	4.1		E	E	2.7
Annual	108	*	43	12	1	4	5.1		W	E	3.2

*¹ Greenwich mean time. * = <1.

TABLE LXV

CLIMATIC TABLE FOR MANGALORE

Latitude 12°52′N, longitude 74°51′E, elevation 22 m

Month	Mean stat. press. (mbar)	Temperature (°C)				Relative humidity (%)		Precipitation (mm)		Mean evap. (mm)
		daily max.	daily min.	extreme		03 h*¹	12 h	mean	max. in 24 h	
				max.	min.					
Jan.	1009.1	31.4	21.7	36.3	16.7	71	61	4.7	40.6	
Feb.	1008.5	31.1	22.8	37.8	16.7	75	66	1.9	36.1	
Mar.	1007.5	31.7	24.5	37.3	18.3	75	67	8.9	82.8	
Apr.	1006.3	32.4	26.1	36.6	20.0	73	69	40.0	117.1	
May	1004.8	32.1	26.0	36.7	18.9	77	72	232.7	360.9	
June	1004.9	29.4	23.9	34.4	19.6	89	85	981.6	252.0	
July	1005.3	28.5	23.5	35.6	19.2	91	88	1058.6	268.2	
Aug.	1005.7	28.5	23.6	32.2	19.4	91	87	576.9	232.4	
Sep.	1006.5	28.7	23.5	31.7	19.0	89	83	267.0	184.7	
Oct.	1007.0	29.8	23.8	35.0	18.8	85	79	205.9	181.6	
Nov.	1007.9	31.1	23.2	35.6	18.0	77	69	70.6	112.1	
Dec.	1008.9	31.7	21.9	35.0	16.7	69	62	18.2	153.2	
Annual	1006.9	30.5	23.7	37.8	16.7	80	74	3467.0	360.9	

Month	Number of days with						Mean cloudi-ness (oktas)	Mean sun-shine hours	Wind		mean speed (m/sec)
	precip. ⩾0.3 mm	hail	thunder-storm	fog	dust-storm	squall			preval. direct.		
									03 h*¹	12 h	
Jan.	*	0	*	*	0	*	2.2		E	NW	2.4
Feb.	*	0	*	0	0	0	2.2		E	NW	2.4
Mar.	1	0	1	*	0	0	2.5		E	NW	2.4
Apr.	4	0	5	*	0	*	4.1		E	NW	2.5
May	12	*	8	0	0	1	5.5		E	NW	2.7
June	27	0	5	0	0	1	7.1		E	SW	2.6
July	30	0	2	*	0	2	7.5		W	W	2.6
Aug.	28	0	1	*	0	*	7.1		W	NW	2.2
Sep.	22	0	1	1	0	*	6.3		E	NW	1.9
Oct.	15	0	7	1	0	*	5.7		E	NW	2.0
Nov.	6	0	3	*	0	*	4.3		E	NW	2.0
Dec.	1	0	1	0	0	*	2.9		E	NW	2.3
Annual	147	*	35	3	0	5	4.7		E	NW	2.3

*¹ Greenwich mean time. * = <1.

TABLE LXVI

CLIMATIC TABLE FOR TIRUCHIRAPALLI

Latitude 10°46′N, longitude 78°43′E, elevation 88 m

Month	Mean stat. press. (mbar)	Temperature (°C) daily max.	daily min.	extreme max.	min.	Relative humidity (%) 03 h*1	12 h	Precipitation (mm) mean	max. in 24 h	Mean evap. (mm)
Jan.	1002.7	30.1	20.6	35.6	14.4	79	54	18.4	114.8	
Feb.	1001.6	32.7	21.3	40.0	13.9	78	43	7.5	137.9	
Mar.	999.9	35.1	22.9	40.0	15.6	76	38	8.4	80.8	
Apr.	997.7	36.7	25.8	42.8	18.3	73	42	70.1	160.5	
May	995.0	37.1	26.4	43.3	19.4	67	41	79.8	183.1	
June	1004.9	36.4	26.5	43.9	20.0	59	43	33.9	75.2	
July	995.5	35.5	25.9	41.1	21.1	61	43	40.5	94.7	
Aug.	995.9	35.1	25.4	40.6	20.6	65	47	104.6	109.7	
Sep.	996.6	34.2	24.9	40.6	20.6	70	51	107.6	84.3	
Oct.	998.5	32.3	23.9	38.9	18.9	79	63	170.0	319.0	
Nov.	1000.5	29.9	22.7	36.7	16.7	81	67	156.2	298.2	
Dec.	1002.3	29.3	21.3	35.6	14.4	79	65	70.6	135.6	
Annual	998.4	33.7	24.0	43.9	13.9	72	50	867.6	319.0	

Month	Number of days with precip. ≥0.3 mm	hail	thunder- storm	fog	dust- storm	squall	Mean cloudi- ness (oktas)	Mean sun- shine hours	Wind preval. direct. 03 h*1	12 h	mean speed (m/sec)
Jan.	3	0	*	0	0	0	3.5	8.5	NE	NE	2.8
Feb.	1	0	1	*	0	0	3.1	9.3	NE	E	2.1
Mar.	1	0	1	*	0	*	2.8	9.6	NE	E	2.4
Apr.	5	0	5	0	0	*	4.0	9.1	W	E	2.7
May	6	0	8	0	*	*	4.7	8.2	W	W	4.8
June	3	0	3	0	0	*	5.9	7.3	W	W	8.0
July	4	0	3	0	*	0	6.3	5.6	W	W	8.7
Aug.	7	0	7	0	0	*	5.9	6.4	W	W	7.2
Sep.	8	0	6	0	0	0	4.7	7.6	W	W	5.4
Oct.	13	0	9	0	0	*	5.7	6.7	W	NE	3.0
Nov.	11	0	3	0	0	0	5.3	6.3	NE	NE	2.4
Dec.	6	0	*	0	0	0	4.7	7.1	NE	NE	3.1
Annual	69	0	46	*	*	1	4.7	7.6	W	E	4.4

*1 Greenwich mean time. * = <1.

TABLE LXVII

CLIMATIC TABLE FOR JAFFNA

Latitude 9°39′N, longitude 80°01′E, elevation 4 m

Month	Mean stat. press. (mbar)	Temperature (°C)				Relative humidity (%)		Precipitation (mm)		Mean evap. (mm)
		daily max.	daily min.	extreme		03 h*1	12 h	mean	max. in 24 h	
				max.	min.					
Jan.	1012.9	28.4	22.3	31.6	18.1	78	69	96.5	138.9	
Feb.	1012.2	29.8	22.4	32.9	16.3	77	65	36.8	146.6	
Mar.	1010.9	31.6	24.3	35.2	16.6	77	63	30.0	92.5	
Apr.	1009.2	32.1	26.8	36.7	21.9	80	68	70.1	240.3	
May	1006.8	31.3	27.6	35.0	21.2	80	77	62.7	222.0	
June	1006.6	30.4	27.2	35.3	21.1	79	79	16.3	125.2	
July	1007.1	30.1	26.6	35.1	21.8	81	78	16.5	96.0	
Aug.	1007.3	30.1	26.3	35.1	21.3	82	78	31.5	112.5	
Sep.	1008.1	30.2	26.4	34.2	21.5	81	78	47.5	134.6	
Oct.	1009.6	29.9	25.4	34.1	20.0	83	78	243.6	252.0	
Nov.	1010.8	28.9	23.8	32.2	19.2	81	76	411.2	520.2	
Dec.	1012.2	28.1	22.9	30.8	18.2	80	75	266.7	282.4	
Annual	1009.5	30.1	25.2	36.7	16.3	80	74	1329.4	520.2	

Month	Number of days with						Mean cloudi-ness (oktas)	Mean sun-shine hours	Wind		
	precip. ⩾0.3 mm	hail	thunder-storm	fog	dust-storm	squall			preval. direct.		mean speed (m/sec)
									03 h*1	12 h	
Jan.	8		0	0			4.4		NE	NE	2.0
Feb.	3		0	0			3.7		NE	NE	2.1
Mar.	3		1	0			3.2		SE	NE	2.6
Apr.	7		6	0			4.8		SE	SW	4.1
May	4		3	0			5.4		SW	SW	6.7
June	1		1	0			5.8		SW	SW	7.4
July	2		1	0			6.0		SW	SW	6.5
Aug.	4		3	0			5.8		SW	SW	6.4
Sep.	3		3	0			5.4		SW	SW	6.4
Oct.	13		6	0			5.8		SW	SW	4.1
Nov.	18		4	0			5.7		NE	NE	2.3
Dec.	14		1	0			5.2		NE	NE	2.1
Annual	80		29	0			5.1				4.4

*1 Greenwich mean time.

TABLE LXVIII

CLIMATIC TABLE FOR TRINCOMALEE

Latitude 08°35′N, longitude 81°15′E, elevation 3 m

Month	Mean stat. press. (mbar)	Temperature (°C)				Relative humidity (%)		Precipitation (mm)		Mean evap. (mm)
		daily max.	daily min.	extreme		03 h*1	12 h	mean	max. in 24 h	
				max.	min.					
Jan.	1012.2	27.0	24.2	31.7	19.6	78	76	210.6	208.5	
Feb.	1011.6	28.1	24.3	34.3	18.3	77	72	95.2	227.8	
Mar.	1010.5	29.9	24.8	37.4	19.8	78	71	48.3	239.0	
Apr.	1008.9	32.0	25.4	38.6	20.9	77	72	76.7	141.0	
May	1006.4	33.6	26.1	38.7	19.4	72	65	67.8	271.5	
June	1006.2	33.7	26.2	38.3	20.6	70	57	18.5	112.3	
July	1006.6	33.7	25.6	38.4	21.2	71	59	54.1	99.8	
Aug.	1006.7	33.5	25.3	37.8	20.8	71	62	102.9	107.4	
Sep.	1007.6	33.5	25.1	37.6	21.1	71	65	88.9	128.5	
Oct.	1009.0	31.3	24.3	37.6	21.0	77	73	234.7	154.2	
Nov.	1010.2	28.7	23.8	35.9	18.7	81	77	355.1	264.7	
Dec.	1011.4	27.3	24.0	32.6	19.4	80	78	373.9	322.8	
Annual	1008.9	31.0	24.9	38.7	18.3	75	69	1726.7	322.8	

Month	Number of days with						Mean cloudi- ness (oktas)	Mean sun- shine hours	Wind		
	precip. ≥0.3 mm	hail	thunder- storm	fog	dust- storm	squall			preval. direct.		mean speed (m/sec)
									03 h*1	12 h	
Jan.	13	0	0				5.6	6.6	NE	NE	5.3
Feb.	6	1	0				4.8	7.5	NE	NE	4.0
Mar.	5	3	0				4.0	8.8	NE	NE	2.9
Apr.	7	8	0				4.8	8.9	SW	E	2.9
May	6	6	0				5.6	8.0	SW	SW	4.7
June	2	2	0				6.0	7.7	SW	SW	6.1
July	4	4	0				6.1	7.3	SW	SW	5.5
Aug.	7	8	0				5.8	8.2	SW	SW	5.1
Sep.	6	8	0				5.8	7.8	SW	SW	4.6
Oct.	16	10	0				5.9	6.9	SW	NE	3.8
Nov.	19	4	0				5.8	5.6	SW	NNE	3.9
Dec.	10	2	0				6.0	5.5	N	NNE	5.2
Annual	109	56	0				5.5	7.4			4.5

*1 Greenwich mean time.

TABLE LXIX

CLIMATIC TABLE FOR TRIVANDRUM

Latitude 08°29'N, longitude 76°57'E, elevation 64 m

Month	Mean stat. press. (mbar)	Temperature (°C)				Relative humidity (%)		Precipitation (mm)		Mean evap. (mm)
		daily max.	daily min.	extreme		03 h*¹	12 h	mean	max. in 24 h	
				max.	min.					
Jan.	1003.8	31.3	23.3	35.5	17.7	77	63	20.1	52.1	4.8
Feb.	1003.5	31.7	22.9	35.0	18.1	78	63	20.3	88.1	5.5
Mar.	1002.6	32.5	24.2	36.2	20.6	80	66	43.5	80.0	5.6
Apr.	1001.5	32.4	25.1	35.3	20.3	81	73	122.1	129.8	4.9
May	1000.5	31.6	25.0	35.2	21.1	84	77	248.6	277.9	4.5
June	1001.2	29.4	23.6	34.4	20.0	90	82	331.2	154.7	3.7
July	1001.5	29.1	23.2	32.4	20.6	89	81	215.4	151.6	3.5
Aug.	1001.5	29.4	23.3	32.8	19.9	88	78	164.0	102.4	3.8
Sep.	1001.9	29.9	23.3	33.3	20.8	86	77	122.9	125.5	4.1
Oct.	1002.7	29.9	23.4	33.4	20.6	87	80	271.2	215.9	3.7
Nov.	1003.1	30.1	23.1	34.3	18.9	87	78	206.9	162.8	3.5
Dec.	1003.6	30.9	22.5	34.4	18.2	80	69	73.1	148.8	3.9
Annual	1002.3	30.7	23.5	36.2	17.7	84	74	1839.3	277.9	4.3

Month	Number of days with						Mean cloudi- ness (oktas)	Mean sun- shine hours	Wind		mean speed (m/sec)
	precip. ⩾0.3 mm	hail	thunder- storm	fog	dust- storm	squall			preval. direct.		
									03 h*¹	12 h	
Jan.	3	0	1	0	0	0	3.1	8.3	NE	SW	1.4
Feb.	3	0	3	0	0	0	2.9	8.5	NE	SW	1.6
Mar.	5	0	8	0	0	0	3.3	8.3	NE	SW	1.8
Apr.	10	0	16	0	0	0	4.9	6.7	N	W	2.2
May	15	0	12	0	0	0	5.7	5.9	N	NW	2.6
June	24	0	3	0	0	*	6.6	3.7	NW	NW	2.7
July	21	0	1	*	0	*	6.7	4.1	NW	NW	3.0
Aug.	17	0	1	0	0	0	5.9	5.3	NW	NW	3.1
Sep.	14	0	3	0	0	*	5.3	5.9	NW	NW	2.9
Oct.	16	0	9	*	0	0	5.9	5.5	N	NW	2.0
Nov.	15	0	9	1	0	0	5.3	5.9	NE	SW	1.5
Dec.	6	0	3	0	0	0	3.9	7.4	NE	SW	1.3
Annual	149	0	70	1	0	*	5.0	6.3	N	NW	2.2

*¹ Greenwich mean time. * = <1.

TABLE LXX

CLIMATIC TABLE FOR MINICOY

Latitude 08°18′N, longitude 73°00′E, elevation 2 m

Month	Mean stat. press. (mbar)	Temperature (°C) daily max.	daily min.	extreme max.	min.	Relative humidity (%) 03 h*¹	12 h	Precipitation (mm) mean	max. in 24 h	Mean evap. (mm)
Jan.	1011.1	29.6	22.7	32.8	17.8	79	72	35.0	126.2	3.8
Feb.	1010.8	29.8	23.5	33.1	17.2	77	72	25.4	57.1	4.1
Mar.	1010.0	30.5	24.7	33.4	19.1	74	72	17.0	54.6	4.5
Apr.	1009.1	31.1	26.2	35.6	20.7	75	74	53.5	121.2	4.5
May	1007.9	31.3	26.3	36.7	19.7	77	77	199.8	213.1	4.4
June	1008.3	30.0	25.3	33.9	20.4	80	80	293.6	148.6	3.9
July	1008.8	29.5	25.1	32.5	19.7	81	80	217.6	154.9	3.6
Aug.	1009.1	29.4	25.1	31.7	19.7	81	80	199.8	200.7	3.9
Sep.	1009.7	29.5	25.1	32.2	20.3	79	79	144.1	107.7	4.0
Oct.	1009.9	29.6	24.6	33.3	19.4	79	78	185.1	128.3	3.8
Nov.	1010.1	29.2	23.6	32.8	17.2	79	77	141.4	132.1	3.4
Dec.	1010.7	29.7	23.3	33.3	16.7	77	74	75.7	187.5	3.3
Annual	1009.6	29.9	24.6	36.7	16.7	78	76	1588.0	213.1	3.9

Month	Number of days with precip. ≥0.3 mm	hail	thunder-storm	fog	dust-storm	squall	Mean cloudi-ness (oktas)	Mean sun-shine hours	Wind preval. direct. 03 h*¹	12 h	mean speed (m/sec)
Jan.	4	0	1	0	0	0	3.8		NE	NNE	1.8
Feb.	3	0	1	0	0	0	3.9		N	N	1.9
Mar.	3	0	2	0	0	*	3.7		N	N	1.9
Apr.	6	*	4	0	0	*	4.5		NW	NW	2.1
May	14	0	6	0	1	1	5.7		NW	NW	3.1
June	22	0	3	0	0	1	6.9		W	W	4.7
July	21	0	2	0	0	*	6.5		W	W	4.5
Aug.	18	0	1	0	0	0	6.4		W	W	4.1
Sep.	17	0	1	0	0	1	5.7		NW	NW	3.6
Oct.	15	0	1	0	0	0	5.4		W	NW	2.8
Nov.	11	0	2	0	0	*	4.9		NW	NW	1.9
Dec.	6	0	1	0	0	0	4.2		NE	NE	1.7
Annual	140	*	24	0	1	2	5.1		W	NW	2.9

*¹ Greenwich mean time. * = <1.

TABEL LXXI

CLIMATIC TABLE FOR NUWARA ELIYA

Latitude 6°58′N, longitude 80°46′E, elevation 1,895 m

Month	Mean stat. press. (mbar)	Temperature (°C) daily max.	daily min.	extreme max.	min.	Relative humidity (%) 03 h*1	12 h	Precipitation (mm) mean	max. in 24 h	Mean evap. (mm)
Jan.	813.8	19.9	8.7	25.9	−2.6	83	80	145.0	106.4	
Feb.	813.8	20.9	7.7	25.6	−2.7	76	74	75.9	120.4	
Mar.	814.0	21.8	7.9	27.1	−1.7	72	72	96.5	96.5	
Apr.	813.4	21.9	10.0	27.2	1.0	82	82	153.7	118.9	
May	812.6	21.3	12.1	27.3	0.8	83	82	236.7	209.6	
June	812.4	18.7	13.3	26.4	6.9	86	85	266.2	175.8	
July	812.2	18.5	12.8	24.4	6.7	87	84	222.5	173.7	
Aug.	812.3	19.0	12.6	24.7	6.2	87	85	179.6	244.9	
Sep.	812.8	19.4	11.9	25.4	5.2	85	84	165.1	231.4	
Oct.	813.2	19.8	11.3	24.4	3.1	85	87	222.2	207.5	
Nov.	813.7	19.9	10.8	24.6	0.8	85	85	208.5	102.1	
Dec.	813.4	19.8	9.7	24.1	−1.1	87	85	190.0	203.7	
Annual	813.1	20.1	10.7	27.3	−2.7	83	82	2161.9	244.9	

Month	Number of days with precip. ≥0.3 mm	hail	thunder-storm	fog	dust-storm	squall	Mean cloudi-ness (oktas)	Mean sun-shine hours	Wind preval. direct. 03 h*1	12 h	mean speed (m/sec)
Jan.	13	1	3				5.4	5.4	E	E	2.6
Feb.	9	2	2				4.8	5.8	E	E	2.8
Mar.	11	7	1				4.4	6.4	E	E	2.6
Apr.	16	18	1				5.7	5.2	E	E	1.9
May	17	8	1				6.2	3.3	W	W	2.5
June	24	3	2				7.0	2.8	W	W	3.9
July	22	2	0				6.9	2.2	WNW	WNW	3.7
Aug.	22	5	3				7.0	2.4	W	W	3.5
Sep.	20	5	1				6.6	2.9	NW	W	3.1
Oct.	21	8	1				6.4	3.8	Calm		2.3
Nov.	21	6	1				6.0	4.1	E	Calm	2.3
Dec.	17	2	2				5.8	4.6	E	E	2.6
Annual	213	67	18				6.0	4.1			2.8

*1 Greenwich mean time.

180

TABLE LXXII

CLIMATIC TABLE FOR COLOMBO

Latitude 6°54′N, longitude 79°52′E, elevation 7 m

Month	Mean stat. press. (mbar)	Temperature (°C) daily max.	daily min.	extreme max.	min.	Relative humidity (%) 03 h*1	12 h	Precipitation (mm) mean	max. in 24 h	Mean evap. (mm)
Jan.	1011.1	30.3	22.2	34.4	15.2	81	70	87.9	124.7	
Feb.	1010.8	30.6	22.3	36.2	16.0	82	72	96.0	132.6	
Mar.	1010.2	31.0	23.3	36.1	17.7	83	72	117.6	126.5	
Apr.	1009.5	31.1	24.3	35.3	21.0	84	74	259.8	263.1	
May	1008.5	30.6	25.3	33.8	20.6	83	78	352.6	302.3	
June	1009.1	29.6	25.2	33.0	21.4	82	78	211.6	164.1	
July	1009.4	29.3	24.9	31.7	21.2	82	78	139.7	293.6	
Aug.	1009.4	29.4	25.0	32.3	21.0	81	77	123.7	126.0	
Sep.	1010.0	29.6	24.7	32.6	21.3	81	77	153.4	180.3	
Oct.	1010.4	29.4	23.8	32.7	20.2	83	78	354.1	256.3	
Nov.	1010.4	29.6	22.9	33.5	17.8	83	77	324.4	210.3	
Dec.	1010.8	29.8	22.4	33.5	17.3	81	74	174.8	161.8	
Annual	1010.0	30.0	23.9	36.2	15.2	82	75	2395.6	302.3	

Month	Number of days with precip. ≥0.3 mm	hail	thunder-storm	fog	dust-storm	squall	Mean cloudi-ness (oktas)	Mean sun-shine hours	Wind preval. direct. 03 h*1	12 h	mean speed (m/sec)
Jan.	8		3	0.1			4.8	7.5	NE	N	2.5
Feb.	7		5	0			4.6	8.2	NE	NW	2.1
Mar.	11		11	0			4.3	8.8	E	W	1.9
Apr.	18		19	0.1			5.8	7.9	E	SW	2.1
May	23		9	0			6.6	6.2	SW	SW	2.8
June	22		2	0			6.7	6.0	SW	SW	3.0
July	15		1	0.1			6.6	6.1	WSW	SW	2.8
Aug.	15		2	0.1			6.6	6.5	SW	SW	2.9
Sep.	17		2	0			6.4	6.4	SW	SW	2.8
Oct.	21		8	0.1			6.4	6.2	Calm	SW	2.2
Nov.	19		10	0.1			5.8	6.8	NE	NW	1.9
Dec.	12		8	0.1			5.2	6.9	NE	N	2.4
Annual	188		80	0.7			5.8	7.0			2.5

*1 Greenwich mean time.

TABLE LXXIII

CLIMATIC TABLE FOR GALLE

Latitude 6°2′N, longitude 80°13′E, elevation 12 m

Month	Mean stat. press. (mbar)	Temperature (°C) daily max.	daily min.	extreme max.	extreme min.	Relative humidity (%) 03 h*1	12 h	Precipitation (mm) mean	max. in 24 h	Mean evap. (mm)
Jan.	₁010.9	28.8	22.8	32.1	18.7	82	74	113.0	114.3	
Feb.	1010.6	29.5	23.2	33.1	18.1	80	74	115.8	110.7	
Mar.	1010.1	30.2	23.9	34.1	20.8	78	73	116.6	142.0	
Apr.	1009.4	30.1	24.8	33.7	18.3	80	76	252.5	195.8	
May	1008.4	29.4	25.6	32.2	20.4	82	80	202.3	319.8	
June	1008.9	28.6	25.1	30.6	20.7	83	81	220.2	173.5	
July	1009.1	28.2	24.8	30.4	20.9	83	81	170.7	132.8	
Aug.	1009.2	28.1	24.9	30.3	20.9	83	81	178.8	117.9	
Sep.	1009.8	28.2	24.9	30.7	20.4	83	80	179.3	170.2	
Oct.	1010.2	28.3	24.2	30.7	20.8	82	80	356.1	204.7	
Nov.	1010.2	28.7	23.4	32.1	20.4	82	78	322.3	218.2	
Dec.	1010.6	28.7	23.0	31.9	18.8	83	77	185.7	237.0	
Annual	1009.8	28.9	24.2	34.1	18.1	82	78	2513.3	319.8	

Month	Number of days with precip. ≥0.3 mm	hail	thunder-storm	fog	dust-storm	squall	Mean cloudi-ness (oktas)	Mean sun-shine hours	Wind preval. direct. 03 h*1	12 h	mean speed (m/sec)
Jan.	11		5	0			5.0		NE	W	2.4
Feb.	9		5	0			4.7		NE	W	2.5
Mar.	11		11	0			4.4		NE	W	2.5
Apr.	16		11	0			5.4		NE	W	3.1
May	21		6	0			5.8		W	W	4.6
June	22		1	0			5.9		W	W	5.4
July	19		1	0			5.7		W	W	5.2
Aug.	19		0	0			5.7		W	W	5.3
Sep.	19		0	0			5.6		W	W	5.0
Oct.	21		4	0			5.7		W	W	4.2
Nov.	19		8	0			5.4		E	W	3.1
Dec.	14		6	0			5.2		NE	W	2.5
Annual	201		58	0			5.4				3.8

*1 Greenwich mean time.

Chapter 3

The Climate of the Near East

M. F. TAHA, S. A. HARB, M. K. NAGIB and A. H. TANTAWY

Introduction

The Near East or the southwestern part of Asia comprises the territories of 12 countries, viz. Turkey, Iran, Cyprus, Iraq, Kuwait, Syria, Lebanon, Jordan, Israel, Saudi Arabia, Yemen and South Yemen not taking into account other Arabian territories on the Arabian Gulf. The entire area extends approximately from 27°E to 60°E and from 41°N to 12°N. Besides the Gulf of Aden and the Arabian Gulf, the area is bordered by six seas, the Aegean, the Mediterranean, the Red, the Arabian, the Caspian and the Black seas. The influence of the surrounding seas, however, is restricted by the prevailing pressure patterns and the mountainous ranges which emerge in the different plateaus of the region so that aridity and continentality characterize most of the area.

In general, lack of meteorological observations forms a difficulty in clarifying the climatic picture, while short-period observations, broken climatological series and non-homogeneous periods of observations serve little in defining different climatic regimes.

The climate of southwest Asia has been discussed in general terms in many books of climatology as part of the climatic regimes of Asia but has not yet been treated as an integrated subject and has not been described climatologically before in a detailed manner. Therefore, it was difficult for the present authors to find adequate references. However, the assistance and continued collaboration of all meteorological services of the entire region in providing climatological information, summaries, periodical reports and bulletins, made the present work possible.

Before discussing the climate of each country in the region separately, the following items which are related to the climate of the whole region will be discussed: location and topography; the climatic classification of southwest Asia; the general factors affecting the climate of the region including the major air masses; and the general circulation (tropospherical) over the region during the four seasons.

In the discussion of the climate of each country in the region a general description of the location, topography, and specified climatic districts of the region is first given, followed by a complete discussion of the following climatic elements: temperature; humidity; precipitation; surface wind; and weather phenomena.

Considerable climatological information in tabular form, with a selected series of climatic charts, is given as an Appendix at the end of this chapter. It serves as a reference as well as representing the climate data.

Location and topography

Southwest Asia as a whole can be defined on the map as being bordered by eight vast water surfaces: the Aegean, the Mediterranean, the Red, the Arabian, the Caspian and the Black seas, the Gulf of Aden and the Arabian Gulf. It extends from the middle latitudes to the tropical latitudes.

Concerning topography, southwest Asia is mainly comprised of the following 3 major plateaus.

(*a*) The Iranian plateau with the Elburz ranges bordering it in the north and the Zagros ranges which form part of its western border.

(*b*) The Asia Minor (Anatolia) plateau with the Septus and the Pentus ranges bordering it in the north and the Toros bordering it in the south.

(*c*) The plateau of the Arabian Peninsula with maximum height on its southwestern edge. The height decreases gradually to the east with the exception of the small plateau of Oman.

The region is characterized by vast desert areas of which the Arabian, the Syrian and the Iranian deserts are most important.

Moreover, the region is characterized by coastal plain strips which vary in width in different areas.

Climatic classification of southwest Asia

Southwest Asia is a region of diverse climates. According to the world climatic classi-fication following Köppen (GLENN, 1954), the entire area can be divided broadly into three main climatic types: the dry climate, the low latitudes' semi-arid climate and the warm temperate rainy climate. Taking into consideration the seasonal distribution of rainfall, the degree of dryness or coldness of the season, the whole area following the Köppen classification is subdivided into the following eight climatic types.

The warm temperate rainy climate with dry and hot summers (CSa)

This climatic type is found in the Black Sea area, the Aegean and the Mediterranean coastal strips of Turkey, Cyprus and the Mediterranean coastal strips of Syria, Lebanon and Israel. It is rainy, characterized by mild winters and dry and hot summers.

The warm temperate rainy climate with no definite dry season and warm summer (Cfb)

This climate is typical for the interior parts of the Anatolian plateau.

The warm temperate rainy climate with dry and warm summers (CSb)

This climate is typical for the eastern parts of Turkey and for the northern and western parts of Iran.

The warm temperate rainy climate with dry winters (CW)

This climate is typical for the mountain ranges in the southwest corner of the Arabian Peninsula (Yemen territory).

Low-latitude arid climate with dry winters (BWhw)

This climate is typical for the desert areas of the Arabian Peninsula and Iran with altitudes below 500 m. This region is bordered in the northwest by the deserts of Syria and Iraq.

Low-latitude semi-arid climate (BSh)

This climate is typical for the high plateau of the Arabian Peninsula whose altitudes range from 500 to 1,500 m above mean sea level. This high plateau roughly includes the western half of the peninsula and Oman plateau with the exception of the narrow coastal strips. In the northwest it is bordered by the interior parts of Jordan, Syria and Israel.

Low-latitude steppe climate with dry summer (BShs)

This climate is typical for the southern parts of Iraq (south of 32.5°N) and Iran (south of 32.5°N) excluding the narrow coastal strip of the Indian Ocean and the Iran desert whose climates are of the BWhs type.

Middle-latitude steppe climate with dry summer (BSKS)

This climate is typical for the northern parts of Iraq (north of 32.5°N).
The climate of the whole area is affected by the following factors.

(1) The dominating pressure systems. The most significant pressure systems are: *(a)* the cold Siberian anticyclone in winter over central Asia (GLENN, 1954, fig.2.4); *(b)* the monsoon Asiatic low in summer over India (GLENN, 1954, fig.2.5); *(c)* the travelling secondary depressions through the Mediterranean and adjacent areas from west to east during non-summer seasons along their preferred tracks (KENDREW, 1961, fig.112).
It should be noted, however, that in the north the climate is controlled mainly by synoptic scale disturbances of the extra-tropical type, while in the south, seasonal factors as well as small-scale thermodynamic factors are more important.
Satellite cloud pictures, however, have frequently indicated that frontal cloud patterns could extend from middle latitudes to the southern-most edge of the peninsula particularly in non-summer seasons. The potentiality of such systems in producing rainfall decreases generally from north to south and from west to east.
In summer time active fronts in their usual sense are infrequent. In addition some fronts, like discontinuities, might appear delineating the limits of sea breezes in daytime.
In the extreme south there is a possibility for another discontinuity where the moist summer southwest monsoon current merges with the hot dry air blowing from the desert. It is this discontinuity over Africa which is called "intertropical front".

(2) *The vast and largely extended water surfaces surrounding the whole area* act as sources of moisture and heat.

(3) *Orography* (nature and altitude of the land surface) largely determines the climate of some areas of the whole region but in other areas its effect on climate is limited.

Air masses

Air masses over southwest Asia have different source regions of which the central part of Asia and North Africa are regarded as the main sources for, respectively, polar continental and tropical continental air masses in winter.

The characteristic air masses (HAURWITZ and AUSTIN, 1944; KENDREW, 1961) of the region are: (*a*) in winter cPW–cPK and cT; (*b*) in summer cTKu and mTu.

Polar continental air (cP)

Polar continental air masses are formed over central Asia in winter. They are characterized by extremely low temperatures in the shallow surface layers and by a significant inversion reaching to about 1,500 m. The cP air masses are generally stable in their source regions; when they enter the southwest Asia region, they are modified. For example, in case of the cP advection over a warm moist surface (e.g., Caspian Sea or Arabian Gulf) the surface inversion is destroyed by heating from below, while widespread cloudiness occurs as a result of moisture absorption.

During summer, the southwest Asia region is beyond the influence of cP air masses as their source is shifted to the north of 50°N.

Tropical maritime air (mT)

During winter the source region of tropical maritime air masses is over the southern parts of oceans; accordingly mT invasions over south Asia are rare.

During summer the modified tropical maritime air (mTKu) prevails over India and eastern and southern parts of Asia. It is moist and convectively unstable up to an altitude of 3 km.

When this air circulates in the Asiatic monsoonal trough it loses its moisture in the form of heavy monsoonal rainfall over northern India and is thus modified into a very hot and dry air mass which infrequently invades the eastern parts of the region.

Polar maritime air (mP)

During winter mP air masses invade Asia Minor and the Black Sea occasionally and are considerably modified because of the long land trajectory. They are generally more humid than cP air masses and they follow the transitory Atlantic low-pressure systems through Europe. Much of the precipitable water content is shed when the air masses are lifted by the mountain ranges.

During summer mP air masses do not invade the region.

Tropical continental air (cT)

These air masses are formed over the dry land surface of central Asia and the African Sahara in summer, late spring and early autumn when surface heating is pronounced. These air masses invade the western parts of the region in advance of the transitory Khamsin or desert disturbances which occur in late spring (April, May) or early autumn (September). They are very dry and hot and frequently proceed eastwards through the Sinai or over the northern Red Sea.

General circulation and its influence over southwest Asia

The mean tropospherical charts for January (GLENN, 1954) and July (HAURWITZ and AUSTIN, 1944), which are published and amended from time to time in the literature, illustrate the principal features of the general circulation. Particularly in southwest Asia observations are rare in vast areas, pressure gradients over mountainous parts do not give a true picture of the mean circulation. Consequently such charts are inaccurate in certain parts but for the purpose of obtaining characteristic features they are satisfactory. Also these charts show summaries of main pressure systems within a certain period with no indication of their life history or relative changes. Moreover, small-scale or local disturbances do not show on these charts. In this part special reference will be given to the predominant pressure systems and a more detailed synoptic description will be given in this connection.

From the monthly mean tropospherical charts, reference will be specific and confined to the January (winter) and July (summer) charts as they are supposed to give extreme limits of the annual variations of various meteorological elements. The mean tropospherical charts for April (spring) and October (autumn) are considered transitional from one regime to the other.

The following is a complete discussion of the winter and summer regimes of the general circulation.

Winter circulation (January)

The outstanding features of the mean sea level pressure map for January (GLENN, 1954, fig.2.4) can be summarized as follows.

(1) Anticyclonic ridge over the region

The northern part of this ridge is an extension of the Siberian anticyclone, while the southern part may be related to the subtropical high-pressure belt which extends over the southern half of the Arabian Peninsula. This ridge causes the predominance of northeast surface winds over the Black and Caspian seas but in the mountainous regions of Iran, Turkey and Iraq surface winds may be quite incompatible with the pressure pattern.

The southern subtropical ridge is associated with an easterly current which may be related to the trade easterlies or to the northeast monsoons which prevail over the Indian subcontinent in winter.

187

These ridges, however, are subject to rather periodic changes of intensity due to the frequent passage of migratory winter depressions from west to east.

(2) The Mediterranean low pressure system

The mean sea level pressure chart for January shows a low-pressure system over the Mediterranean and another over the Arabian Gulf; each is relatively warm compared with the surrounding land areas.

The daily series of mean sea level pressure charts during winter show families of low pressure systems traversing the Mediterranean and adjacent land areas along their preferred tracks (see KENDREW, 1961, fig.112) from west to east. Some of these depressions migrate from the Atlantic to the western Mediterranean through northern Spain (*op. cit.*, fig.112, track 2), southern France (track 1), or through the Strait of Gibraltar (track 3), but others develop over the warm Mediterranean, especially in the Gulf of Genoa, and in the Adriatic and Cyprus regions. Orographic depressions also develop in the lee sides of the Alps and Atlas mountains over North Africa and start their courses along their usual tracks affecting the weather in the Mediterranean and adjacent areas. North-African secondary depressions follow track 10.

The Asia Minor plateau is a natural barrier in the tracks of Mediterranean depressions originating in the Adriatic Sea. Some of these depressions advance northeastwards (track 8) along the Marmara and Black seas while others proceed eastwards (track 9) towards Cyprus. Most Mediterranean depressions follow both tracks more frequently in winter than in the transitional periods. Depressions following track 9 in autumn are much less frequent than in spring, while those originating over the west and central Mediterranean follow track 3 mainly in the winter season.

The Mediterranean winter depressions that invade the east Mediterranean and its surroundings follow tracks 8, 9 and 3. These depressions control the surrounding coastal areas in general. Their influences do not reach farther than 20–30 miles inland, but sometimes they are very vigorous and show marked effects up to the 200-mbar level and proceed eastwards assisted by the warm Caspian Sea and Arabian Gulf. One of the tracks crosses Iraq, Iran and Afghanistan, and winter depressions following this track give these territories their major supply of rain.

The Mediterranean winter depressions produce a large percentage of the annual rainfall in many localities in the northern half of the region. Rainfall, however, varies with locality and season; it is particularly heavy in mountainous coastal areas of the Black, Mediterranean and Caspian seas.

(3) Upper air structure

The mean tropospheric charts for January are more simplified than the mean sea level pressure charts. As an example, the mean 700-mbar chart (GLENN, 1954, fig.2.8) shows a marked upper low over northeastern Siberia with a broad belt of prevailing westerlies over southwest Asia. Between latitudes 20°N and 50°N, the easterly surface winds change to westerlies at the altitude of 2 km or more, in response to the pronounced meridional temperature gradient. In lower latitudes where the temperature gradient is diminished, easterly winds extend to higher levels. As an example, easterly winds predominate to 8 km at latitudes 10° to 12°N (KENDREW, 1961).

On the average, the westerly winds definitely increase in speed with elevation until they reach jet stream velocities in the upper troposphere around the 200-mbar level.

The predominant pattern of upper westerlies over the whole area is disturbed by the consecutive passage of westerly troughs from west to east or by the meridional oscillations (5°–10° latitude) of jet streams in the upper troposphere.

The subtropical jet stream

The axis of that broad belt of strong upper westerlies normally denoted as the subtropical jet stream crosses the peninsula generally round 23°–27°N with its centre at approximately the 200-mbar level (GLENN, 1954, fig.44). It is noticeable that the eastern Mediterranean and northern Arabia are dominated during winter by one of the most intense phases of the subtropical jet stream (KRISHNAMURTY, 1959). The current is normally very broad, covering usually more than 5°, and is subject to meridional oscillations in a range of approximately 10° latitude. It is also subject to changes in intensity similar to zonal index oscillations in middle latitudes.

The upper tropospheric divergence around the subtropical jet stream, in association with lower convergence, has been considered an important factor in the development of the convective activity observed in non-summer seasons. Although the prevailing dryness and the consequent high cloud bases do not allow the falling rain to reach the earth's surface over most of Saudi Arabia, strong gusts, thunderstorms, dust- or sandstorms are still observed. Also the superposition of upper tropospheric divergence and lower convergence is considered an important factor for the maintenance of small-scale systems such as desert depressions (TANTAWY, 1961a, b).

The polar jet

In addition to the subtropical jet stream, which is a permanent feature of the upper tropospheric flow pattern in non-summer seasons, the area is frequently invaded by active branches of the polar jet streams in association with the passage of Mediterranean depressions. Frequent amalgamations of the two types of jet streams occur in periods of active disturbances which cause the formation of a rather complicated wind field in the upper half of the troposphere.

In general the interaction of the two jet streams, and the subsequent changes in the upper tropospheric wind field, is thought to be a major cause of hydrodynamic instability, which leads to the deepening of winter depressions in the east Mediterranean and adjacent areas. On such occasions active cold air outbreaks penetrate deep throughout the Arabian Peninsula to the extreme south, where it is subject to quick modifications by surface heating and active subsidence.

Accordingly, the majority of southwest Asia is dominated by westerlies in the upper troposphere.

Summer circulation (July)

The outstanding features of the mean sea level pressure map (GLENN, 1954, fig.2.5) for July, over the southwest Asia region can be summarized as follows.

(1) Western arm of the Indian monsoon trough

During summer, the southwest Asia region is dominated by the western extension of the permanent thermal depression over north India and Pakistan which is centered at approximately 30°N. The daily series of the mean sea level pressure charts show this feature as a permanent trough of low pressure over the Arabian Gulf, which is mostly a complex monsoon trough that dominates the region. Analysis of the daily synoptic charts shows that this complex monsoon trough experiences zonal oscillations and undergoes relative and occasional deepening, thus extending its influence to the east Mediterranean and vicinities. In such pressure distributions warm modified subtropical air advects over a continental course through Asia Minor, probably extending over Greece and Italy or westwards through the east Mediterranean causing excessive heat waves over these territories. It has been found that some of these summer heat waves are related to the latitudinal elongation of the monsoon trough over the Arabian Gulf or to a relative deepening of that trough.

(2) Quasi stationary surfaces

The summer distribution is outstandingly stable without any appreciable variation from one day to another. Such steadiness allows for the appearance of quasi stationary surfaces of separation between different air streams; of these, two surfaces are known and described in the literature (SOLIMAN, 1958; TANTAWY, 1970.)

The northern surface of separation, normally called the subtropical front, extends from the Iranian plateau to run parallel to the eastern coasts of the Mediterranean. The nature and structure of this separation surface is not clear from the hydrodynamical point of view, but is considered to be the northern limit of the cyclonic circulation around the Asiatic monsoon trough. To the north of the subtropical front the air mass, which circulates in the subtropical anticyclone, is characterized by moderate temperature and moisture content. To the south of that front very hot and dry air, which circulates in the monsoon trough, prevails over almost all of the Arabian Peninsula, perhaps with the exception of a very narrow coastal strip in the extreme south.

The second separation surface, usually called the intertropical front extends in the extreme south. In fact, there are many contradicting ideas concerning the nature and significance of that front, even over Africa where it is most distinct. It seems that the existence of the Red Sea basin, with its characteristic topography, behaves in a manner that causes a strong meridional flow in the lower layers, thus separating the circulation of the Asiatic monsoon low from that of the African trough, and pushing the intertropical convergence zone southwards so that it runs almost along the southern coastal boundary of the Arabian Peninsula.

(3) Upper distribution

Over southwest Asia the cyclonic circulation extends up to about the 700-mbar level, or a little higher. From the 500-mbar level to the top of the troposphere the area is dominated by an anticyclonic belt with an axis running at approximately 25°N on the 500-mbar level and inclining northwards to approximately 30°N on the 100-mbar. The intense middle tropospheric warming associated with that anticyclonic belt indicates an active large-scale subsidence.

In the middle troposphere steady easterlies prevail over the southern half of the region. These easterlies spread northwards and increase their strength with height until they reach maximum intensity in the upper troposphere at levels ranging from 14 to 17 km. The axis of this strong upper tropospheric easterly current, denoted in the literature by "tropical easterly jet stream" and described by several authors (KOTESWARAM, 1958; TANTAWY, 1963; FLOHN, 1964), extends from India to east Africa traversing the region off the southern coasts of the Arabian Peninsula where this jet reaches its maximum intensity. It is interesting to note that the appearance of a strong solid easterly flow in the upper troposphere, where the wind reaches jet stream velocities, is almost restricted to that part of the world where orography is thought to be a major cause of such a formation.

On the northern flank of the middle tropospheric ridge, which covers the northern part of the entire region, a belt of upper westerlies prevail; they increase further north since the axis of the subtropical jet stream is at approximately 40°N in summer.

Transitional seasons

Spring and autumn represent transitional periods. The distribution during either of these two seasons could be partly related to the winter pattern and partly to the summer one, with a pronounced change taking place in a short period of time. In consequence, we find that mean tropospherical distributions have little significance from the climatological point of view. Nevertheless, the most significant feature of such distribution patterns is the tendency for the surface subtropical anticyclone to be situated at approximately 50°N, which is also the average location of the subtropical jet stream axis (JENKINSON, 1955).

One of the most important factors, which have a great influence on the climate of the Middle East region in general, is the marked variation of static stability in the lowermost layers of the atmosphere. The rapid heating of the continental surfaces in spring and the rapid cooling in autumn cause weak stability conditions in spring and relatively large stability in autumn (TANTAWY, 1961a). The stability parameter definitely contributes to the formation of desert depressions which frequently affect the northern part of the Arabian Peninsula.

Most of the quick changes, which divert the circulation from the summer to the winter pattern and vice versa, take place in two or three weeks between May–June and October–November. It is postulated that the seasonal warming of the air above northern India, the Himalayas, the Iranian plateau and northern Arabia, causes a reversal of the meridional thermal gradient during the first period. Such a reversal takes place simultaneously with the first summer pattern of the mean sea level pressure distribution described above, as well as the reversal of the upper tropospheric flow pattern from westerlies to easterlies since the subtropical jet is displaced to approximately 40°N (SUTCLIFFE and BANNON, 1954).

Spring

Spring is characterized by the frequent formation of desert depressions, which originate in the extreme northwest of Africa, and move eastwards parallel to the north coast of Africa until they reach the northern region of the Arabian Peninsula.

In many cases, the depression may die out before reaching northern Arabia, but the frequency with which these depressions affect this region in spring is quite remarkable. A desert depression is a shallow phenomenon with a slightly subsynoptic scale which usually develops below the 700-mbar level. In advance of a desert depression southerly sand-laden winds cause intense heat waves as well as active sand or dust phenomena (TANTAWY, 1969).

The particular nature of the earth's surface in the desert area in western Iraq and on the western coasts of the Persian Gulf, which is covered with very finely granulated sand or dust particles, increases the intensity of sand or dust phenomena. Sand storms frequently reach the height of 3 km. On certain occasions active advection of moisture from the Arabian Gulf may supply these spring depressions with a sufficient amount of water vapour to cause frequent thunderstorms.

Upper troughs in the subtropical jet stream and the uplift of the air over mountainous regions are considered conducive to spring rainfall.

Autumn

In autumn, the surface layers are more stable and consequently the frequency and intensity of sand phenomena, which associate subsynoptic scale disturbances, decrease considerably. The upper layers, however, are generally more unstable particularly when active cold air intrudes behind the westerly troughs. In vigorous cases such intrusion causes the formation of an upper cut-off low. Such occasions are characterized by active thunderstorms in the northern part of the Red Sea and scattered localities of the Arabian Peninsula and Iraq (TANTAWY, 1961b).

Tropical cyclones

In spring and autumn the Arabian Sea is sometimes affected by tropical disturbances (GLENN, 1954), some of which develop into tropical storms. The number of tropical disturbances of all intensities, observed in the Arabian Sea during the 60-year period ending in 1950, was approximately 140 (KOTESWARAM, 1962). Of this amount, 31 cases were in the pre-monsoon period (May and June), and 61 were in the post-monsoon period (October and November). It was further indicated in the same work, that the number of occasions during which disturbances developed into tropical storms was 21 in May and June and 27 in October and November. These tropical disturbances have a predominating influence on the climate of the southern part of the peninsula. An ordinary disturbance in that region is usually in the form of an easterly trough moving from east to west, and initiating convective activity on the rear side. The development of natural tropical cyclones, which affect the southern part of the peninsula, is a rare event, and, according to statistical results, the maximum frequency occurs in autumn and spring with insignificant proportions in the other two seasons.

Climate of Turkey

Location and topography

The greater part of Turkey, the Asiatic part, includes the Asia Minor Peninsula or Anatolia which, as an approximation takes the form of an irregular rectangle with sides that are multi-curved. It extends approximately meridionally between 41.5°N and 36.5°N and zonally between 26°E and 44°E. Its northern boundaries are the Black Sea and the high Pentus Mountains. Its western boundary is formed by the Aegean Sea. Its southern boundaries are the Mediterranean Sea and the high Toros Mountains, and its eastern boundary is formed by the high Soragos Mountains.

In general, the peninsula is a high plateau (1,000–2,000 m, longitude 27°E) which reaches to Iran and gradually descends westwards through a wide-stepped coastal strip. The northern and southern sides of this high plateau falls rather stiffly towards the Black and Mediterranean seas through rather flat coastal strips. As far as the high plateau extends, many scattered high mountain tops (2,000–3,000 m) emerge of which the two main chains of the Septus and Toros mountains are characteristic for this area. They border most of the northern and southern parts of the peninsula, respectively, and run eastwards with pronounced penetrating tops of 3,000 to 4,000 m height.

General climate

The climate of Turkey is either of the hot and dry summer Mediterranean type (CSA) (CSA) (Black Sea, Aegean and Mediterranean coastal strips) or of the middle latitude dry semi-arid (steppe) of BSK type (interior parts of Turkey). The entire area in Asia Minor can be divided into the following climatic districts.

(a) The northern coast (Black Sea) district

This district includes the Asiatic Black Sea coastal area of Turkey. Its altitude ranges from 1,000 to 5,000 m. It is surrounded in the south by the Septus mountain ranges (2,000–3,000 m) which extend from west to east. It is well-watered, receiving much rain from the Black Sea depressions in autumn and winter and a considerable amount from the strong on-shore winds in spring and summer (southeastern coast of the Black Sea). The prevailing weather is similar to that of the Mediterranean coasts of France and Italy. The eastern part (east of longitude 35°E) is sheltered by the Caucasus Mountains and enjoys the foehn winds, while the western part (west of 35°E) is open to cold northeast winds and with less frequent foehn winds from the plateau. This district is characterized by much vegetation, particularly the Mediterranean figs, olives, tea and the Turkish tobacco.

(b) The interior (plateau) district

This district includes the plateau which ranges in altitude from 1,000 to 2,000 m above mean sea level and is enclosed by a mountainous rim. Its beginning is close behind the forests of the coastal ranges. The dominating climate is continental with very little rain.

193

Weather conditions are severe in winter, particularly when vigorous Black Sea depressions extend over the plateau bringing warm south winds shortly before they approach, and bitter, cold weather as cold fronts, which follow these depressions, degenerate and produce heavy snowfall on the mountain slopes. Dominating weather in summer is hot and dry in general, though the aridity is tempered by occasional thunderstorms. Spring is the rainiest season, but late summer is rather rainless. The existence of the great salt lakes in this district is an example of this aridity.

(c) The south coast (Cilicia) district

This district includes the coastal plain of Cilicia at the foot of the Toros Mountains running from west to east in the south of Anatolia. It is bordered by the Mediterranean in the south and sheltered by the Toros Mountains. The dominating climate is very warm temperate and rainy with a dry summer; it is of the typical CSa type in Köppen's classification. The maximum rainfall is in winter and spring. This plain is particularly rich in agriculture when streams provide irrigation facilities, as is the case near Adalia and Adana. Cotton is an important product of these plains. There is much contrast between this rich coastal strip and the arid steppes of the interior at a distance of a few miles.

(d) The west coast district

This district includes the western coastal plains which are bound from the west by the Aegean Sea and from the east by the mountainous rim of the plateau of Anatolia. Its altitude ranges from 100 to 300 m above mean sea level. The lowlands in this district are wider and extend inland for rather long and variable distances along the valleys running through the eastern parts of the plateau. The dominating climate is again of the humid dry summer Mediterranean type (CSa).

Temperature

The distribution of temperature over Turkey shows marked differences due to the diversity of orography, locality and distance from the surrounding seas. Generally, the temperature minimum is in January and the maximum in August. For the country as a whole the temperature decreases from the coastal areas towards the interior (plateau). Freezing temperatures are frequent in the higher mountain regions. The eastern parts of the Black Sea coastal areas of Turkey, east of longitude 35°E, are sheltered by the Caucasus while the southern coastal areas (Cilicia plains) are sheltered by the Toros Mountains; in both these mountain ranges foehn winds occur in winter.

During winter, the mean temperatures range between 5°C and 7.5°C in the northern district, between 0°C and −5°C in the plateau (interior) district, between 8°C and 10°C in the western and southern districts and is about 1.5°C in the eastern district. Minimum temperatures range from 2.5°C to 5°C in the northern district, from −3.5°C to −13°C in the plateau district, from 4°C to 6°C in the western and southern districts and about −2.5°C in the eastern district. Maximum temperatures are between 8° and 11°C in the northern district, between 0° and 5°C in the plateau district, between 12° and 15°C in the western and southern districts, and about 6.5°C in the eastern district.

The absolute minimum temperature, recorded in February, was $-16.1°C$ in Istanbul (Goztepe), $-34.4°C$ in Sivas, $-8.4°C$ in Izmir. In January it was $-7.1°C$ in Adana and $-24.2°C$ in Diyar Bakir.

During summer, the mean temperature is 23.5°C in the northern district, between 20° and 24°C in the plateau district, 28°C in the western and southern districts and approximately 31°C in the eastern district. The minimum temperature ranges from 18.5° to 20.5°C in the northern district, from 10.5° to 15.5°C in the plateau district, from 21° to 27.5°C in the western and southern districts and approximately 21.0°C in the eastern district. Maximum temperatures range from 27°C to 29°C in the northern district, from 28°C to 30.5°C in the plateau district, from 33°C to 35°C in the western and southern districts and is about 38.5°C in the eastern district.

The absolute maximum temperature, recorded in August, was 39.4°C in Istanbul, 40.0°C in Ankara, 42.7°C in Izmir, 45.6°C in Adana and 45.9°C in Diyar Bakir.

Humidity

Relative humidity is generally high during the whole year over the northern and southern districts where it shows slight variability (5%) throughout the whole year. Elsewhere it is maximum in January (or December) and minimum in August.

Precipitation

Rainfall

Heavy rainfall occurs in winter and large parts of spring and autumn (March and April, September and October, respectively) and to a moderate extent from March to August in the coastal and the eastern districts. As an exception the plateau district receives its maximum rain in May. For the entire country, rainfall is minimal in August.

The average annual rainfall shows marked variations from district to district; it is approximately 650–830 mm in the northern district, 350–500 mm in the plateau district, 700 mm in the western district, 650–1,050 mm in the southern district and 500 mm in the eastern district.

The average maximum monthly rainfall is approximately 75–110 mm in the eastern parts of the northern district in October (autumn in general), 105 mm in the western parts of the northern district during December, 50–75 mm in the plateau district in May, 145 mm in the western district in December, 115–280 mm in the southern district (rainfall in the western parts is more than twice the amount of the rainfall in the eastern parts), and is 75 mm in the eastern district in January. The average monthly rainfall is minimal in August and is 20–30 mm in the northern district, 5–20 mm in the plateau district, 2–5 mm in the western and southern districts and is 1 mm in the eastern district. Maximum daily rainfall attains very high records in general. For example the maximum daily rainfall, recorded in October was 112.5 mm in Florya (Istanbul), 46.3 mm in Erzurum and 231.1 mm in Izmir. Also, in December, it was 69.8 mm in Ankara and 290.7 mm in Anatalya.

The annual number of raindays ($\geqslant 0.1$ mm) averages 85–100 days in the northern district, 65–80 days in the plateau districts, 55–65 days in the western, southern and eastern districts. On the other hand, annual number of days with $\geqslant 10.0$ mm rain averages 20–30 days in the northern, western and southern districts and 10–15 days in the plateau and eastern districts.

Snow

Snowfall over Turkey is confined to the mountain ranges, particularly in the Black Sea and interior plateau districts; it occurs mainly during the winter season and very occasionally in the early spring and late autumn. The annual frequency shows a maximum over the eastern mountains where it ranges between 30 and 50 days. Both the western and southern coastal plains of Turkey can be regarded as snowless areas.

Surface wind

The wind regime over Turkey shows marked variations during the four seasons. It is directly influenced by the major, the transitory and the local pressure systems in the Near East and adjacent areas, particularly the Mediterranean Basin region.

During winter, the prevailing winds are southwesterly over the eastern parts of the northern parts (56.5% in Samson, 46.5% in Trabazon) and the plateau region (40% in Ezzusum and 33% in Sivas) and are northerly over the eastern districts (43% in Florya and Goztepe and 49% in Ankara). Over the western district southeasterly winds prevail with a frequency of 62% while over the southern and eastern districts northerly winds prevail with a frequency of 72%, 62.5% and 43.5% in Anatalya, Adana and Diyar Bakir, respectively.

The prevailing winds are light to moderate in general but occasionally fresh to strong (17 to 27 knots) and may reach gale force in association with vigorous depresssions. Gales during winter ($\geqslant 34$ knots) are most frequent in the Western district (13 days in Izmir) and infrequent in the northern (western parts) and the southern districts where they average 3 days in Florya and Anatalya, respectively.

During summer, the prevailing winds over the country as a whole are northerlies but in the eastern parts of the southern district southerly winds blow frequently (49.5% in Adana). Gales are infrequent in general over the whole country in summer, each summer averages one day in Florya and Izmir. Elsewhere the frequency is approximately one day every two or three years.

During spring the prevailing winds are northerly for the country as a whole, an exception is the western district where westerly winds are more frequent.

During autumn northerly winds show marked prevalence over the entire country with the exception of the eastern parts of the northern and plateau districts where southwesterly winds are more frequent (37.5% in Samson and 45% in Trabazon), and the westerly district where southeasterly winds are more frequent (37% in Izmir).

Gales ($\geqslant 34$ knots) are more frequent in the transitional periods over the western district (e.g. 7 days in spring and 3 days in autumn in Izmir) and less frequent elsewhere where their average occurrence is 1–2 days during either season.

Calms (<1 knot) are frequent in the eastern half of Asia Minor (east of longitude 35°E) where they average 20–50% per month. Sivas shows a marked prevalence of calms in October (56%). In the western half, the average monthly frequency of calms is 5–20%. Anatalya has a very marked low monthly frequency (2%) in January and October.

Weather phenomena

Fog

Visibility is generally good in the western and the southern districts where the annual frequency averages 1 and 2–5 days, respectively. The annual frequency of days with fog is maximal (20–25 days) in the plateau district and moderate (10–15 days) elsewhere. Fog occurs generally in winter, early spring and late autumn and is infrequent in summer. (One day of fog every two or three years.) Izmir is an example of the fogless places in the western district (annual frequency 0.6 day).

Thunderstorms

Thunderstorms develop over the western and Mediterranean districts more frequently with an annual average ranging from 20 to 30 days and a monthly frequency averaging 2 to 4 days. Elsewhere, thunderstorms develop less frequently, with an annual frequency averaging 10 to 20 days and a monthly frequency averaging 1 to 3 days. They are most prevalent during non-winter seasons.

Climate of Cyprus

Location and topography

Cyprus is located in the northeast corner of east Mediterranean Basin. Zonally, it extends approximately between longitudes 32°20′E and 35°5′E, and meridionally between latitudes 35°15′N and 34°35′N.

Cyprus is a rocky island characterized by two mountain ranges running from the east. The southern ranges are the more extensive; they range in altitudes from 500 to 1,000 m and have tops with altitudes of 1,000–2,000 m. In the northern mountains, both ranges are separated by a treeless plain, namely the Messoria, which is between 10 and 30 km in width.

General climate

The climate of Cyprus (AIR MINISTRY METEOROLOGICAL OFFICE, 1936, 1963) is of the Mediterranean type and as a whole it comprises one climatic district, though minor variations occur due to locality and altitude above mean sea level. The dominating climate is mild and rainy in winter (November–March), dry and hot in summer (June–September), and changeable in transitions.

Temperature

Temperature varies with time, locality and altitude; however, detailed information of the mountain regions is limited. The temperature is lowest in January and highest in August. The latitude of Cyprus permits only small differences in temperature in the surrounding coastal strip.

During January, the mean temperature is 12°C on both northern and southern coasts and is 10°C inland at Nicosia. The mean maximum temperature is 16°C on northern, southern and lower inland levels, while extreme maximum temperatures of 23°C, 21°C and 30°C have been recorded at Kyrenia, Nicosia and Limassol, respectively, in January. At Prodromos (1,500 m), the mean monthly temperature in January is 4°C and the extreme maximum temperature is 13°C.

Winter nights in Cyprus are comparatively cold and on clear nights frost is frequent on high ground. The minimum temperature reaches 6°C in general in January and February at lower levels or coastal areas.

During summer the monthly mean temperatures for August range from 23°C to 28°C. The maximum temperature at most low-level stations averages 32°–35°C while it averages 36°C at inland stations. Maximum temperatures exceeding 40°C have recorded in each of the summer months for low-level stations.

The temperature increases slightly, but fairly steadily, from January to July and falls steadily from September to December. The interdiurnal variation in temperature approximates to 1°C during half year of summer and are much larger during the half year of winter as a result of the transits of depressions and their associated fronts.

Humidity

The relative humidity is changeable with time, locality, wind and precipitation. It is maximal in January and minimal in July. The mean relative humidity inland is fairly high in winter and low in summer.

On the coasts, the relative humidity is affected by the directions of the wind; it is high with seawinds and much lower with land winds. In quiet spells, the daily range of relative humidity is much reduced. On the other hand, in winter months with moisture-laden southwesterly-to-westerly wind, the relative humidity becomes very high.

Precipitation

Rainfall

Cyprus has a typical Mediterranean climate, which is obvious from the distribution of rainfall throughout the year. Winter is normally rainy and rainfall varies from year to year. The rainfall is maximal in December and minimal in June. Most of the rain falls between October and March. The annual rainfall ,which varies with locality and orography, is between 400 and 550 mm in general.

Rainfall occurs usually in the form of heavy showers or local thundery showers. However, it may be of the continuous type when depressions are stationary over Cyprus. Rainfall over mountains is greater than at lower levels. The greatest rainfall is on the southern and western slopes of Troödos Ranges and the northern range of the Kyrenia Mountains, where the annual rainfall exceeds 750 mm.

Variability in the monthly rainfall is large in the wet season and ranges between 25 and 175 mm.

Snow

Snow is infrequent at low altitudes in Cyprus. Its average annual frequency is 1–2 days in January or February. At high altitudes (600 m above mean sea level or more) snow falls frequently in winter months and accumulates at great depths on the northern slopes of mountains above 1,000 m.

Frost

The nights during winter in Cyprus may be comparatively very cold, and on clear nights frost is frequent on high ground but seldom occurs at low levels or in coastal areas.

Surface wind

The wind regime over Cyprus is not simple. Marked differences appear between prevailing winds over different parts of the island and, with respect to surface wind regimes, the entire island shows no homogeneity. During winter months (November–March) winds vary in direction as depressions pass either north or south of the island on their eastward tracks through the east Mediterranean. Generally winds are mainly between southwest and northwest or between northeast and southeast. The westerly winds tend to predominate in the western and southern parts of the island and the easterly winds predominate in the northern and eastern parts.

During the summer months (June–September), the westerlies predominate. While in the west of the island there is a bias towards northwesterlies, in the eastern parts the bias is towards southwesterlies.

During the short transitional periods, spring- (April and May) and autumn-winds (October and November) vary much in direction. The surface wind direction and velocity is affected by topography, and in coastal localities winds are influenced by the very marked land and sea breezes; the latter is most pronounced in the half year of summer.

The prevailing winds over Cyprus are generally light to moderate (3–16 knots), but occasionally fresh (17–21 knots) in the afternoon over the coastal areas in summer where the sea breeze is well pronounced and speeds of 17 knots or more are observed. Strong winds (22–27 knots) blow during winter months over Cyprus in association with transitory Mediterranean depressions. They are frequent in unsheltered coastal areas, especially in the northeast. Gales (≥34 knots) also occur mostly in unsheltered areas with a monthly frequency of 1–2 days but are less frequent elsewhere. Both strong winds and gales blow from between southwest and northwest (i.e., westerly) or from north and east.

Weather phenomena

Fog

Visibility is generally good in Cyprus throughout the year although fog occurs occasionally. Fog is confined to the southern coastal parts in summer months. Dense coastal fog with a monthly frequency of 2–3 days occurs at Larnaca in June and July. The northern coast of Cyprus is free from fog. Fog develops late at night or in the early morning but disappears quickly after sunrise. Hill fog is experienced from time to time in the mountains especially during winter months, when it is more persistent.

Thunderstorms

Thunderstorms develop either on frontal systems during winter or in the instability showers that develop in the cold air following winter low-pressure systems through the Mediterranean. They occur with a monthly frequency averaging 1–2 days on the coasts and 2–3 days inland. In summer they are infrequent in the mountains and may be severe and accompanied with hail.

Climate of Syria

Location and topography

The northern boundary of the Syrian Arab Republic is the Asia Minor plateau or Turkey and its western boundaries are the Mediterranean Sea and the Lebanon borders. In the east it is bordered by the Republic of Iraq, and in the south by Jordan and Israel. In general, the whole area of Syria can be regarded as a plateau; the northern part, north of 35°N has altitudes of 300–500 m, while the southern part, south of 35°N, has altitudes of 500–1,000 m. Scattered high mountain tops (1,000–2,000 m) stand along the northern and western parts, and in particular near the eastern borders of Lebanon and the southern borders of Jordan where mountain tops range between 1,500 and 2,500 m. The eastern part of Syria is characterised by the Euphrates valley while most of the southern half is a sandy desert.

General climate

The climate of Syria (AIR MINISTRY METEOROLOGICAL OFFICE, 1936; SYRIAN METEORO-GICAL DEPARTMENT, 1960–70) is of the Mediterranean climatic type in the western parts, the dry-steppe type in the inland parts and the arid-desert type in the desert area. Considering the distribution of temperature and rainfall, Syria can be divided into the following climatic districts.

(a) Coastal district

(*1*) The Mediterranean coastal strip. It sometimes rises gradually and sometimes rapidly towards the western mountains; its altitude varies and ranges between 50 and 500 m.

(*2*) The western chains of the mountain ranges run in an approximate north–south parallel to the Mediterranean coast (500–1,000 m altitude with a few scattered tops of 1,000–1,500 m). Chains of mountains descend gradually from their eastern sides towards the inland parts of Syria. The dominating climate of this district is mild and dry in summer, rainy and very cold in winter, mild and rather cold and rainy in the transitional periods.

(b) The western inland district

This district is at the foot of the western slopes of the coastal district and those bordering east Lebanon, (1,000–2,000 m with scattered tops of 1,500 to 2,000 m). It extends eastwards till longitude 38°E north of latitudes 34°N and as far as 37°E approximately south of this latitude. Its altitude is below 500 m with emerging tops ranging from 500 to 800 m. The dominating climate of this district is rather mild and dry in summer, rainy to light rainy and cold in winter, rather dry and mild in the transitional periods.

(c) Northeastern district

This district includes the northeastern sector, north of latitude 36°N, which borders Turkey. Its altitude is generally below 400 m but reaches 500–700 m near the borders of Turkey. The dominating climate of this district is mild and dry in summer, rainy and intensively cold in winter, mild and light rainy in the transitions.

(d) Eastern district

This district includes the eastern part of Syria east of longitude 38°E, south of latitude 36°N and north of 34°N approximately. It includes the Euphrates, its tributaries and the fertile lands of El Gezeira. Its altitude is generally below 400 m above mean sea level with few penetrating mountains (500–1,000 m with tops of 1,000–1,500 m). The dominating climate of this district is hot and dry in summer, cold and slightly rainy in winter and rather hot and almost rainless in the transitions.

(e) Desert district

It includes the great desert of Syria which is below the 34°N latitude, east of 38°E and extends towards the borders of Iraq and Jordan. It is a desert plateau ranging from 500 to 1,000 m above mean sea level with a few tops in the southwest of 1,000–1,500 m. The dominating climate of this district is excessively hot and dry in summer, rather warm in daytime and intensively cold at nighttime during winter, and hot and rainless in the transitions.

Temperature

The distribution of temperature over Syria is greatly diversified with respect to time, locality and orography.

The temperature in general is lowest in January and highest in July and August.

During winter (January), the mean temperature averages 10.5°C in the coastal district, 6.5°C in the western inland district, 8°C in the eastern district, 6.5°C in the northeastern district and 7.5°C in the desert district. Very low temperatures, ranging from 5°C to 0°C, are experienced on the upper slopes of the mountain ranges and temperatures are freezing on their snow-covered tops.

The maximum temperature ranges between 12°C and 17°C in the coastal district, between 12°C and 7°C in the western hilly and western inland and northeastern districts and between 13°C and 18°C in the eastern and desert districts. Maximum temperatures as low as 3°C are experienced on the upper slopes of mountains.

The minimum temperature ranges from 5°C to 9°C in the coastal district, from 0°C to 5°C in the western hilly, western inland and northeastern districts and from 2°C to 4°C in the eastern and desert districts.

During summer (July and August) the mean temperature averages 26°C in the coastal district, 26°–29°C in the western hilly district, 27°–30°C in the western inland district and 29°–33°C in the northeastern and desert districts.

The mean maximum temperature ranges from 30°C to 35°C in the coastal district, from 25°C to 30°C in the western hilly district, from 32°C to 37°C in the western inland district and from 39°C to 42°C in the northeastern, eastern and desert districts. Several lovely summer resorts are located on the mountain slopes where the maximum temperature does not exceed 30°C. Ice caps crowning high tops melt and supply the down-sloping valleys with fresh water.

Excessive heat waves are experienced during summer when the monsoon over the Arabian Gulf extends westwards or relatively deepens. The circulating warm and dry air is of the Tch type. The mean minimum temperature averages 21°–24°C in the coastal district, 16°–21°C in western hilly and western inland districts and 22°–27°C in the northeastern and desert districts.

Humidity

Relative humidity experiences diurnal and seasonal variations but generally it decreases in the far inland parts, while it is relatively high in the northern and western parts. Its monthly value is lowest in summer (July or August) when it averages 65–75% in the coastal district, 35–45% in the western inland district, 25–30% in the northeastern district, 25–30% in the eastern district and 35% in the desert district.

Precipitation

Rainfall

Rain falls in moderate amounts in winter, and to a lesser extent in the transitions. The maximum rainfall is generally in January and the minimum in July and August. The mechanical lifting of moist air masses up the slopes of the western districts and the northeastern districts favours the increase of rainfall in these regions. For Syria as a whole, the coastal district receives the maximum rainfall, while the western inland and the northeastern districts receive a lower amount and the desert district receives the least amount of rain.

The annual rainfall averages 600–1,000 mm in the coastal district in general but exceeds 1,000 mm in some areas particularly on the upper slopes of the western mountains. The annual rainfall averages 200–600 mm in the western inland and northeastern districts, 100–175 mm in the eastern district and 75–125 mm in the desert districts.

During January the monthly rainfall averages 216 mm in the coastal district, 50–75 mm in the western inland district, 75–100 mm in the northeastern district, 25–35 mm in the eastern district and 25 mm in the desert district.

During May the monthly rainfall averages 20–25 mm in the coastal district, 10–20 mm in the western inland district, 20–40 mm in the northeastern district and 10 mm in the eastern and desert districts.

Rainfall is associated with infrequent thunderstorms during winter and the transitional periods. The monthly frequency of days with thunderstorms is maximum during spring (April and May), moderate in autumn (October and November) and rare in July and August.

Snow

Snowfall over Syria is confined generally to the coastal, western inland and northeastern districts in the wet season. The annual frequency of days with snow averages four days.

Surface wind

Syria is influenced generally by two main wind regimes: the northeast flow in summer; and the southwest to west winds in winter and the transitions.

Those two general wind regimes undergo different changes in direction and velocity depending on locality, altitude and time.

During winter (January), the prevailing winds over the northern parts are east to north-east (frequency 40% in Aleppo and 42% in Kamishly). Elsewhere, west-northwest winds prevail (frequency is 22% in Deir Ezzor, 32% in Abu Kamal and Palmyra) with exception of the coastal district where south to southwest winds prevail.

During spring (April), the prevailing winds are west to southwest in the northern parts (average frequency of 48% in Aleppo) east to northeast in the northeastern parts (frequency of 33% in Kamishly) and south to southwest in the coastal district. Elsewhere, west to northwest winds prevail.

During summer, the prevailing winds are west to southwest in the northern parts, north to northwest in the northeastern parts and southwest in the coastal district.

During autumn (October) west to northwest winds prevail in general.

Calms (<1 knot) show marked high frequency prevalence in the winter and autumn seasons, a lower frequency in spring and the lowest frequency in summer.

The prevailing surface winds are light to moderate (3–16 knots) in general, fresh to strong (17–27 knots) in association with transitory depressions. Gales (≥34 knots) occur in association with vigorous fronts.

Gales (≥34 knots) are infrequent in general, though their monthly frequency averages 2 to 3 days in the western inland district and 1 to 2 days elsewhere.

Weather phenomena

Fog

Fog shows maximum monthly frequency in December and develops rarely in June and July.

Thunderstorms

Thunderstorms develop over Syria most frequently in the transitions, less frequently in winter and rarely in July and August. The annual frequency averages 15–20 days in the northeastern district and 10–15 days elsewhere. The monthly frequency averages 1 to 4 days in spring and 2 to 3 days in October.

Climate of Lebanon

Location and topography

The Lebanon territory includes in general most of the mountainous part of the east Mediterranean coast. Meridionally it extends approximately between 33.2° and 34.7°N and zonally between 35.2° and 36.6°E. In the north and east it is surrounded by the Syrian Arab Republic and in the west it is bordered by the Mediterranean Sea. In addition, it is bordered by Israel and Jordan from the south.

In general, the Lebanon territory stands as a high plateau (500–1,000 m), characterized by two parallel chains of separated high mountains (1,500–2,500 m) all running northeastwards. The precoastal chain (western mountains) descends gradually towards the Mediterranean shore in a stepped narrow coastal strip, while both the eastern and western mountain chains descend gradually in a moderate narrow valley, which separates them. Scattered high tops (2,500–3,500 m) emerge from both ranges.

General climate

The climate of Lebanon (AIR MINISTRY METEOROLOGICAL SERVICE, 1936; LEBANON METEOROLOGICAL SERVICE, 1966a, b) is of the Mediterranean type, characterized by a mild dry summer, a cold and very wet winter and two rather short and mild transitions. Lebanon can be divided into the following climatic districts.

(a) Coastal district

This district comprises the narrow (5–15 km) Mediterranean coastal strip. It rises sometimes gradually and sometimes rapidly in steps towards the western mountainous ranges. Its altitude varies and ranges from 50 to 800 m. The dominating weather of this district is mild and dry in summer, cold and wet in winter and mild with light rainfall in transitions.

(b) The hilly district

This district includes the western chain of mountainous ranges running approximately in a northeasterly direction through the whole territory ranging in altitude from 800 to 1,500 m with some scattered penetrating tops (2,000–2,500 m). These mountainous ranges gradually descend from the east towards the inland parts of Lebanon. The dominating climate of this district is mild and dry in summer, very cold and with heavy rainfall in winter and mild but rather cold and rainy in transitions.

(c) The inland district

This district represents the eastern parts of Lebanon next to the western mountainous ranges. It also includes the other mountain ranges bordering the country from the east. Its width ranges between 10 and 50 km and its altitude ranges between 900 and 1,800 m. The dominating weather is hot and dry in summer, cold and wet in winter and mild and slightly rainy in transitions.

Temperature

The distribution of temperature over Lebanon (LEBANON METEOROLOGICAL SERVICE, 1966a) is complicated due to the diversity of orography and the nature of the earth's surface. Temperature is generally lowest in January and highest in August.

During winter (January) the mean temperature ranges between 10 and 15°C in the coastal district, between 10°C and 5°C on the slopes of the western mountainous and inland districts, a very low temperature ranging between 5°C and 0°C on the upper slopes of the mountainous ranges in both districts and freezing temperatures (below 0°C) on their emerging tops. The mountain tops are covered with snow and popular wintersport resorts have been developed in these places.

The maximum temperature ranges between 13° and 18°C in the coastal district, 8° and 13°C on the slopes of the western mountainous and inland districts and between 3° and 8°C on the upper slopes of the mountain ranges in both districts which have very low temperatures of less than 3°C in the highest regions.

During summer (August) the mean temperature ranges between 25° and 28°C in the coastal district and in the northern part of the inland district. It ranges between 20°C and 25°C on the lower slopes of the western and eastern mountains in the western and inland districts. It ranges between 15° and 25°C on their upper slopes and between 10°C and 15°C on the tops of the mountains that emerge in both districts.

The mean maximum temperature ranges between 27° and 32°C in the coastal district, between 22°C and 27° on the slopes round the mountains of the hilly western and eastern districts and ranges between 32° and 37°C in most of the eastern district particularly its northern and central parts. Numerous lovely summer resorts are on the mountain slopes. As a consequence tourism plays a great part in the economy of the country during summer; this is a direct result of the prevailing mild climate. During summer the tops of mountains have temperatures as low as 0°C and ice caps crowning mountain tops melt and supply the down-sloping valleys with fresh water necessary for summer and autumn agriculture.

Humidity

Relative humidity experiences diurnal and seasonal variation, it varies also with the locality and nature of the earth's surface. Its monthly value is lowest in summer (August) when it averages approximately 65–85% in the coastal district, 40–65% on slopes of the mountains of the hilly western and eastern districts and 45–75% in most of the eastern district particularly its northern and central parts.

Relative humidity is rather lower in the inland district where air is much drier during summer than in the western hilly and coastal districts. The diurnal range of relative humidity is rather large, but due to earth's active radiation during night, it approaches 90% or more on most summer nights; a favourable condition for the formation of dew during night and early morning.

Transits of desert depressions through the Lebanon territory affect the relative humidity which is generally lower under the influence of the southerly and southeasterly winds moving in advance of these depressions.

Precipitation

Rainfall

Rain falls heavily in the winter season, particularly in January which is characterized by the maximum monthly rainfall, and falls to a lesser extent in the transitions (LEBANON METEOROLOGICAL SERVICE, 1966a). It is favoured in most cases by the lifting of the moist air masses which enter the east Mediterranean via the slopes of the mountains of both the western and inland districts. An exception is the northern part of the coastal district; it receives more rain than the mountainous parts in September. On the other hand rain generally decreases over the coastal area from north to south. In July and August it is almost rainless in Lebanon ($\leqslant 0.5$ mm).

The annual rainfall ranges from 800 to 1,000 mm in the coastal district, from 1,000 to 1,400 mm in the western hilly district, from 200 to 600 mm in the northern and central parts of the eastern district and from 600 to 1,000 mm in the southern parts of the eastern district.

During January the monthly rainfall averages 150–200 mm in the coastal district, 200–350 mm in the western hilly district, 50–100 mm in the northern and central parts of the eastern district and 100–200 mm in its southern parts.

During May, the monthly rainfall averages 5–20 mm in the coastal district, 25–50 mm in the western hilly district, 10–20 mm in the northern and central parts of the eastern district and 25–40 mm in its southern parts.

Snow

Snow falls in the hilly districts particularly during winter, early spring and late autumn following the transits of vigorous depressions. The annual frequency averages 5–10 days.

Surface wind

Lebanon is influenced generally by two main wind regimes: the summer northeast winds; and the southwest and west winds.

These two general wind regimes undergo various changes in different localities (LEBANON METEOROLOGICAL SERVICE, 1966b).

During winter, winds blow frequently between southeast and southwest in the coastal district, west to northwest in the hilly district and southwest in the inland district. East and northeast winds blow also with a high frequency after the prevailing winds and average 20–30% in the coastal district, 15–20% in the hilly district and 10–20% in the inland district.

During spring and summer, prevailing winds are southwest in the coastal district and the northern parts of the inland district, and west to northwest elsewhere.

During autumn, east and southeast winds prevail over the coastal district, southwesterly winds over the northern parts of the inland district and west to northwest winds prevail elsewhere.

Winds in Lebanon lessen their force and become calms with considerable frequency throughout the year, with averages of 10–20% in the coastal district, 10–35% in the hilly district and 15–35% in the inland district. In general the frequency of calms over the inland district is greater than over the coastal district.

The prevailing surface winds are light to moderate (3–16 knots) in general, fresh to strong (17–27 knots) in association with transitory disturbances through Lebanon. Gales ($\geqslant$34 knots) blow in association with vigorous disturbances. The monthly frequency of days with gales (surface wind velocity $\geqslant$34 knots) is maximum in the coastal district particularly during winter and spring. In winter it averages 9 days in the coastal district (Beirut) 4 days in the inland district (Rayak).

Weather phenomena

Fog

Fog develops over scattered localities of the hilly and inland districts, particularly during winter, but rarely develops over the coastal district where the monthly frequency is almost nil. Fog formation in the Lebanese interior and hilly parts is influenced by locality, cloudiness and nocturnal radiation. During summer fog in general develops infrequently.

Thunderstorms

Thunderstorms develop frequently over Lebanon during winter and the transitional periods. The annual frequency generally averages 20–30 days, with the exception of the southern parts where it averages 5–10 days. The monthly frequency averages 5 days in general, apart from the southern parts where it averages 1–2 days.

Climate of Jordan

Location and topography

The Jordan territory extends zonally from approximately the world's deepest rift, the Jordan Valley or Ghor, running from north to south and meridionally from 37°E to 37°, 38° and 39°E, depending on latitude. It is bounded in the north by the Syrian Arab Republic, in the east by the Republic of Iraq, in the south by the Saudi Arabian Kingdom and in the west by Israel.

Apart from the Great Rift, which is 200–390 m below mean sea level, the whole area is regarded generally as a plateau of approximately 500–1,000 m height. Scattered mountain tops of about 1,000–1,500 m height stand in the western, central and southern parts. Pronounced high tops ranging from 1,500 to 2,000 m emerge, especially in the west and southeast ranges. Apart from the narrow fringe of low lands (valleys) created by rain-fed floods running off the mountains during winter all over the western borders, the whole area is mainly of the desert type.

General climate

The climate of Jordan (JORDAN METEOROLOGICAL SERVICE, 1968) is of the Mediterranean type in the western hilly district, which is characterized by a hot dry summer, a cool wet winter and two rather short transitions. Elsewhere it is of the middle latitude steppe type or the arid type. Jordan can be divided into the following climatic districts.

(a) The hilly district

This district comprises both the western and eastern hilly sides of the Jordan Valley, and its altitude ranges from 500 to 1,000 m with tops reaching approximately 1,500 m in the southeast. The dominating climate is cool and dry in summer, cold and wet in winter. It is the most populated district of Jordan and a predominant characteristic is its summer resorts (Rannallah and Ajlum about 870 and 760 m above mean sea level, respectively).

(b) The Ghor district (area below mean sea level)

This district comprises the Jordan River and the Dead Sea. Its altitude ranges from 192 to 392 m below mean sea level. Characteristic for the Ghor district are its winter resorts (e.g., Jericho).

(c) The desert district

This district forms a part of the Syrian Desert, it is characterized by a hot, dry summer a cold, rainless winter and mostly clear skies.

Temperature

The distribution of temperature over Jordan is rather complicated due to the diversity of orography and the nature of the earth's surface. The temperature is generally lowest in January and highest during August. It is difficult to place elements of temperature within specified classes.

During winter, the temperature is pleasant in the Ghor district with a daily mean of approximately 15°C and in the hilly and desert districts the temperature is relatively low with a daily mean of 8°–12°C. The minimum temperature averages 9°C in the Ghor, 4°–8°C in the hilly district and 2°–4°C in the desert district. When cold air penetrates, the minimum temperature may on rare occasions fall to 0°C in the Ghor Valley, but falls several degrees below 0°C in both the hilly and desert districts.

During summer, the temperature is high in the hilly district with a daily mean of approximately 25°C and the temperature is high in the Ghor and desert districts with a daily mean of approximately 31° and 29°C, respectively. The maximum temperature averages approximately 32°C in the hilly district, approximately 39°C in the Ghor district and approximately 35°C in the desert district.

Humidity

Relative humidity experiences diurnal and seasonal variations. Its monthly value is lowest in summer, then it averages round 50% in the western hills and round 22% in the desert district, while it is maximal in winter when it ranges between 60 and 77% in all districts with the exception of Aqaba at the Red Sea and the desert district where it ranges between 47% and 53%.

During winter, the hourly relative humidity is above 80% most of the night and early morning periods and on many days it reaches 95 or 100% for many hours.

During summer, the relative humidity is fairly low. The air in the eastern hills is much drier than the air in the western hills. The diurnal range is large, but still the relative humidity reaches 90% or more during the night. Dew is formed during most summer nights, but its formation in the eastern hills is less frequent than in the western hills.

During the transits of desert depressions in spring the relative humitidy is very low. It has decreased on some occasions, to 10% in Jerusalem and 6% in Amman and has been even much lower in arid areas.

Precipitation

Rainfall

Rain falls mainly in the winter season, with heavy rainfall in January and February, but it starts in October or early in September and usualy ends around mid May although sometimes the rainfall period ends at an early date; this depends entirely on the weather conditions and transitory Mediterranean depressions which influence the rainfall over the east Mediterranean coasts. Studies of the whole distribution over the Jordan territory reveal the following outstanding features:

As we go from west to east and from north to south rainfall decreases to approximately nil in the southeast corner of the country.

The western hills receive the highest rainfall (600–800 mm), the Ghor receives little rainfall (150–250 mm), the eastern hills have good rainfall (500–600 mm) and the eastern desert is almost arid and receives very little rainfall (≤100 mm).

Snowfall

Snowfall is rare, it falls on high terrain during the wet season for 1 to 2 days per year on the average. The maximum number of days with snowfall in one month was 6 and 7 in Jerusalem and Amman, respectively. At Shoubak (1,500 m altitude) the snowcover may remain for several days.

Ground frost

Ground frost forms frequently during winter in the hilly and desert districts and may occur during November, March and April also.

Surface wind

Two wind regimes show marked prevalence over Jordan; they are as follows: the summer northwest flow; and the southwest to west wind in the non-summer season.

During winter, the southwest to west winds vary in velocity with respect to time and locality in general. Their velocities average from 12 to 18 knots with occasional gales ($\geqslant$34 knots) in the hilly district, from 8 to 15 knots and occasionally fresh to strong (22–28 knots) in the Jordan Valley district and from 10 to 15 knots with occasional gales in the desert district. South to southeast winds may blow in advance of transitory depressions and cause dust-rising or dust-storms. Winds drop to calms or grow weaker with a frequency averaging 20–45% during the night and early morning.

During the summer season prevailing surface wind velocities range from 8 to 15 knots in general and are southwest to northwest in the hilly district, northerly in the Jordan Valley district and west to north in the desert region. In the afternoon surface winds become generally stronger (20–25 knots).

As an exception, the prevailing surface winds over Aqaba are northerly (10–20 knots) and are generally associated with dust. There are very few gales and southerly winds occur occasionally in advance of active cold fronts or depressions.

Weather phenomena

Fog and mist

Fog and mist develop mainly in the hilly district with an average annual frequency of 5–20 days.

Thunderstorms

Thunderstorms, a vigorous instability feature of the atmosphere, occur in winter and to a lesser extent in transitions. Their annual frequency over the western hills and the Jordan valley is rather greater than their frequency over the eastern hills and the desert districts. It averages about 10 days per year in the former and approximately 5 days per year in the latter, respectively.

Dust-rising and dust-storms

These two phenomena are very frequent in the Ghor and desert districts, especially in summer, with an average frequency of 25–60 days annually.

Climate of Israel

Location and topography

Israel is bounded in the east by Syria and Jordan, in the southwest by the Sinai Peninsula, in the west by the Mediterranean Sea and in the north by Lebanon.

A long-stretched, broad and high plateau of about 500–1,000 m height extends from the north to the south in the middle parts between the Great Rift and the Mediterranean coast. It descends gradually from both the west and the east side towards the Mediterranean coast in a wide-stepped coastal strip and towards the Great Rift track in a rather narrow strip. The great desert of Negev extends south of latitude 31°N.

General climate

The climate of the Mediterranean coastal strips (north of latitude 31°N) is of the Mediterranean type. It is characterized by a mild, dry summer, a cold and rainy winter and mild and light rainy transitions. South of latitude 31°N, the climate is of the middle latitude arid type characterized by a very hot, dry summer, a rather cold and light rainy winter and rather hot transitions. Israel can be divided into the following districts.

(a) The coastal district

This district comprises the relatively wide-stepped Mediterranean coastal strip (15–35 km). Eastwards it rises gradually in irregular steps towards the mountain ranges. Its altitude is variable depending on locality and ranges from 50 to 300 m. The dominating weather of this district is mild and dry in summer, cold and rainy in winter and mild with light rainfall in transitions.

(b) The hilly district

This district includes the mountain ranges running north to south through the whole territory; its altitude is from 400 to 800 m, with emerging scattered tops from 800 to 1,200 m. These ranges stand as a mountain barrier between the Mediterranean and the Great Rift. The dominating climate of this district is mild and dry in summer, rainy and very cold in winter and mild to cold with light rainfall in the transitions. As an exception, south of latitude 31°N there is little rainfall and the temperature shows a pronounced rise; rather stable climatic conditions dominate.

(c) The Great Rift (eastern) district

This district represents the eastern district and it comprises a part of the Great Rift which runs from north to south with an altitude varying from 190 m below mean sea level in the north, to 400 m below mean sea level near the Dead Sea and in the south to a few decametres above mean sea level (Aqaba Gulf and Red Sea). Its dominant feature is that it is enclosed between two high mountain ranges from east and west. The dominating climate is very hot and dry in summer, rather cold with light rainfall in winter and rather warm with light rainfall in transitions. The area is characterized by transitional and even winter floods, as a result of the drainage of cumulated rainfall down the slopes of the mountainous regions on both sides.

Temperature

The temperature over Israel shows marked variations with respect to time, orography and locality. In general it is lowest in January and highest in July and August.

During winter (January), mean temperatures range from 10° to 14°C in the coastal district and from 1° to 6°C in the hilly district. In the inland district it shows great variability from the northern parts (10°–14°C) through the middle parts (15°–18°C) to the southern parts (13°–17°C).

Maximum temperatures range from 14° to 19°C in the coastal district and from 14° to 9°C in the hilly district. In the inland district it averages 13°C to 18°C in the northern parts and 20°C to 24°C in the southern parts. The minimum temperature averages 11° to 7°C in the coastal district and 8° to 4°C in the hilly district. In the inland district it averages 5° to 10°C in the northern parts and 10° to 14°C in the middle and southern parts.

During summer (July and August) the mean temperature ranges from 29° to 25°C in the coastal district and from 25° to 20°C in the hilly district. In the inland district it varies from 30° to 26°C in the northern parts and from 30° to 34°C in the middle and southern parts.

The maximum temperature averages 33° to 29°C in the coastal district and 29° to 24°C in the hilly district. Over the inland district it averages 36° to 32°C in the northern parts, 40° to 36°C in the middle parts and 42° to 38°C in the southern parts.

The mean temperature averages 9° to 14°C in the coastal district and 14° to 19°C in the hilly district. Over the inland district it averages 14° to 19°C in the northern parts, 19° to 24°C in the middle parts and 22° to 26°C in the southern parts. It is important to emphasize that the Dead Sea, and the lakes of Tiberias and Hulle impose local variations on the temperature of the surroundings. The sea breeze (on-shore currents) and the hot and dry katabatic flowing down the mountain ranges play a great part in dominating the climate throughout the year. In the inland district in particular there are at least three variant climates and it can be divided accordingly into sub-districts. The northern part is more or less similar to the coastal district climate. The middle parts comprise a large part of the Ghor which experiences a climate similar to that of south Jordan, while the southern parts experience a climate similar to the north of the Red Sea.

It is noteworthy that, although in summer the temperature rises very often to 40°C in the Hulle Valley, in winter temperatures at night may drop to 0°C or below. Due to these freezing temperatures, orange trees do not survive; for this reason the orange trees have been replaced by apple trees.

Humidity

The relative humidity experience diurnal and seasonal variations dependent on locality and nature of the earth's surface. It shows a marked decrease down the eastern slopes of mountainous regions while it shows a marked increase in the eastern district in the surroundings of the Dead Sea and the lakes of Tiberias and Hulle. The monthly value is generally lowest in summer (July or June), when it averages 78 to 68% in the coastal district and 50% in the hilly district. In the inland district, relative humidity averages 65 to 45% in the northern parts and 45 to 25% in the middle and southern parts.

Precipitation

Rainfall

Most of the rainfall in Israel comes from the Mediterranean with the cyclonic centres usually located between Cyprus and Beirut leaving the Negev and Arava on the extreme outer circumference of the vortex which is slightly affected by condensation and precipitation associated with these lows. Floods in the southern parts (Arava) are associated with southern depressions located over the Red Sea. Troughs of low pressure extend northwards and induce the interaction of air masses from the Mediterranean with air masses from the Red Sea and modified tropical air from the south. A result of interaction of these air masses is the development of heavy clouds and huge thundery cells over the Negev, trans-Jordan and the Syrian Desert; in these areas flash floods are particularly common. Every year, usually at the beginning of the rainy season or at its end, such lows occur, although there have been cases of midwinter floods too. Heavy rains fall not only in the desert regions of the country but also in the rainy parts with unusual intensity.

The annual rainfall averages 400–800 mm in the coastal district and 600–1,000 mm in the hilly district. In the inland district it averages 300–600 mm in the northern parts including the Jordan Valley, 50–200 mm in the middle parts and 20–50 mm in the southern parts.

The rainfall decreases from north to south in the coastal district, it varies from 600 to 800 mm north of latitude 32°N, decreases to 400 mm north of 31°30′N and drops to 200 mm more southwards. Also the rain decreases in the Ghor as winds descend the mountainous hills of the hilly district and a pronounced dryness is experienced over the landscape west of the Dead Sea which is more desert. In the northeastern parts of the Hulle Valley winds are forced to rise over the Hermon Mountain (2,880 m), the Golan Heights and old volcanoes (800–1,300 m) and the northern hills (500 m) causing more precipitation which averages 400–750 mm. In particular over the Hermon precipitation is high. It annually averages 1,500 mm (mostly in the form of snow) and over Upper Galilee it averages more than 1,200 mm.

Snow

Snow falls sometimes in the hilly district of Israel, particularly in winter after transits of vigorous depressions. The annual frequency is not more than a few days, for example at Jerusalem it averages 1 to 2 days.

Surface wind

Israel is influenced generally by three main wind regimes: the summer northeast winds;

the southwest to west winds; and the shallow depressions developing over the northern parts of the Red Sea in transitions (particularly in autumn); these may produce a subsiding wind regime.

The sea breezes coming from the vast water areas (Dead Sea, Tiberias and Hulle lakes) affect the surrounding land areas to an average distance of 5–15 km while the rather hot and dry katabic downslope winds in the inland district play a pronounced part in the wind regime and accordingly in dominating the climate of this district.

During winter the prevailing winds are blowing from the southeast and southwest over the southern and middle parts of the coastal district and from west to southwest in its northern parts. Over the hilly district west to northwest winds show marked prevalence while southeasterlies are the next frequent.

Over the inland district northerly winds prevail while southerly winds are the next frequent in northern and middle parts.

During spring, the prevailing winds are west to northwest in the coastal and hilly districts with the exception of the coastal middle parts where southwesterly winds prevail. Northerly winds prevail over the middle and southern parts of the inland (Rift) district but are mostly from south and west over its northern part.

During summer the prevailing winds are west to northwest over the coastal and hilly districts with the exception of the middle parts of the coastal district where west to southwest winds prevail. Northerly winds prevail over the inland district with the exception of its northern parts where east to southeast winds are more frequent.

During autumn, the prevailing winds are west to northwest over the hilly district and northerly over the inland district where southerly winds are next frequent over its northern and middle parts. Over the coastal district prevailing winds are either westerly or easterly over its northern parts, southwesterly over its middle parts and north to northwest over its southern parts where the southeast winds are next frequent.

The prevailing winds are light to moderate in general, occasionally fresh in the afternoon over the coastal and hilly districts, strong in association with the transitory Mediterranean or desert disturbances and their cold fronts. Gales develop in general from November, are most frequent in January and February and are rare in summer. They either blow from south or west over the south Mediterranean coast and are northerly or easterly over its northern part. Gales are generally preceded by southeasterly winds and are followed by north and northeast breezes.

On-shore winds (sea breezes) blowing from the Dead Sea to the surrounding land have higher velocity in the northern part of the region than in the southern part as a result of differences in the environment. They are a vital biological factor on the Dead Sea shores, characterized by low temperatures and high humidity values until the late afternoon. The night winds (land breezes) are very light and variable, and summer nights near the Dead Sea are mostly unpleasant and it is extremely difficult to sleep then.

In the northern parts of the inland district, winds rise in the afternoon and continue until 22h00. These strong and continuous winds, which blow for several hours every afternoon for several months, affect the growth of trees in general and sensitive fruit trees in particular. Therefore, wind breaks are necessary for the protection of trees.

Weather phenomena

Dew

Formation of dew in Israel shows generally marked variations with time and locality, The mean dew amount reaches its maximum during summer in the coastal district. during spring in the inland district and during autumn in the hilly district. Besides those two factors of general distribution of temperature and rainfall, influences of topography, soil characteristics, vegetational cover and irrigation practices play a great part in dew formation.

Fog

Fog develops in different districts in Israel. It is of the typical radiation type in the coastal district, with maximum frequency during May. The maximum frequency of foggy days occurs during April in the Emeq Valley (7.9 days), during August in the northern Negev (7.6 days) and during winter in the hilly district (4.2 days in Jerusalem in February, and 9.7 days in Har Kenaan in January). In the mean, fog lasts from 2 to 4 h, but in individual cases it may last for 12 h.

Thunderstorms

Thunderstorms develop frequently over Israel during winter and transitions. The annual frequency averages 5–10 days in general, while the monthly frequency averages 2–4 days during winter and 1–2 days at other times.

Climate of the Arabian Peninsula: Saudi Arabia, The Yemen Arab Republic (North Yemen), The People's Democratic Republic of Yemen (South Yemen), Oman

In this part it was found difficult to deal with each country separately, because the lack of climatological data makes detailed analysis on a smaller scale impossible. Consequently we shall deal here with the entire Arabian Peninsula extending roughly south of 30°N.

Location and topography

The Arabian Peninsula is a large elongated rectangle with the long sides running from north-northwest to south-southeast. In the west it is bounded by the Red Sea, in the east by the Arabian Gulf and the Gulf of Oman and in the south by the Arabian Sea. In the north it is bounded by Jordan, Iraq and Kuwait. The peninsula extends meridionally from 35° to approximately 60°E and zonally from approximately 12° to 30°N. From the topographic point of view we can divide the area into the following zones.
(*1*) In the west a coastal strip which extends along the Red Sea with varying breadth. It is broadest around 25°N and fairly narrow around 20°N. In the north the region is irregular and with frequent intervals of high land, while in the south it is mainly com-

posed of short valleys running from the mountains of Yemen and southwestern Saudi Arabia, towards the Red Sea.

(*2*) A triangular high plateau in the extreme southwest with one side parallel to the Red Sea coast and another side parallel to the Gulf of Aden. This zone extends from the Asir Mountains in Saudi Arabia to include most of the territory of North Yemen. The region is generally mountainous with individual mountain blocks extending from southwest to northeast. These are joined in the middle and separated in the west and east by narrow valleys running towards the Red Sea or to the interior of the peninsula. The height of the main plateau is approximately 3,000 m above mean sea level with several peaks reaching the height of approximately 3,800 m. The height decreases sharply on the western side and more gradually on the eastern side.

(*3*) An elongated plateau extends in the western part of the peninsula from the extreme north to the extreme south where it spreads westwards approximately parallel to the coasts of the Arabian Sea, with on the west the triangular high plateau described above. This zone is broad at approximately 25°N, where it covers about half the breadth of the whole peninsula, and it is relatively narrow in the extreme north. With the exception of a few peaks in the north, the height ranges from 1,000 to 1,500 m.

(*4*) A vast desert plain extends on the eastern side of the peninsula, covering the majority of the area, with the exception of the small Oman Plateau. This zone extends from Rubalkhali in the south to the southern borders of Iraq and the eastern borders of Jordan. The height above mean sea level does not exceed 200 m in general.

(*5*) The Oman Plateau stands in the extreme east with an average height of approximately 1,500 m above mean sea level.

General climate

The prevailing climates in the Arabian Peninsula are mainly of the dry types. They firmly depend on the proximity of water surfaces and topography. Following the Köppen classification the whole area can be divided into three different climatic types:

(*a*) The hot desert climate (BWh) prevails in two regions.

(*1*) The narrow coastal strip on the eastern side of the Red Sea extending from approximately 25°N to the Gulf of Aden. This region is rather hot and dry in summer and warm with some rain in winter and spring. Temperatures generally increase from north to south with a maximum in July and a minimum in January. Local circulations influence to a large extent the climate of that zone, particularly in spring and summer when the temperature difference between land and sea is most pronounced.

(*2*) The vast desert plain extending parallel to the coasts of the Arabian Gulf covers more than half the area of the peninsula. The climate is very hot and dry in summer and relatively cool in winter with small amounts of rainfall in spring and winter. The annual range of temperature is very large, approximately 20°C.

(*b*) The hot steppe climate (BSh) prevails in two regions.

(*1*) The elongated plateau extending parallel to the Red Sea from the extreme north to the extreme south, including Gebel Shefa Al Qusseim, Neged, Tarique, and Hadramaut. This region is generally mild with some rains in winter and spring and hot and dry in summer and autumn. The effect of the topography on the temperature, as well as the rainfall distribution, is very distinct.

(2) The Oman Plateau extending along the western coast of the Gulf of Oman. The central peak, called Jabal Akhdar, reaches a height of about 1,700 m above mean sea level while the height of most of the area ranges from 500–1,000 m. No representative climatological records could be obtained for this region since the nearest stations available, Muscat and Sharjah, are both coastal.

(c) The warm temperate rainy climate with a dry winter (CW) prevails in the high triangular plateau of the southwestern corner of the Arabian Peninsula. This zone includes the Asir Mountains of Saudi Arabia, most of the territory of North Yemen and the norther part of South Yemen.

Orography and local circulations play a very important role in the distribution of various climatological elements. Rain falls in all seasons of the year with maxima in spring and summer and a total annual yield ranging from 200 to 500 mm. The temperature is generally lower than that of the surrounding areas as a result of the high elevation; the average annual temperature range does not exceed 10°C. High peaks are covered with ice caps in the cold season. These melt in spring and supply the numerous valleys running east and west.

Temperature

The temperature distribution throughout the year shows a diversity of patterns which closely follow the orographic features described above.

In the western coastal zone of the Red Sea the daily mean temperature in winter (January) increases rather steeply from about 29°C in the north to about 33°C in the extreme south. The summer distribution, however, does not show a regular change from north to south as it does in winter. In both seasons the daily mean temperature decreases eastwards following the increase of elevation, and causes an east–west temperature gradient during the entire year throughout that coastal effect zone. The coastal zone reduces the annual range of temperature to about 10°–15°C, as well as the diurnal range of temperature, which varies between 15° and 19°C during various seasons.

In the broad eastern area of plain lands the daily mean temperature in winter (January) shows little variation from north to south, with a cold "tongue" covering the lowlands in Rebalkhali and higher temperatures in the coastal stations on the Arabian Gulf, Gulf of Oman and Arabian Sea. In general the temperature increases from west to east with decreasing altitude. In summer the temperature in the eastern belt of lowlands is extremely high. It is considered one of the hottest areas in the world in the summer season. Daily mean temperatures in July reach about 36°C, decreasing gradually towards the coasts of the Arabian Gulf and decreasing more rapidly southwards and westwards.

The diurnal variation, as well as the annual range, of daily mean temperature are very large in the interior of that zone, decreasing towards the coast and towards the western plateau.

A peculiar feature of this area is the climatological discontinuity in the monthly mean temperatures. It usually increases very rapidly by about 5°C between May and June and between September and October, with a more gradual variation between the other months. Such climatological discontinuities were shown to exist also for some other climatological elements, and are considered to be the result of a change in the distribution pattern from one regime to another.

In the main plateau extending from the extreme north to the extreme south, the temperature distribution follows the orographic features closely with a belt of low temperatures extending along the plateau in all seasons.

In winter the mean minimum temperatures fall below 0°C in scattered localities, especially on the higher peaks. However, the height of this region tends to decrease the diurnal, as well as the annual ranges of temperature, more than the corresponding ranges in the neighbouring lower region.

In the triangular area of the high plateau in the southwest of the peninsula there are relatively low temperatures in all seasons. Several peaks are covered with snow in the cold season. The annual range of temperature does not exceed 10°C.

Precipitation

The peninsula is generally dry with the exception of the southwestern plateau. The small amounts of rainfall in the greater part of the area have definite distribution patterns in space and time. These patterns are quite consistent with the synoptic systems which prevail in different seasons.

In winter, the Mediterranean cyclones migrating from west to east, in association with upper troughs and active phases of the subtropical jet as well as the polar front jet, are considered the main rain-producing synoptic systems. Their potential for producing rain decreases generally from north to south over the entire peninsula, except for mountainous areas where the uplift acts as an exterior factor. Consequently, the distribution of winter rainfall in the peninsula shows maximum values in the northern part of the main plateau and gradually decreases in the lowlands on the eastern and western sides. Winter rainfall also decreases towards the south, except over the high mountains in the extreme southwest where the influence of topography is predominant. Coastal stations and lower altitude stations in the extreme south have no, or insignificant, values of winter rainfall.

Spring is an important season for rainfall, during which troughs in the upper westerlies, in association with the depressions migrating from west to east, still affect the northern part of the peninsula. Moreover, spring is characterized by a weak stability in the lowermost atmospheric layers as well as large daytime temperature differences between land and water surfaces. Such conditions stimulate active local circulations between land and sea and between mountains and valleys. These local circulations, as rainfall producing systems, are most pronounced in the high plateau in the extreme southwest.

As a result of these factors the spring rainfall is maximum in the southwestern high plateau and the southern Red Sea coasts, with another minor concentration in the northern part of the main plateau.

In summer the whole circulation pattern alters so that middle latitude disturbances of the extratropical type do not affect the area any more. The thermal monsoon trough is established across the peninsula with southwesterlies blowing over the southern edge of the peninsula. This current is conditionally unstable and its potential for rainfall production depends on the topography as well as the nature of the underlying surface. The Arabian Sea off the southern coasts of the peninsula is characterized by active upwelling. The cooling of the sea surface which results acts as a stabilizing factor which decreases the amount of summer rainfall in the southern coastal stations. The summer rainfall is consequently almost restricted to the southwestern plateau, with insignificant amounts over the southern coastal strip. The northern part of the peninsula is quite dry. In autumn, middle latitude disturbances begin to affect the northern part of the peninsula, but the local circulations are considerably weakened due to the stability of the lower atmospheric layers. Consequently, the distribution pattern shows a minor concentration in the northern part of the main plateau where mean monthly values for October are approximately 10 mm. Over the southwestern plateau the rainfall values are much smaller than the corresponding values in summer.

Over this entire area snowfall is a rare phenomenon and is climatologically insignificant.

Surface wind

Prevailing wind patterns differ from one part to another in the peninsula. These patterns are generally affected by the prevailing pressure distributions as well as by the surrounding orographic features. The influence of each of these two main factors varies from one locality to another.

In the northern part of the western coastal strip, where the channel effect is thought to be a major influence, the prevailing wind blows from the sector between north and west throughout the year; in autumn and early winter other wind directions form a small percentage of the total frequency. In the southern section of the western coasts the prevailing winds are between west and northwest with a small percentage of southwesterlies in summer months. In autumn and early winter the winds are between south and southwest. In spring they are usually westerlies.

In the northern part of the main plateau seasonal variations are more pronounced, particularly at greater distances from the coastal regions. In the summer months the winds are between north and west. This is quite consistent with the mean summer location of the monsoon trough which extends from northwestern India to the Arabian Gulf. In non-summer months the wind blows from different directions although easterlies and southerlies are most frequent.

On the southern coasts the most frequent winds blow from between northeast and southeast in winter, between east and south in spring and autumn, and between southeast and southwest in summer.

In the eastern coasts the prevailing wind blows steadily from the northwest in winter and from the sector between north and east with varying degrees of steadiness in the three other seasons.

Thunderstorms

The frequency of thunderstorms in the Arabian Peninsula is well correlated with rainfall distribution. In the northern parts, where rainfall and thunderstorms are attributed to synoptic scale disturbances, thunderstorms have maximum frequency in spring and autumn, moderate frequency in winter and they are practically absent in summer. The available data shows that the frequency distribution of thunderstorms depends to a large extent on the orographic features and the proximity of water surfaces, with higher frequencies near the eastern and western coasts in the neighbourhood of high land and lower frequencies in the lower interior areas.

In the southern sector of the peninsula thunderstorms are generally more frequent, especially over the high plateau in the extreme southwest where the maximum frequencies are in spring and summer.

Climate of Iraq

Location and topography

Iraq occupies most of the area between 40° and 48°E, and 30° and 37°N. It is mainly a flat, little above sea level, in which Euphrates and Tigris rivers run from northwest to southeast. To the east, the high Zagros Mountains of Iran act as a barrier, while the west is a desert with mountains near the Syrian border ranging between 500 and 1,500 m. To the north, mountains range between 200 and 800 m increasing to 3,000 m in a small area in the extreme northeast. The Arabian Gulf lies in the south.

General climate

The country in general can be considered semi-arid. In winter it is mostly covered by an extension of the cold high pressure region of central Asia, while the Mediterranean is a region of low pressure, but without direct influence. Only the deep depressions can advance to the country. They are fed by the warmth and humidity of the Arabian Gulf and produce rain which is sometimes accompanied by thunder.

In the transitional seasons, Mediterranean depressions which pass eastwards are similar to the winter depressions but have less frequency and intensity. Especially in the north thunderstorms are expected more in these seasons than in winter. Storms of very fine particles of sand and dust generally blow in advance of winter and transitional depressions. Particularly in spring the air carrying these particles is generally very dry and hot. In summer, due to the absence of travelling depressions, weather conditions are settled. However, the central and southern parts are subject, during day time hours, to heavy dust and sandstorms which sometimes reach heights over 3 km; these have been observed by aircraft pilots flying over these regions.

Climatic types

According to Köppens' classification for world climates, Iraq comprises three climatic types: (*1*) CS; warm, temperate, rainy with a dry summer in a relatively small area in the north; (*2*) BSh; dry, hot steppe covering the central and southern parts; and (*3*) BWh; dry, hot desert in the west.

Temperature

Extreme continentality characterizes the climate of Iraq. The mean annual range of temperature is very large. It reaches its maximum value (27°C) in the north, gradually decreases to 24°C in the western and central parts and attains its minimum value (21°C) in the Gulf area in the extreme south. Temperature extremes generally occur in January and July. For instance, in Mosul where the temperature sometimes falls below 0°C in winter, it sometimes rises to more than 50°C in summer.

The mean daily winter temperature ranges from 6°C in the extreme north to 12°C in the Gulf area. The absolute minimum at Mosul is below −10°C while Basra recorded −5°C. In summer the entire country is generally very hot. The mean daily temperature ranges from 34° and 36°C everywhere except in the western desert where it is reduced by altitude in some parts to about 30°C. Excluding these high parts, absolute maxima of 50°C have been experienced several times. Even higher absolute maxima have been recorded in the central and southern districts on some occasions. Near the Gulf, due to high humidity, the weather is very uncomfortable especially with light or calm winds. An interesting feature of some buildings is that they have underground chambers to provide relief from such high temperatures during day. The diurnal range of temperature is also large, from 15° to 20°C everywhere, and the nights are hot.

Humidity

The air over Iraq is dry in general. Due to the great rise in temperature in spring and summer seasons the mean relative humidity drops to about 26% in these two seasons, except in the Gulf area where it remains about twice that value. In the western desert relative humidity values of 0% are reported several times.

Precipitation

Rainfall

The Mediterranean depressions are the source of all the precipitation of Iraq. The rainy season begins normally in October and continues till May. The months from June to September can be considered rainless. Most of the precipitation occurs in winter and spring. Its amount is relatively slight. The mean annual total is less than 250 mm.

The distribution for both the amounts and the number of rainy days shows great variability with locality. Large amounts fall in a relatively small area in the extreme northeast of the country where the annual total is about 700 mm and exceeds 1,000 mm in some localities. Southwestwards (south of 36°N and west of 45°E) the rainfall considerably decreases. The mean annual total in the extreme southwest is less than 100 mm. The mean annual number of rainy days is over 60 in the extreme northeast and decreases to less than 10 in the extreme southwest.

Sometimes rain is accompanied with thunder, but in spring most of the rainfall occurs with thunderstorms which are, in certain situations, violent in the north.

Snow

Snow falls heavily in winter on the mountains in the north. Tigris and Euphrates reach their highest flood level in spring, mainly from the melting snow. In the central plains snow sometimes falls but usually melts immediately. In the south and southwest, snow is very rare.

Surface wind

The prevailing winds in the wet season are mainly northwest, but they shift to the southeast in advance of the travelling low pressure systems. In the Gulf region these southeast winds are followed sometimes by southwest winds for some hours after the passage of the trough. These winds are called "suhaili". They are sometimes strong enough to be a danger for vessels in the Arabian Gulf. The winds behind the trough are usually not strong, but they are cold and dry. In the western desert the winds are mostly variable. In summer the main air stream is northwest blowing between the Azor anticyclone and the Indian monsoon low down Mesopotamia under the influence of the thermal low pressure of south Asia. These air streams are well-known under the name of "shamal" which is the Arabic word for north. They are remarkably persistent due to the absence of cyclonic disturbances. However, although they may be very strong and carry clouds of dust and sand during the day, their strength decreases and is almost calm by night.

Sandstorms

Sand and dust storms are the most marked weather phenomena in Iraq. They occur mostly in the central and southern parts and can be expected in any month, but the maximum frequency is in July and the minimum frequency in December and January. In the transitional periods sandstorms cover most of the central and southern parts while in summer sandstorms are mainly confined to the southern area near the Gulf.
The annual frequency is about 20 days in the central parts and it increases to 30 days in the south, of which 16 days in the summer.

Thunderstorms and hail

Thunderstorms occur in association with the vigorous Mediterranean depressions which pass over the country in the wet season. Their maximum frequency occurs in spring, averaging 3 times per month. In winter and autumn, they occur with less frequency. Its maximum occurrence is in the north, due to the high mountain ranges, as well as in the southern Gulf area where humidity is high. The central and western parts have the smallest frequency. In summer they are very rare.
Hail falls in the wet season in the north with an annual frequency of about 2 days, decreasing to 1 day in the central and southern regions.

Fog

Fog is generally expected in autumn and winter in the Tigris and Euphrates valleys. Its maximum frequency of 6 days occurs in the north in January; this frequency decreases to 3 days in the Gulf area. In March it also occurs, with a frequency of 1 day. The annual frequency is about 20 days decreasing to 10 days in the Gulf area. Fog is very rare in the western desert. Also, it seldom occurs in late spring and summer.

Climate of Kuwait

Location and topography

Kuwait occupies a relatively small and mainly desert area at the northwestern part of the Arabian Gulf south of Iraq. The eastern part is mainly a flat rising no more than 100 m above sea level. The western parts are hills sloping from approximately 300 m in the west to approximately 100 m above sea level in the central parts.

General climate

Kuwait can be considered as a natural southward extension of Iraq, and therefore its climate is extremely continental and very similar to that of southern Iraq. In winter and transitions, only the vigorous Mediterranean depressions advance to it producing rain accompanied sometimes by thunder.

In summer, the shamal winds raise a great amount of dust and sand during the day hours.

Climatic types

The Kuwait climate is in general of the BWh-type: a dry, hot desert-type climate in Köppen's classification.

Temperature

The mean annual range is very large; approximately 24°C. Temperature extremes occur in January and July. The mean daily temperature in winter is about 14°C, although −6°C has been registered. An absolute maximum of 50°C has been observed, while the mean daily summer temperature is 36°C. The daily range is approximately 12°C in winter and increases to approximately 18°C in summer.

Humidity

The littoral is characterized by high humidity especially in late spring, summer and early autumn.

The relative humidity reaches a maximum in winter with a mean daily value of 60% inland. It falls to approximately half this value in summer. Near the Gulf these values are increased by about 15%.

Precipitation

Since Kuwait is mainly a flat desert, the rainfall amounts are generally slight and even less than those of southern Iraq. The rainy season generally lasts from November to May. The mean annual total is less than 100 mm. The Arabian Gulf is the source of the moisture for the rain-bearing depressions. No precipitation is expected in summer. In the transitional seasons, when upper troughs drive cold air from the northwest over warm air in the lower layers, the vertical temperature gradient in some upper air patterns is favourable for violent thunderstorms with hail and heavy rainfall.

Rain is expected more in the east than in the west and in autumn more than in spring. A maximum of 65.9 mm was registered at Ahmadia in November 1965; of this amount 33 mm fell on the 5th. Most of the annual total falls in winter. The annual mean of rainy days is 24, half of it occurs in winter.

Snow is practically unknown in Kuwait.

Surface wind

The prevailing winds are mainly northwest, but they turn to southeast and southwest in advance of the travelling Mediterranean depressions in winter and in the transitions, carrying large amounts of very fine dust particles and causing dust-storms which are extensive at times. Here also the southwest winds, as in Iraq, are sometimes strong and hazardous for vessels in the Arabian Gulf.

In summer the changes in weather are very few and the wind is persistently northwest. As in Iraq, the winds sometimes cause heavy sandstorms during the day hours; these reduce visibility considerably. The local name for these winds is toz.

Sandstorms

They are expected in late spring, summer and early autumn. The maximum occurrence is in summer with the largest frequency of 5 days in July. In spring and autumn the frequency drops to only one day per month.

Thunderstorms

They occur in association with the vigorous Mediterranean depressions in winter and in the transitions. The maximum frequencies occur in winter and spring (1 to 2 days per month). The annual frequency is about 8 days. Thunderstorms are expected more in the east than the west due to the increased humidity of the Gulf area.

Fog

Fog is expected in winter and autumn with a frequency of about one day per month. The annual frequency ranges from 5 to 10 days. It is mainly of the radiation type and dissipates shortly after sunrise.

Climate of Iran

Location and topography

Iran is a land area extending approximately between 46° and 62°E and 26° and 38°N, in southwest Asia, with the Caspian Sea in the north, the Arabian Gulf in the west and the Gulf of Oman the south. The main features of the topography can be summarized as a great plateau between two mountain ranges. In the north, the Elburz Mountains have long east–west ranges of more than 2 km height; these reach more than 6 km height in

224

some places. In the west and southwest, the Zagros Mountains extend over a very long distance. Most of this area is over 1,800 m, much of it is over 3 km and many summits exceed 3,600 m. The great plateau, rising 1 km above sea level, occupies most of the country. Some parts, such as the Dasht-e-Lut and Seistan, are only about 500 m above sea level. South of the Caspian Sea there are some narrow lands about 20 m below sea level.

General climate

The country is mainly arid or semi-arid. Except for the littoral, the climate is extremely continental.

The average frequency of the travelling depressions in winter does not exceed 3 per month. In summer, as in Iraq and Kuwait, not many changes occur from day to day over most of the country. Hot and dry weather prevails in general.

Climatic types

Three climatic types are found in Iran: (*1*) CS: warm, temperate, rainy with dry summer in a narrow strip in the north; (*2*) BWh: dry, hot desert in the central plateau; (*3*) BSh: dry, hot steppe covering the rest of the country.

Temperature

Apart from the littoral, the temperature in Iran is extremely continental. The annual range is great, from 22° to 26°C. The daily range is also large. Temperature extremes occur in January and July.

Winters are cold, especially in the north. The January mean temperature at Mashhad (985 m above sea level) is 2°C and the absolute minimum is −25°C. Nights are generally very cold especially in the northeast. On the plateau it is less cold than in the Elburz Mountains.

In summer, as in Iraq and Kuwait, weather is settled in most of the country. Hot weather generally prevails, especially in the lowlands and enclosed valleys such as those of Khuzestan and Luristan where the daily maxima often exceed 44°C. They are known to be like furnaces. More than 50°C has been recorded at Abadan. On the great plateau the low humidity makes the heat bearable. Although the mean daily range there is about 17°C, nights remain hot. In the higher places the weather is generally mild and pleasant. At the coasts, where the daily range of temperature is not so great as more inland, the weather is very unpleasant due to the excess of moisture and increased heat. The July mean temperature at Bandar Abbas is 35°C, while the absolute maximum is only 46°C.

Humidity

Since Iran is for the greater part situated at a considerable height above sea level, the humidity is generally low except for the coastal regions where it is particularly high especially in summer due to intense heating. In Bushehr, on the Arabian Gulf, the mean relative humidity in the dry season is about 60% while at Kerman, which is far inland, it sometimes is as low as 8 %. For this reason the summer weather is very sultry in the Gulf area where both temperature and humidity are high.

Precipitation

Rainfall

The rainy period in most of the country lasts from November to May. The average annual rainfall total is about 240 mm. Maximum amounts fall on the Elburz and Zagros slopes facing north. The mean annual rainfall total in these regions is more than 1,200 mm. Rasht, in the northwest near the Caspian shore, has more than 100 days with rain every year. All the months there are rainy and the mean annual total is 1,240 mm of which approximately 200 mm fall in October. However, rain begins earlier in the northwest on the Caspian littoral with maximum amounts in autumn. On the other hand, going inland, the amount of precipitation decreases to less than 120 mm annually. Seistan has less than 70 mm. The amounts vary considerably with the topography. In the northern and western mountains the annual mean precipitation is more than 480 mm; snow forms most precipitation. The plateau has most of its rainfall in spring, while on the western and southern coasts most of the rain falls in winter. In the Caspian littoral, where the rain falls earlier, the rainfall is maximal in autumn.

In the dry period, between May and October, rain is rare in most of the country.

Snow

Snow is a very well-known phenomenon in Iran in winter. Maximum amounts fall on the mountains in the north and west. The highest summits are snow-covered most of the year. Snow falls also on the plateau, mainly in winter, but it does not stay for long. The snowfall is maximum in the northeast, and minimum on the south and southeast shores.

Surface wind

Mountain ranges in the north and west tend to deflect the general wind pattern.

In winter northerly winds prevail generally. Calms are frequent due to the small frequency of depressions.

In summer the prevailing winds, due to the fall of pressure over southern Asia, are mainly north and northwest in the northern and central parts. These winds are persistent, hot and dry. Frequently they are strong in most places for long periods. These winds which blow in the Gulf area are similar to the shamal in Iraq. Among the best examples for persistence are the Seistan winds which blow continuously from June to September in the region of Seistan in the east and are known as "the winds of 120 days". They are very hot, dry, often exceed gale force and carry great amounts of dust and sand.

In the south, south of the axis of the monsoon low, the prevailing winds are west and southwest.

Thunderstorms

Thunderstorms occur in the wet season in connection with the travelling depressions. They are most frequent in the extreme north near the Caspian Sea.

Frost

The distribution of humidity and temperature in the wet season in the lower regions of Iran is suitable for the formation of frost over most of the country. Maximum occurrence takes place in winter over the highlands far from the shores. The average annual number of days with frost exceeds 100 in the extreme northwest and decreases to approximately 65 in the Zagros Mountains in the west. The south and southwest littoral are frost-free places.
In the dry season frost is not reported generally.

Acknowledgements

The authors of this chapter wish to express their deep gratitude to the Directors of the Meteorological Services who have kindly supplied them with the data, reprints and reports on the basis of which this part of the World Survey of Climatology could be written.
The authors are particularly indebted to Dr. H. E. Landsberg, Research Professor of the Institute of Physical Science and Technology, Maryland University, for his kind assistance in providing them with many valuable references and tables.

References

AIR MINISTRY METEOROLOGICAL OFFICE, 1936. *Weather in the Mediterranean, Vol. II, Parts 7–12.* H.M. Stationary Office, London.

AIR MINISTRY METEOROLOGICAL OFFICE, 1958, 1966. *Tables of Temperature, Relative Humidity and Precipitation of the World, Part V. Asia.* H.M. Stationary Office, London.

AIR MINISTRY METEOROLOGICAL OFFICE, 1963. *Report No. 12, Cyprus, Parts I, II.* H.M. Stationary Office, London.

ASHBEL, D., 1964. *Climate of the Great Rift Arava Dead Sea Jordan Valley.* Hebrew University, Jerusalem.

ASHBEL, D., 1969. *Frequencies of Temperature Threshold, and Maximum Minimum Graphs.* Hebrew University, Jerusalem.

CYPRUS METEOROLOGICAL SERVICE, 1970. *Monthly Climatological Data.* Supplied on request, unpublished, Nicosia.

EL BASHAN, D., 1968. *Frequency of Calm Days in Israel.* Israel Meteorological Service. Bet Dagan.

FLOHN, H., 1964. The tropical easterly jet and other regional anomalies of the tropical circulation. *Proc. Symp. Tropical Meteorol., Rotorua.* Wellington, pp.160–171.

GILEAD, M. and ROSENAN, N. 1954. Ten years of dew observation in Israel. *Israel Explor. J.* IV (2): 120–123.

GLENN, T., 1954. *An Introduction to Climate.* McGraw-Hill, New York, Toronto, London, 3rd ed.,

HAURWITZ, B. and AUSTIN, T. M., 1944. *Climatology.* McGraw-Hill, New York and London. Plates I, II, IV, V, IX, XVI, XVII, pp.269–324.

IRANIAN METEOROLOGICAL DEPARTMENT, 1965–1969. *Meteorological Year Book.* Publication Section of IMD, Tehran.

IRANIAN METEOROLOGICAL DEPARTMENT, 1970. *Climatological Data for Selected Stations.* Supplied on request, Unpublished.

IRAQ METEOROLOGICAL DEPARTMENT, 1945. *Climatological Atlas for Iraq.* Publ. No. 13, Survey Press, Bagdad.

IRAQ METEOROLOGICAL DEPARTMENT, 1970. *Climatological Data for Selected Stations in Iraq.* Supplied on request, unpublished.

JENKINSON, A. F., 1955. Average Vector Wind Distribution of Upper Air in Temperate and Tropical Latitudes. *Meteorol. Mag.,* 84: 140–147.

JORDAN METEOROLOGICAL SERVICE, 1968. *Jordan Climatological Data Hand Book.* Jordan Meteorological Service, Amman.

JORDAN METEOROLOGICAL SERVICE, 1970. *Monthly Climatological Data for Amman Airport, Mafraq and Jerusalem.* Supplied on request, unpublished.

KENDREW, W. G., 1961. *The Climate of the Continents.* Oxford University Press, New York, 5th ed.

KOTESWARAM, P., 1958. The easterly jet stream in the tropics. *Tellus,* 10(1): 43–57.

KOTESWARAM, P., 1962. Origin of tropical storms over the Indian Ocean. *Proc. WMO Inter-Regional Seminar Tropical Cyclones, JMA ,Tokyo,* pp.69–71.

KRISHNAMURTY, T. N., 1959. *The Subtropical Jet Stream of Winter.* Res. Rep., Contr. No.N60RI–02036 NR082–120.

KUWAIT METEOROLOGICAL SERVICE, 1956–1967. *Annual Climatological Report.* Kuwait Government Printing Press, Kuwait.

KUWAIT METEOROLOGICAL SERVICE, 1970. *Climatological Data for Kuwait.* Supplied on request, unpublished.

LEBANON METEOROLOGICAL SERVICE, 1966a. *Atlas Climatique du Liban, Tome I.* Meteorological Service, Lebanon and Ksara Observatory, Beyrouth.

LEBANON METEOROLOGICAL SERVICE, 1966b. *Atlas Climatique du Liban, Tome III.* Meteorological Service, Lebanon and Ksara Observatory, Beyrouth.

LEBANON METEOROLOGICAL SERVICE, 1970. *Climatological Data for Beyrouth, Tripoli El Mina and Rayak.* Supplied on request, unpublished.

LEVI, M., 1967. Fog in Israel, Ser. D (contributions) No. 21. *Israel J. Earth Sci.,* 16: 7–21.

SAUDI ARABIAN METEOROLOGICAL DEPARTMENT, 1971. *Climatological Data for Selected Stations.* Supplied on request, unpublished.

SOLIMAN, K. H., 1958. On the Intertropical Front and Intertropical Convergence Zone over Africa and Adjacent Areas. *Symposium on Monsoons of the World, Indian Meteorological Department, New Delhi,* pp.135–142.

SUTCLIFFE, R. C. and BANNON, J. K., 1954. Sci. Proc. IAM Rome.

SYRIAN METEOROLOGICAL DEPARTMENT, 1960–1970. *Monthly Climatological Data Bulletins.* Syrian Meteorological Department, Damascus.

SYRIAN METEOROLOGICAL DEPARTMENT, 1970. *Climatological Data for Selected Stations.* Supplied on request, unpublished.

TANTAWY, A. H. I., 1961a. The role of the jet stream in the formation of desert depressions in the Middle East. *WHO Tech. Note,* 64.

TANTAWY, A. H. I., 1961b. The role of jet streams in the development of autumn thunderstorms in the Middle East. *WHO Tech. Note,* 64.

TANTAWY, A. H. I., 1963. *The Tropical Easterly Jet Stream over Africa.* Meteorological Department, Cairo.

TANTAWY, A. H. I., 1969. On the genesis and structure of spring desert depressions. *Meteorol. Res. Bull.,* 1(1): 69–106. Meteorological Department, Cairo.

TANTAWY, A. H. I., 1970. On the kinematic and dynamic structure of the African Intertropical Convergence Zone in relation to summer rainfall distribution. *Meteorol. Res. Bull.,* 2(2): 131–148. Meteorological Department, Cairo.

TURKISH METEOROLOGICAL DEPARTMENT, 1967. *Mean and extreme meteorological bulletin.* Ankara Basimve Cilithia, Ankara (in Turkish).

TURKISH METEOROLOGICAL DEPARTMENT, 1970. *Climatological Data for Selected Stations in Turkey.* Supplied on request, unpublished.

UNITED ARAB REPUBLIC, 1970. *Climatological Data for Gaza.* Unpublished.

YEMEN PEOPLE'S DEMOCRATIC REPUBLIC METEOROLOGICAL SERVICE, 1971. *Climatological Data for Selected Stations.* Supplied on request, unpublished.

Appendix—Climatic data in tables and charts

TABLE I

LIST OF STATIONS

Station	Lat.°N	Long.°E	Elev. (m)
Turkey			
Samsun	41°17′	36°20′	44
Trabzon	41°00′	39°43′	35
Florya (Ist.)	40°59′	28°48′	34
Goztepe (Ist.)	37°05′	37°22′	855
Sivas	39°45′	37°01′	1,285
Ankara	39°57′	32°53′	902
Izmir	38°26′	27°10′	25
Antalya	36°42′	30°44′	43
Adana	37°00′	35°25′	66
Erzurum	39°55′	41°16′	1,829
Diyarbakir	37°53′	40°11′	677
Cyprus			
Kyrenia	35°20′	33°19′	13
Famagusta	35°07′	33°57′	22
Nicosia	35°09′	33°17′	220
Limassol	34°41′	33°03′	16
Marphoubay	35°18′	32°57′	14
Syria			
Safita	34°49′	36°08′	369
Aleppo	36°11′	37°13′	395
Damascus	33°29′	36°14′	729
Kameshli	37°03′	41°33′	455
El Haseke	36°32′	40°44′	295
Deir ez Zor	35°20′	40°08′	212
Palmyra	34°33′	38°18′	404
Abu Kamal	34°25′	40°55′	182
Lebanon			
Tripoli—El Mina	34°27′	35°48′	10
Beirut	33°49′	35°29′	16
Cedres	34°15′	36°03′	1,915
Dahr el Baidar	33°49′	35°46′	1,524
Riyaq	33°52′	36°00′	921
Ksara	33°50′	35°53′	918
Marjayoun	33°22′	35°35′	773
Jordan			
Wadi Ziglab	32°31′	35°36′	−150
Wadi Yabis	32°24′	35°35′	−200
Deir Alla	32°12′	35°37′	−224
Shouneh South	31°54′	35°37′	−230
Wadi Husban	31°49′	35°39′	−185
Aqaba Airport	29°33′	35°00′	8
Amman Airport	31°59′	35°59′	766
Shoubak	30°30′	35°32′	1,365
H.4	32°30′	38°12′	686

TABLE I *(continued)*

Station	Lat.°N	Long.°E	Elev. (m)
Jordan			
Mafraq	32°22'	36°15'	686
Shoumary	31°48'	36°49'	533
Ma'an Airport	30°10'	35°47'	1,069
West Bank and Gaza Strip			
Gaza	31°30'	34°27'	16
Nablus	32°13'	35°16'	580
Jerusalem	31°52'	35°13'	755
El Aroub	31°36'	33°07'	960
Jericho	31°52'	35°30'	−276
Dead Sea north	31°47'	35°31'	−387
Israel			
Tel Aviv	32°06'	34°47'	49
Haifa	32°49'	35°00'	8
Lod	32°00'	34°54'	49
Beer Sheba	31°14'	34°47'	275
Dead Sea south	—	—	−390
Eilat	29°33'	34°57'	34
Arabian Peninsula			
Al Wajh	26°13'	36°27'	20
Tabuk	28°24'	36°35'	2,400
Jiddah	21°30'	39°12'	11
Jizan	12°54'	42°33'	
Hodeida	14°48'	42°25'	00
Kamaran Island	15°20'	42°37'	6
Medina	24°39'	39°39'	594
Hail	27°31'	41°44'	914
Taif	21°29'	40°32'	1,395
Sana	15°23'	44°12'	2,200
Taizz	13°35'	43°57'	1,100
Riyadh	24°42'	46°43'	594
Dhahran	26°17'	50°09'	72
Sharjah	25°20'	55°24'	5
Muscat	23°37'	58°35'	4
Misrah Island	20°41'	58°54'	16
Salalah	17°03'	54°06'	10
Riyan (Mukala)	14°39'	49°23'	25
Khormaksar (Aden)	12°50'	45°01'	6
Beihan	14°53'	45°42'	1,097
Dhala	13°42'	44°44'	1,395
Mukayras	14°00'	45°42'	2,041
Iraq			
Mosul	36°19'	43°09'	222
Kirkuk	35°28'	44°24'	330
Khanaqin	34°18'	45°26'	201
Rutbah	33°02'	40°17'	615
Habbaniya	33°22'	43°34'	43
Baghdad	33°20'	44°24'	34
Kut-al-Hai	32°30'	45°45'	14
Diwaniya	31°59'	44°59'	20
Nasiriyah	31°01'	46°14'	3
Basrah	30°34'	47°47'	2

TABLE I *(continued)*

Station	Lat.°N	Long.°E	Elev. (m)
Kuwait			
Kuwait	29°14′	47°59′	55
Iran			
Tabriz	38°08′	46°15′	1,362
Mashhad	36°16′	59°38′	985
Tehran	35°41′	51°19′	1,191
Abadan	30°22′	48°15′	3
Bushehr	28°59′	50°50′	14
Shiraz	29°36′	52°32′	1,491
Kerman	30°15′	56°58′	1,749
Esfahan	32°37′	51°40′	1,598

Note to Tables II–XII

Recording periods for these tables are as follows:

Turkey	1951–1965
Cyprus	1931–1965
Syria	1951–1969
Lebanon	1931–1970
Jordan	1952–1965
Israel	1952–1965
Arabian Peninsula	1941–1969, with interruptions; for the former Southern Arabian Protectorates the recording period is 1922–1955, with interruptions; for North and South Yemen 1948–1970, with interruptions.
Iraq	1941–1970
Kuwait	1956–1967
Iran	1963–1967

TABLE II

MEAN MINIMUM TEMPERATURE (°C)

	Jan.	Feb.	Mar.	Apr.	May	June	July	Aug.	Sep.	Oct.	Nov.	Dec.
Turkey												
Samsun	3.8	3.5	4.2	7.5	11.9	15.8	18.7	19.2	16.1	12.7	9.6	6.4
Trabzon	4.6	4.2	5.0	8.3	12.8	17.0	19.9	20.4	17.4	13.8	10.4	6.9
Florya (Ist.)	2.5	3.0	3.5	7.1	11.6	15.8	18.3	18.6	15.7	12.1	8.9	5.4
Goztepe (Ist.)	2.7	2.4	3.4	1.0	11.6	15.5	18.0	18.4	15.2	11.9	8.6	5.1
Erzurum	−12.9	−11.2	−7.4	0.2	5.0	8.1	11.8	12.0	7.7	2.7	−2.3	−8.8
Sivas	−8.1	−6.9	−3.2	2.1	6.1	8.5	10.4	10.5	7.0	3.3	−0.5	−4.9
Ankara	−3.7	−3.2	−0.3	4.5	9.4	12.5	15.2	15.4	11.1	6.6	2.7	−1.1
Izmir	5.3	5.3	6.6	9.9	14.2	18.3	21.0	21.0	17.4	13.7	10.5	7.1
Antalya	6.2	6.5	8.1	11.2	15.1	19.5	22.7	22.5	19.3	15.3	11.2	8.0
Adana	4.6	5.6	7.6	11.0	14.9	18.9	21.9	22.4	18.9	14.6	10.4	6.6
Diyarbakir	−2.6	−1.1	1.8	6.7	10.9	15.8	21.5	20.8	15.6	9.4	4.3	−0.4
Cyprus												
Kyrenia	8.8	9.0	9.6	11.9	15.7	19.5	22.2	22.5	20.6	17.4	14.2	10.9
Famagusta	6.4	6.3	7.4	10.5	14.7	18.9	21.8	21.8	19.1	15.1	11.6	8.4
Nicosia	5.3	5.4	7.4	10.0	14.2	18.4	21.2	21.1	18.2	14.3	10.4	7.2
Limassol	7.4	7.1	7.7	10.2	13.8	17.3	19.3	19.9	18.0	15.4	12.4	9.2
Syria												
Safita	7.4	7.1	9.3	12.8	16.0	18.8	20.8	21.7	19.8	17.6	14.2	9.7
Aleppo	1.8	2.4	4.7	8.4	12.6	17.4	20.0	20.1	16.5	11.6	6.2	3.6
Damascus	2.5	3.3	5.2	8.6	12.6	16.1	17.2	17.5	15.3	12.4	7.7	4.1
Kameshli	2.8	3.6	6.0	10.1	14.6	19.7	23.1	23.1	19.2	14.8	8.9	4.8
El Haseke	1.3	2.0	4.9	9.1	14.0	18.7	21.8	21.2	16.1	10.6	5.5	2.6
Deir ez Zor	2.5	3.9	6.7	11.6	16.8	21.9	25.1	24.8	20.2	14.1	7.6	3.5
Palmyra	2.6	3.7	6.7	11.2	15.6	19.6	21.1	21.4	18.5	13.9	7.8	3.9
Abu Kamal	2.3	3.2	6.6	11.4	16.3	20.5	23.2	22.8	18.0	13.4	7.3	3.4
Lebanon												
Tripoli–El Mina	9.3	9.7	11.1	13.6	17.0	19.9	22.0	22.7	21.4	18.4	14.9	11.0
Beirut	9.1	9.2	10.7	12.9	15.9	19.4	21.3	22.2	20.8	17.6	13.9	11.3
Cedres	−2.9	−3.5	−1.7	2.7	6.9	9.5	11.8	12.7	10.4	7.5	3.8	0.2
Dahr el Baidar	−0.8	0.0	1.4	4.5	6.7	13.8	15.4	17.4	15.1	11.7	8.7	2.8
Riyaq	0.4	1.1	2.6	5.3	8.4	11.3	13.3	13.9	11.9	9.0	5.4	2.1
Ksara	1.9	2.0	4.0	7.0	10.4	13.1	15.3	15.6	13.0	10.0	6.5	3.1
Marjayoun	5.9	6.1	7.9	10.7	14.1	16.6	18.2	19.1	17.8	16.1	12.3	7.9

TABLE II *(continued)*

	Jan.	Feb.	Mar.	Apr.	May	June	July	Aug.	Sep.	Oct.	Nov.	Dec.
Jordan												
Wadi Ziglab	9.8	10.3	11.9	14.7	19.7	22.2	24.3	25.2	24.2	20.6	15.3	11.9
Wadi Yabis	6.8	8.4	9.2	12.5	17.5	19.0	21.5	23.0	21.2	16.2	12.3	8.8
Deir Alla	10.1	10.6	12.2	14.7	17.6	20.9	22.4	23.6	22.4	20.2	16.1	12.6
Shouneh South	11.5	12.2	14.3	17.6	21.9	24.3	26.2	26.5	25.7	22.8	17.9	13.9
Wadi Husban	10.5	11.1	12.6	15.8	19.3	21.5	23.2	23.9	23.9	21.0	17.7	13.4
Aqaba airport	10.0	10.3	14.0	17.5	21.6	24.0	25.5	25.4	24.3	20.8	16.4	12.0
Amman airport	3.7	4.3	6.2	9.3	13.4	16.3	18.1	18.4	16.2	13.7	9.6	5.4
Shoubak	0.3	0.3	2.7	5.4	7.3	10.1	12.6	13.5	10.0	8.4	5.2	1.4
H.4	2.9	4.3	7.5	10.7	15.6	17.8	19.1	19.8	17.3	13.3	7.8	4.3
Mafraq	2.2	2.7	5.4	8.3	11.8	13.8	15.3	16.0	14.3	11.4	7.0	3.7
Shoumary	2.9	4.4	6.9	10.7	14.2	18.5	18.4	19.0	17.0	11.4	8.1	4.7
Ma'an airport	2.7	3.6	6.7	10.1	14.0	16.5	17.6	18.3	16.4	12.8	8.3	4.2
West Bank and Gaza Strip												
Gaza	9.4	10.0	11.6	13.8	16.4	19.5	21.4	22.2	20.5	17.7	14.5	11.6
Jerusalem	4.6	4.8	6.6	9.1	12.4	15.5	17.2	17.8	16.5	14.1	10.4	6.6
El Aroub	4.4	4.8	6.3	8.1	12.3	14.7	15.9	16.2	14.4	12.1	9.6	6.4
Jericho	9.3	10.0	12.0	15.9	20.0	22.4	24.0	24.8	23.6	20.2	15.0	11.2
Dead Sea north	10.2	9.9	14.1	17.0	21.5	23.0	25.0	25.8	24.4	18.9	16.3	12.3
Israel												
Tel Aviv	9.3	9.9	11.2	14.0	16.6	18.7	22.2	23.3	21.2	17.8	14.1	10.3
Haifa	8.9	10.0	10.9	13.5	17.3	20.5	22.8	23.6	22.3	19.3	14.9	11.0
Beersheba	5.4	6.1	7.4	8.7	13.1	15.8	17.6	17.8	16.5	14.2	10.9	7.4
Dead Sea south	11.4	13.3	15.6	18.5	24.1	26.7	28.6	29.3	27.1	23.9	17.8	12.3
Arabian Peninsula												
Al Wajh	9.8	10.3	12.0	14.1	17.0	20.5	22.2	22.6	20.6	18.3	14.2	10.8
Tabuk	−1.1	0.0	2.7	4.8	11.7	16.8	18.2	19.2	15.5	8.9	5.1	1.8
Jiddah	15.8	15.9	17.1	18.3	21.1	23.4	24.7	25.7	23.5	21.7	19.8	17.6
Jizan	19.8	20.2	20.1	21.6	24.4	27.0	26.6	27.4	25.3	24.2	20.7	19.3
Hodeida	20.3	21.7	23.1	25.6	26.7	28.3	28.4	28.2	27.2	25.0	22.6	20.0
Kamaran Island	23.3	23.0	25.0	26.1	27.8	28.9	29.4	29.4	28.9	27.8	25.5	23.7
Medina	7.6	9.8	12.4	15.0	20.2	25.0	24.6	26.2	23.8	19.0	13.1	9.3
Hail	3.9	3.9	7.8	11.1	17.2	21.1	22.8	22.8	19.4	16.1	12.2	5.6
Taif	4.2	5.3	6.8	8.9	13.3	17.2	17.9	18.4	16.0	10.6	8.1	4.5
Sana	3.0	6.0	6.6	9.5	9.5	10.9	12.2	12.0	9.0	4.3	3.5	2.4
Taizz	14.9	16.7	18.2	19.7	20.1	20.8	21.5	20.7	19.5	18.3	16.9	14.8
Riyadh	3.0	3.8	8.4	12.0	17.5	21.4	23.3	23.7	19.4	14.3	8.5	3.3
Dhahran	5.9	7.8	11.2	15.3	19.7	23.9	25.4	25.6	21.9	17.9	10.9	7.1
Sharjah	12.2	13.9	15.5	18.3	22.2	25.0	27.8	27.8	25.0	21.6	17.8	14.4
Muscat	18.9	19.4	22.2	25.5	30.0	31.1	30.5	28.9	28.3	26.6	22.8	20.0
Misrah Island	23.9	23.9	25.0	26.1	28.3	29.4	30.0	30.0	28.9	27.2	25.5	24.4
Salalah	17.8	18.9	20.5	22.8	25.0	26.1	23.9	23.3	23.3	20.5	20.0	19.4
Riyan (Mukala)	19.6	20.0	21.3	23.5	25.3	26.8	25.3	24.7	25.8	22.7	21.2	19.6
Khormaksar (Aden)	22.2	23.2	23.9	25.1	27.7	28.9	28.5	27.9	28.1	24.7	22.8	22.6
Beihan	8.3	11.0	14.8	16.5	18.2	19.2	19.3	21.2	18.1	12.9	9.6	7.3
Dhala	10.6	11.4	12.7	14.8	17.1	18.9	18.4	16.7	16.1	13.1	12.8	10.3
Mukayras	3.4	5.6	7.3	9.6	11.0	12.8	14.9	14.1	11.8	8.1	6.1	4.5
Iraq												
Mosul	2.5	3.5	6.3	10.2	15.0	19.5	22.9	21.8	16.6	11.4	7.0	3.3
Kirkuk	4.5	5.5	8.5	13.0	18.8	23.7	26.7	26.5	22.4	17.2	11.3	6.8

TABLE II *(continued)*

	Jan.	Feb.	Mar.	Apr.	May	June	July	Aug.	Sep.	Oct.	Nov.	Dec.
Iraq												
Khanaqin	4.6	5.7	8.5	13.6	18.8	22.4	25.3	24.8	20.7	15.9	10.5	5.8
Rutbah	1.7	2.9	6.0	10.7	15.8	19.3	21.7	21.4	17.6	12.9	7.1	3.2
Habbaniya	4.4	6.0	9.5	14.9	20.1	23.7	25.8	25.0	21.2	16.2	10.5	5.7
Baghdad	4.3	5.9	9.6	14.6	20.0	23.4	25.3	24.6	21.0	16.2	10.3	5.5
Kut-al-Hai	5.5	6.9	10.7	15.6	21.0	24.4	26.4	25.9	22.5	17.4	11.8	7.0
Diwaniya	4.1	5.8	9.6	14.5	20.0	23.1	24.3	23.6	20.3	15.8	10.4	6.4
Nasiriyah	5.9	7.7	11.5	16.6	22.4	25.3	26.1	25.2	22.0	17.1	12.0	7.5
Basrah	7.0	8.7	12.6	18.0	23.7	26.9	27.7	26.3	22.6	18.3	13.2	8.0
Kuwait												
Kuwait	8.4	10.2	13.8	18.4	24.2	27.8	29.6	29.0	25.5	20.7	14.5	9.4
Iran												
Tabriz	−4.9	−5.5	−0.1	3.6	10.3	14.7	18.6	19.9	13.7	7.5	2.0	−3.4
Mashhad	−4.3	−0.8	2.0	7.1	12.1	16.1	18.2	15.6	12.4	7.0	1.9	−2.9
Tehran	−0.3	0.6	4.0	8.9	15.7	19.8	22.7	22.8	18.0	11.9	6.3	1.7
Abadan	7.5	8.3	11.7	16.0	23.2	25.7	27.2	27.6	23.7	20.1	13.6	7.5
Bushehr	10.4	10.9	13.1	16.6	22.6	25.6	27.2	27.2	24.3	21.4	22.5	10.4
Shiraz	−0.1	1.3	3.4	6.7	12.6	15.5	19.4	17.8	13.8	9.1	2.7	1.8
Kerman	−2.7	0.4	2.5	7.4	13.4	15.8	17.5	13.2	10.0	7.0	2.1	−6.1
Esfahan	−1.8	−0.1	3.3	7.5	3.4	18.2	21.1	18.9	15.2	9.7	3.7	−1.8

TABLE III

MEAN DAILY TEMPERATURE (°C)

	Jan.	Feb.	Mar.	Apr.	May	June	July	Aug.	Sep.	Oct.	Nov.	Dec.
Turkey												
Samsun	6.8	6.7	7.6	10.8	15.5	20.0	23.0	23.3	19.8	16.2	12.8	9.4
Trabzon	7.2	7.0	7.9	11.3	15.7	20.0	22.7	23.2	20.1	16.5	13.1	9.6
Florya (Ist.)	5.2	5.4	6.6	10.7	15.6	20.6	23.2	23.3	19.6	15.4	11.7	8.0
Goztepe (Ist.)	5.5	5.3	6.9	11.3	16.2	20.7	23.2	23.4	19.6	15.6	11.8	8.0
Erzurum	−8.6	−6.9	−3.1	5.0	10.9	15.0	19.1	19.6	15.0	7.9	1.8	−5.2
Sivas	−4.0	−2.7	1.7	8.3	13.2	16.6	19.4	19.7	15.5	10.5	4.7	−1.2
Ankara	−0.1	0.9	5.0	11.1	16.0	20.0	23.2	23.3	18.4	12.9	7.7	2.5
Izmir	8.5	9.0	11.1	15.4	20.2	24.9	27.6	27.3	23.1	18.5	14.3	10.5
Antalya	10.0	10.6	12.7	16.3	20.4	25.0	28.2	28.1	24.9	20.3	15.5	11.8
Adana	9.2	10.2	12.9	16.9	21.2	25.1	27.6	28.1	25.2	20.8	15.7	11.1
Diyarbakir	1.6	3.7	8.0	13.8	19.3	25.8	31.0	30.5	24.9	17.2	10.0	4.1
Cyprus												
Kyrenia	12.6	12.9	14.1	16.9	20.7	24.7	27.4	27.9	25.6	22.2	18.4	14.6
Famagusta	11.4	11.6	13.1	16.7	21.2	25.8	28.0	28.1	25.7	21.8	17.3	13.4
Nicosia	10.2	10.7	12.7	17.1	21.7	26.0	28.8	28.9	25.7	21.3	16.4	12.1
Limassol	12.1	12.2	13.3	16.5	20.3	23.9	26.1	26.6	24.6	21.7	17.8	14.1
Syria												
Safita	10.3	11.2	13.0	16.7	20.4	23.4	25.0	25.8	24.2	21.9	17.8	12.7
Aleppo	6.1	7.6	10.7	15.4	20.6	25.6	28.2	28.4	24.8	19.5	12.7	7.9
Damascus	7.4	8.7	11.5	15.7	20.5	24.9	26.5	26.8	23.9	19.7	13.7	9.1
Kameshli	6.8	8.1	11.1	15.9	21.7	27.9	31.7	31.7	27.3	21.7	14.2	9.0
El Haseke	6.2	7.7	11.5	16.3	22.3	27.7	31.1	30.7	25.8	19.5	12.9	8.1
Deir ez Zor	7.7	9.7	13.1	18.5	24.1	29.7	32.7	32.5	27.9	21.9	14.3	9.1
Palmyra	7.5	9.4	13.1	18.1	22.9	27.7	29.5	29.8	26.3	21.1	14.1	9.1
Abu Kamal	8.1	9.7	13.9	18.9	24.5	29.3	31.8	31.5	27.1	21.5	14.8	9.8
Lebanon												
Tripoli–El Mina	12.8	13.4	14.9	17.4	21.1	24.0	26.0	26.8	25.5	22.6	18.7	14.6
Beirut	12.7	13.3	15.1	17.5	20.5	23.9	25.5	26.1	24.9	22.9	18.3	15.1
Cedres	0.5	0.4	2.2	6.9	11.3	14.1	17.0	17.9	15.3	12.0	7.6	2.8
Dahr el Baidar	2.0	2.8	4.8	8.1	10.8	17.4	18.8	21.0	18.8	15.0	12.7	5.4
Riyaq	5.1	6.2	8.8	12.8	17.3	21.4	23.8	24.0	21.3	17.3	11.9	6.9
Ksara	6.8	7.0	9.7	13.9	18.0	21.3	23.8	24.2	21.3	17.8	11.4	8.2
Marjayoun	9.0	8.9	11.2	14.5	18.3	21.0	22.5	23.4	21.9	20.0	15.5	10.7
Jordan												
Wadi Ziglab	13.5	14.2	16.8	20.8	26.0	28.7	30.2	31.0	29.4	26.3	21.1	15.8
Wadi Yabis	12.8	13.9	16.3	20.5	26.0	28.1	30.3	30.7	29.2	24.8	20.3	15.1
Deir Alla	14.6	15.5	17.9	21.8	26.3	29.0	30.5	31.1	29.8	26.6	22.3	16.9
Shouneh South	15.1	15.9	18.8	23.3	27.8	30.6	31.9	32.7	30.7	27.7	22.3	17.4
Wadi Husban	14.5	15.6	17.6	22.2	26.3	28.8	30.2	30.9	29.6	26.9	22.5	17.3
Aqaba Airport	15.6	17.0	20.1	24.3	28.4	31.8	32.5	33.0	30.4	27.1	22.1	17.2
Amman Airport	8.1	9.0	11.8	16.0	20.7	23.7	25.1	25.6	23.5	20.6	15.3	10.0
Shoubak	4.7	5.2	8.5	12.7	14.5	18.1	20.7	20.7	17.7	15.3	11.1	6.8
H.4	8.6	10.5	14.2	18.5	23.8	26.7	28.7	28.8	25.6	21.4	15.6	10.7
Mafraq	7.4	8.7	11.6	15.9	20.2	22.7	24.0	24.4	22.2	19.3	14.0	9.3
Shoumary	9.1	10.2	13.8	18.5	22.7	27.2	27.9	27.7	25.2	21.0	15.9	10.5
Ma'an Airport	8.5	9.7	13.0	17.2	21.5	24.6	25.9	26.4	23.8	20.2	14.6	10.2
West Bank and Gaza Strip												
Gaza	13.7	14.0	16.0	18.2	21.0	24.0	25.5	26.5	25.1	22.5	19.3	15.9
Jerusalem	8.3	8.8	11.3	15.0	19.3	21.9	23.2	23.7	22.3	19.8	15.4	10.6
El Aroub	8.0	8.5	11.2	14.9	19.0	21.3	22.6	23.2	21.5	19.2	14.7	10.2
Jericho	14.2	15.2	18.1	22.7	27.1	29.9	31.0	31.3	29.7	26.8	21.7	16.7
Dead Sea North	15.2	15.6	19.1	22.3	27.5	29.9	31.2	31.6	29.6	25.6	22.4	17.1

TABLE III *(continued)*

	Jan.	Feb.	Mar.	Apr.	May	June	July	Aug.	Sep.	Oct.	Nov.	Dec.
Israel												
Tel Aviv	13.8	14.6	17.0	19.3	21.6	23.5	26.7	27.7	26.1	23.5	19.3	15.0
Haifa	13.6	14.6	16.4	19.1	22.7	25.1	26.9	27.7	26.7	24.2	20.2	15.6
Lod	12.3	12.7	14.3	17.4	21.1	23.9	25.1	26.2	24.6	22.0	18.0	14.3
Beersheba	12.0	12.6	15.5	18.4	22.7	24.8	26.1	26.3	25.0	22.9	18.7	14.0
Eilat	15.8	17.5	20.3	24.5	28.9	32.0	33.5	33.7	30.7	27.4	21.6	17.0
Dead Sea South	16.0	17.1	20.0	22.8	29.0	32.4	33.6	34.3	32.0	28.6	22.6	16.9
Arabian Peninsula												
Al Wajh	19.1	20.9	21.4	23.8	27.0	31.5	28.0	29.2	27.5	26.2	24.0	19.9
Tabuk	12.2	14.1	17.7	20.5	25.5	28.0	29.6	31.0	28.0	22.2	17.5	14.5
Jiddah	23.5	23.9	25.2	26.9	32.1	32.1	32.3	32.4	31.0	28.2	29.2	24.9
Jizan	25.1	26.2	27.0	28.7	32.0	32.6	32.2	32.8	32.8	30.6	27.8	25.6
Hodeida	24.5	25.4	26.7	29.1	30.5	32.1	32.5	32.3	31.4	29.3	27.0	24.5
Kamaran Island	25.6	25.8	27.5	28.9	31.4	32.5	33.1	32.8	32.5	30.8	28.1	26.1
Medina	18.0	20.0	23.6	26.0	31.1	34.3	34.5	35.3	33.4	28.5	23.2	19.1
Hail	12.2	12.9	16.7	20.1	25.5	28.8	30.1	30.6	27.6	23.4	18.5	12.3
Taif	15.3	16.7	19.3	19.9	24.7	27.3	28.1	28.0	26.0	22.2	18.0	15.7
Sana	12.5	15.0	16.0	16.9	17.8	19.2	20.0	19.3	17.6	14.1	12.8	11.6
Taizz	20.9	22.4	24.4	25.5	26.5	27.1	27.0	26.4	25.3	24.6	22.9	20.5
Riyadh	15.1	16.7	21.5	25.0	29.0	32.9	34.4	33.6	30.8	26.0	21.3	15.6
Dhahran	16.0	19.0	22.7	26.1	31.4	34.7	35.5	35.7	33.6	28.7	23.2	17.5
Sharjah	17.8	18.9	21.1	24.1	28.0	30.5	32.8	33.6	31.1	27.5	24.1	20.0
Muscat	22.0	22.2	25.3	28.9	33.3	34.5	33.3	31.1	31.1	30.3	26.7	23.1
Misrah Island	21.9	22.5	22.5	28.3	30.5	30.0	27.2	26.6	26.1	26.6	25.0	23.3
Salalah	22.5	23.4	25.3	27.0	28.6	28.9	25.8	25.3	26.0	25.6	25.0	23.9
Riyan (Mukala)	23.5	24.1	25.3	27.4	29.1	30.6	29.3	28.7	28.9	26.8	25.7	24.1
Khormaksar (Aden)	25.1	25.9	27.0	28.4	30.9	32.8	32.5	31.8	31.9	28.9	26.5	25.6
Beihan	17.4	20.4	23.8	25.3	27.0	28.4	28.3	29.9	26.3	22.7	19.5	16.3
Dhala	17.7	18.1	19.8	21.8	24.4	26.2	25.7	23.9	23.3	21.9	20.2	17.5
Mukayras	10.4	12.4	14.2	16.0	18.1	19.8	20.3	19.1	17.9	15.4	13.0	11.3
Iraq												
Mosul	6.9	8.8	12.4	17.5	24.1	30.4	33.9	32.7	27.2	20.3	13.3	8.2
Kirkuk	8.7	10.3	13.6	19.4	26.4	32.4	35.3	34.9	30.5	24.2	16.6	10.7
Khanaqin	10.0	12.5	15.3	21.1	28.3	33.6	36.4	35.7	30.2	25.2	17.6	11.9
Rutbah	7.0	9.1	12.8	18.4	23.8	27.4	30.4	30.2	26.7	21.2	14.0	8.6
Habbaniya	9.5	12.0	16.1	21.7	28.1	32.7	34.8	34.2	30.0	24.1	16.8	11.0
Baghdad	10.0	12.3	16.1	22.1	28.4	33.0	34.8	34.4	30.6	24.6	17.1	11.3
Kut-al-Hai	11.3	13.3	17.3	22.9	30.8	34.3	35.6	35.3	31.8	26.3	18.1	12.7
Diwaniya	10.6	13.0	16.9	22.8	29.6	32.3	34.2	33.8	31.5	25.7	17.7	12.3
Nasiriyah	11.8	13.9	17.5	23.7	29.7	32.9	34.4	34.4	31.6	26.1	18.8	12.8
Basrah	12.2	14.1	18.6	24.0	29.4	32.6	33.9	33.5	31.1	25.9	19.3	13.7
Kuwait												
Kuwait	13.6	15.6	19.7	24.7	30.9	36.5	37.1	36.1	33.4	27.9	19.9	14.3
Iran												
Tabriz	2.3	1.2	5.3	9.7	16.2	21.9	25.7	26.1	20.9	13.5	8.6	1.1
Mashhad	1.3	5.0	8.7	13.6	19.1	24.2	26.2	24.4	19.7	19.2	8.7	2.6
Tehran	4.5	6.5	10.4	15.1	22.1	25.1	30.0	29.5	24.8	17.8	11.2	6.3
Abadan	12.7	15.1	19.0	23.8	30.2	34.6	35.9	35.8	32.5	24.1	18.9	13.2
Bushehr	14.4	16.5	19.3	23.1	28.2	30.7	32.1	31.9	29.9	26.3	20.7	15.1
Shiraz	5.8	8.3	11.7	14.9	21.2	25.7	26.3	27.5	23.4	17.8	11.3	5.8
Kerman	3.1	8.1	11.3	15.5	21.1	24.4	26.9	24.3	20.4	15.6	8.9	3.5
Esfahan	3.6	6.9	10.7	14.9	20.8	26.3	29.4	27.7	23.2	16.3	10.2	3.6

TABLE IV

MEAN MAXIMUM TEMPERATURE (°C)

	Jan.	Feb.	Mar.	Apr.	May	June	July	Aug.	Sep.	Oct.	Nov.	Dec.
Turkey												
Samsun	10.4	10.6	11.8	15.2	19.1	23.3	26.3	26.9	23.7	20.5	17.0	13.2
Trabzon	10.5	10.5	11.5	14.8	19.0	22.8	25.8	26.3	23.3	19.9	16.6	13.0
Florya (Ist.)	8.0	8.4	10.4	15.4	20.5	25.5	28.7	28.8	25.0	19.8	15.2	10.8
Goztepe (Ist.)	8.5	8.7	11.0	16.2	21.3	25.9	28.5	28.9	25.0	20.4	15.6	12.2
Erzurum	−4.0	−2.3	1.4	9.9	16.6	21.1	25.8	26.6	22.0	14.8	6.8	0.7
Sivas	0.3	1.7	6.9	14.5	19.7	23.6	27.2	28.1	23.9	18.2	10.8	3.3
Ankara	4.1	5.4	10.7	17.4	22.4	26.8	30.1	30.4	25.7	20.1	13.4	6.4
Izmir	12.1	13.1	15.9	20.8	25.8	30.4	33.1	33.2	29.1	24.2	18.8	14.2
Antalya	14.9	15.4	17.7	21.0	24.9	30.0	33.7	33.7	30.8	26.4	21.5	16.9
Adana	14.2	15.7	19.0	23.3	28.1	31.9	33.9	35.0	33.0	28.9	22.9	16.8
Diyarbakir	6.5	9.1	14.1	20.4	26.5	33.3	38.2	38.3	33.2	25.4	16.7	9.3
Cyprus												
Kyrenia	16.4	16.8	18.6	21.8	25.7	29.9	32.6	33.2	30.7	27.1	22.6	18.2
Syria												
Safita	13.2	15.4	16.8	20.7	24.9	28.0	29.1	30.0	28.7	26.2	21.5	15.6
Aleppo	10.4	12.8	16.6	22.4	28.6	33.8	36.4	36.7	33.1	27.3	19.1	12.1
Damascus	12.3	14.1	17.8	22.8	28.5	33.6	35.8	36.1	32.4	27.1	19.8	14.0
Kameshli	10.8	12.7	16.1	21.8	28.9	36.0	40.3	40.3	35.4	28.5	19.5	13.2
El Haseke	11.1	13.5	18.0	23.6	30.7	36.7	40.3	40.1	35.5	28.5	20.4	13.7
Deir ez Zor	12.8	15.5	19.4	25.4	31.5	37.4	40.4	40.3	35.7	29.6	21.1	14.7
Palmyra	12.4	15.1	19.5	25.1	30.3	35.7	38.0	38.2	34.1	28.3	20.5	14.4
Abu Kamal	14.0	16.3	21.1	26.5	32.8	38.0	40.4	40.3	36.1	29.7	22.3	16.2
Lebanon												
Tripoli—El Mina	16.4	16.8	18.6	21.1	24.4	27.1	29.0	29.8	29.1	26.7	23.3	18.7
Beirut	16.7	17.6	19.6	22.2	24.6	27.9	29.1	29.8	28.8	26.5	23.0	19.4
Cedres	4.0	4.3	6.2	11.1	15.8	18.7	22.3	23.1	20.3	16.6	11.3	7.3
Dahr el Baidar	4.9	5.7	8.3	11.7	15.0	21.1	22.3	24.7	22.6	18.4	15.4	8.0
Riyaq	10.1	11.4	15.0	19.9	25.2	29.7	32.3	33.0	30.1	25.6	19.5	12.9
Ksara	11.6	12.1	15.3	20.8	25.7	29.4	32.2	32.7	29.6	25.5	19.3	13.4
Marjayoun	12.2	11.8	14.4	18.4	22.6	25.5	26.8	27.8	26.0	23.9	18.8	13.4
Jordan												
Wadi Ziglab	17.2	18.3	22.0	26.7	32.3	35.5	35.5	35.9	34.8	31.9	26.2	19.4
Wadi Yabis	18.7	20.0	24.2	29.1	34.2	36.8	38.4	38.3	36.5	33.1	27.5	21.0
Deir Alla	18.9	19.9	23.7	28.7	34.1	37.0	38.3	39.2	36.8	33.3	27.7	21.2
Shouneh South	18.5	19.4	23.8	29.0	34.2	36.7	38.1	38.5	35.8	32.4	36.9	21.2
Wadi Husban	18.7	20.1	23.0	28.7	33.5	36.5	37.8	37.3	35.4	32.0	36.9	21.2
Aqaba Airport	21.2	22.9	26.5	30.8	35.3	38.8	40.0	40.6	36.3	33.2	28.2	23.0
Amman Airport	12.5	13.7	17.6	22.6	28.0	31.0	32.1	32.8	30.9	27.5	21.0	14.8
Shoubak	9.1	10.3	14.3	19.0	22.5	26.5	27.5	28.3	25.9	22.7	17.1	12.2
H.4	14.1	16.6	21.0	25.7	32.0	35.6	38.1	38.0	34.5	29.2	22.4	16.9
Mafraq	12.3	13.9	18.8	23.2	28.6	31.6	32.8	33.2	30.9	27.3	21.1	14.8
Shoumary	15.0	16.4	20.5	25.8	31.4	35.6	37.0	37.1	34.3	29.8	23.4	16.9
Ma'an Airport	13.8	15.4	19.2	23.8	28.7	32.3	34.2	34.0	32.0	27.7	21.0	16.0
West Bank and Gaza Strip												
Gaza	17.9	17.9	20.1	22.0	24.8	27.1	28.6	29.7	28.2	26.9	24.0	20.4
Jerusalem	11.9	12.8	16.3	20.6	25.7	28.2	29.1	29.7	28.0	25.4	20.1	14.1
El Aroub	12.3	13.0	16.5	20.9	25.7	28.5	29.6	30.0	28.4	25.7	20.4	14.7
Jericho	19.0	20.6	24.4	29.5	34.4	37.0	38.6	37.9	35.8	32.7	28.1	21.4
Dead Sea North	19.6	21.1	24.4	28.4	33.9	36.5	38.2	38.0	35.7	32.0	28.3	22.4

TABLE IV *(continued)*

	Jan.	Feb.	Mar.	Apr.	May	June	July	Aug.	Sep.	Oct.	Nov.	Dec.
Israel												
Tel Aviv	18.3	19.4	22.9	24.7	26.7	28.3	31.3	32.1	31.1	29.1	24.6	19.7
Haifa	18.4	19.2	21.9	24.7	28.1	29.7	31.0	31.9	31.2	29.3	25.5	20.3
Beersheba	15.8	19.1	23.6	28.1	32.3	33.9	34.7	34.8	33.6	31.6	26.5	20.7
Dead Sea South	20.6	20.9	24.5	27.1	34.0	38.2	38.6	39.4	37.0	33.3	27.4	21.6
Arabian Peninsula												
Al Wajh	28.3	29.5	30.5	33.3	37.0	42.3	33.8	35.8	34.2	34.0	33.7	30.5
Tabuk	25.6	28.7	32.7	36.3	39.2	41.0	42.7	42.3	40.5	35.4	29.9	27.1
Jiddah	31.5	31.9	33.3	35.7	38.8	40.7	39.5	39.1	38.7	37.1	35.1	32.3
Jizan	30.5	32.2	33.8	36.0	39.0	38.2	39.4	38.3	38.2	36.9	34.9	31.7
Hodeida	28.8	29.2	30.4	32.6	34.3	35.9	36.7	36.4	35.7	33.7	31.4	29.1
Kamaran Island	27.8	28.3	30.0	31.6	35.0	36.1	36.6	36.1	36.1	33.9	30.5	28.3
Medina	28.4	30.1	34.8	37.8	42.0	43.9	44.3	44.3	43.1	39.1	33.2	28.8
Hail	16.5	18.9	23.9	27.8	33.4	38.4	38.4	38.9	36.6	32.8	23.9	17.3
Taif	26.6	28.1	31.5	32.2	36.1	37.2	38.2	37.6	32.2	32.7	28.0	25.5
Sana	22.1	23.9	25.4	24.4	26.2	27.5	27.8	26.7	26.3	24.0	22.1	20.8
Taizz	27.0	28.2	30.6	31.4	32.9	33.5	32.6	32.1	32.1	30.9	28.9	26.2
Riyadh	27.5	30.0	35.0	38.2	42.2	44.1	44.8	45.0	42.9	38.1	32.6	28.1
Dhahran	29.5	28.0	35.1	36.8	43.1	45.3	45.2	45.6	44.2	39.5	33.4	27.7
Sharjah	23.3	23.9	26.6	30.0	33.9	36.1	37.8	39.4	37.2	33.3	30.5	25.5
Muscat	25.0	25.0	28.9	32.2	36.6	37.8	36.1	33.3	33.9	33.9	30.0	26.1
Misrah Island	25.5	26.1	29.4	33.3	35.5	34.4	31.1	30.0	30.0	31.1	28.9	26.6
Salalah	27.2	27.8	30.0	31.1	32.2	31.6	27.8	27.2	28.9	30.5	30.0	28.3
Riyan (Mukala)	27.5	28.1	29.2	31.3	32.9	34.4	33.4	32.7	32.1	30.9	30.2	28.5
Khormaksar (Aden)	28.1	28.6	30.1	31.7	34.1	36.7	36.4	35.7	35.7	33.1	30.3	28.6
Beihan	26.5	29.8	32.9	34.1	35.8	37.6	37.4	38.7	35.5	32.4	29.4	25.4
Dhala	24.8	24.8	26.9	28.9	31.8	33.5	33.1	31.1	30.6	30.7	27.6	25.2
Mukayras	17.5	19.2	21.1	22.4	25.2	26.9	25.7	24.1	24.1	22.7	19.9	18.1
Iraq												
Mosul	12.8	15.3	19.0	25.4	32.9	39.6	43.4	43.0	38.7	31.2	22.3	15.0
Kirkuk	13.7	15.5	19.3	25.5	32.9	39.2	42.8	42.5	38.3	31.6	23.0	16.1
Khanaqin	15.7	17.6	21.1	27.6	36.2	39.3	43.7	43.3	39.3	32.9	24.6	17.9
Rutbah	13.8	16.1	19.8	25.4	31.5	36.1	38.5	38.7	35.5	29.6	21.6	15.4
Habbaniya	15.6	18.6	23.0	28.9	36.0	40.9	43.4	43.3	39.5	33.0	24.2	17.6
Baghdad	15.8	18.7	22.7	28.7	35.8	41.0	43.4	43.3	39.8	33.4	24.6	17.7
Kut-al-Hai	17.5	20.0	24.0	29.8	36.9	41.5	43.5	43.8	40.8	35.0	26.2	19.2
Diwaniya	17.1	19.7	24.4	30.1	34.8	41.0	42.7	42.9	40.1	34.5	25.6	18.6
Nasiriyah	17.8	20.4	24.9	30.7	36.9	40.2	42.8	43.6	41.3	35.4	26.0	18.9
Basrah	18.6	21.0	25.3	30.8	36.1	38.8	40.5	41.3	39.7	35.0	26.7	20.0
Kuwait												
Kuwait	18.5	20.9	25.6	30.9	37.7	43.3	44.6	43.3	41.4	35.3	26.4	20.1
Iran												
Tabriz	3.3	3.4	10.3	15.0	23.6	28.8	32.3	33.9	27.5	18.2	13.3	6.9
Mashhad	9.2	10.1	14.1	19.6	26.6	31.9	33.8	33.0	28.1	21.2	16.3	11.6
Tehran	9.4	11.0	15.8	20.9	30.2	34.2	36.6	36.3	31.0	23.2	16.8	11.9
Abadan	19.5	19.7	25.4	30.7	39.0	42.9	44.1	44.5	42.0	35.1	27.0	20.6
Bushehr	19.3	19.6	23.4	27.4	34.4	35.4	36.9	36.9	36.5	33.1	26.6	20.9
Shiraz	12.9	14.9	19.5	21.9	31.1	35.3	37.4	36.1	33.4	27.5	20.3	15.1
Kerman	13.8	14.8	18.5	21.7	29.9	33.7	35.5	33.4	31.2	25.8	19.1	14.5
Esfahan	10.1	12.1	16.8	20.9	28.7	33.9	36.8	35.4	31.3	23.8	17.4	11.7

TABLE V

TOTAL AMOUNT OF RAINFALL (mm)

	Jan.	Feb.	Mar.	Apr.	May	June	July	Aug.	Sep.	Oct.	Nov.	Dec.	Year
Turkey													
Samsun	80.6	72.5	73.5	54.2	43.7	39.4	35.9	29.5	56.8	70.3	83.6	79.4	719.4
Trabzon	90.4	69.6	59.9	56.4	52.7	50.7	36.9	46.3	78.3	109.3	100.6	79.8	830.8
Florya (Ist.)	84.2	75.0	63.5	40.6	29.5	22.9	21.0	19.8	42.9	62.3	86.3	103.0	651.1
Goztepe (Ist.)	83.8	82.3	62.4	39.4	30.7	23.7	24.3	18.8	51.3	66.5	85.0	104.3	672.5
Erzurum	26.1	31.8	41.3	56.8	75.6	55.0	30.4	18.9	26.8	47.3	36.9	23.9	470.8
Sivas	42.4	42.2	42.1	53.3	58.3	32.6	7.2	4.8	17.3	30.3	38.9	42.1	411.4
Ankara	34.9	38.2	35.9	33.6	50.0	30.6	12.7	8.4	18.6	22.0	27.9	45.7	358.5
Izmir	135.8	107.1	72.3	45.4	38.0	8.8	2.6	2.4	15.8	48.7	85.1	142.3	704.4
Antalya	246.3	159.8	89.1	42.6	32.1	11.0	2.0	2.6	13.2	51.0	103.8	277.0	1030.5
Adana	102.3	103.9	65.4	46.3	51.2	20.9	4.2	5.1	15.1	39.2	57.4	113.9	625.0
Diyarbakir	75.2	68.4	59.8	70.6	42.4	6.7	0.9	0.5	3.1	28.3	55.7	68.5	481.0
Cyprus													
Kyrenia	117	79	60	20	13	2	1	1	5	37	68	133	534
Famagusta	90	52	36	16	13	4	1	1	3	33	49	106	403
Nicosia	75.4	54.4	31.7	20.3	29.0	8.9	1.5	2.3	8.4	25.1	33.0	79.3	369.3
Limassol	111.3	79.8	40.9	19.8	10.4	3.3	NIL	NIL	1.5	26.9	48.0	115.1	457.0
Syria													
Safita	215.1	216.7	159.5	108.1	21.8	0.3	0.4	5.1	39.7	75.4	135.2	239.2	1216.5
Aleppo	67.6	54.4	42.4	36.0	19.3	2.6	0.1	0.9	1.4	18.7	25.3	72.1	340.8
Damascus	58.2	37.6	23.9	13.9	8.3	0.2	0.0	0.0	0.1	9.8	27.7	51.9	231.6
Kameshli	97.9	76.2	76.1	73.8	36.5	1.3	0.7	0.3	0.5	19.7	36.1	72.5	491.6
El Haseke	67.0	38.6	38.4	48.9	21.0	0.6	0.0	0.0	1.3	14.0	19.6	41.1	290.5
Deir ez Zor	34.4	27.2	24.8	21.9	9.5	0.7	0.0	0.0	0.4	7.3	11.4	28.8	166.4
Palmyra	24.4	13.6	13.2	16.2	10.7	0.7	0.0	0.0	0.1	8.9	14.2	20.1	122.1
Abu Kamal	22.7	13.8	9.1	18.7	10.2	0.0	0.0	0.0	0.1	8.0	5.7	16.7	96.0
Lebanon													
Tripoli-El Mina	214.1	160.1	110.1	54.3	18.4	0.7	0.0	1.5	14.9	64.9	129.1	186.6	954.7
Beirut	193.7	107.0	95.7	33.7	18.6	0.9	0.4	0.6	5.2	34.5	99.8	153.3	743.4
Cedres	239.0	207.0	145.0	68.0	32.0	5.0	0.5	0.5	5.0	32.0	98.0	158.0	990.0
Dahr el Baidar	281.0	258.0	222.0	86.0	33.0	3.0	0.5	0.5	7.0	38.0	134.0	237.0	1300.0
Riyaq	160.2	120.6	79.2	33.5	16.5	0.5	0.1	0.1	0.8	20.9	62.4	122.8	617.6
Ksara	156.0	138.0	71.0	41.0	14.0	1.0	0.5	0.5	1.0	19.0	62.0	126.0	630.0
Marjayoun	193.0	181.0	129.0	73.0	26.0	1.0	0.5	0.5	3.0	24.0	91.0	162.0	884.0
Jordan													
Wadi Ziglab	92.6	59.9	49.6	25.6	10.6	0.0	0.0	0.0	0.0	11.4	43.3	97.1	390.1
Deir Alla	63.4	59.4	37.5	19.0	5.9	0.0	0.0	0.0	0.4	5.2	38.4	62.2	291.4
Shouneh South	48.3	27.7	22.6	8.0	3.0	0.0	0.0	0.0	0.0	3.1	13.4	41.6	158.1
Aqaba Airport	3.4	4.6	4.4	4.7	0.1	trace	0.0	0.0	trace	0.4	3.1	8.2	28.9
Amman Airport	63.7	67.3	36.8	14.7	4.3	trace	trace	trace	0.5	5.1	30.0	48.5	271.9
Shounak	72.2	70.2	66.1	18.4	6.5	0.0	0.0	0.0	0.0	1.9	22.8	76.1	325.2
H.4	8.7	12.9	10.5	11.3	7.5	0.0	0.0	0.0	0.0	2.9	9.7	13.9	77.4
Mafraq	35.5	33.4	20.5	8.6	3.4	0.0	0.0	0.0	0.4	3.1	16.2	29.1	150.2
West Bank and Gaza Strip													
Gaza	79.9	53.0	32.3	8.5	3.0	0.1	0.0	trace	0.2	15.8	55.2	64.7	312.7
Nablus	164.1	136.7	90.2	23.9	5.9	0.0	0.0	0.0	0.0	18.2	76.8	136.6	652.4
Jerusalem	142.9	129.1	102.1	25.2	5.8	trace	0.0	0.0	0.5	9.2	74.6	137.7	627.1
El Aroub	151.8	140.9	105.1	34.0	6.4	0.0	0.0	0.0	0.0	9.6	78.2	114.7	640.7
Israel													
Haifa	181	144	24	18	3	1	1	0	0	13	69	171	625.0
Lod	131	93	58	15	4	0	0	0	2	18	79	129	529.0
Eilat	2	5	5	3	0.3	0	0	0	0	0.2	2	9	26.5

TABLE V *(continued)*

	Jan.	Feb.	Mar.	Apr.	May	June	July	Aug.	Sep.	Oct.	Nov.	Dec.	Year
Arabian Peninsula													
Al Wajh	3.8	5.4	0.0	3.8	7.0	0.0	0.0	0.0	0.0	0.0	10.9	6.6	37.5
Tabuk	15.7	0.2	10.2	1.1	8.8	0.0	0.0	8.2	6.6	1.9	18.4	5.0	76.1
Jiddah	14.3	3.2	2.6	12.2	2.0	0.5	0.6	0.1	0.0	0.0	13.3	14.4	63.2
Jizan	0.0	4.4	0.0	18.0	0.0	0.0	4.4	0.5	4.9	4.4	2.0	0.0	38.6
Hodei	0.1	17.6	4.7	4.5	7.3	2.8	18.2	20.5	4.8	3.0	7.9	9.6	101.0
Kamaran Island	2.1	2.1	2.5	2.5	2.5	2.5	12.6	17.7	2.5	2.5	10.1	22.8	82.4
Medina	7.8	0.7	4.4	4.6	4.1	0.3	0.1	0.1	0.0	0.5	9.8	4.0	35.5
Hail	9.0	8.4	6.4	11.1	7.0	0.0	0.0	0.0	0.0	10.9	21.9	2.6	77.3
Taif	10.0	7.4	9.5	31.3	18.7	6.6	4.6	11.3	5.6	8.0	48.5	4.8	166.3
Riyadh	20.4	9.8	16.7	20.8	14.3	0.0	0.1	0.0	0.0	0.0	11.5	11.6	105.2
Dhahran	28.1	19.1	4.2	15.7	2.6	1.0	0.0	0.0	0.0	1.6	8.1	4.6	85.0
Sharjah	22.8	22.8	10.1	5.1	0.0	0.0	0.0	0.0	0.0	0.0	10.1	35.4	106.3
Muscat	27.8	17.7	10.1	10.1	2.5	2.5	2.5	2.5	0.0	2.5	10.1	17.7	106.0
Misrah Island	2.5	2.5	5.1	0.0	2.5	2.5	2.5	2.5	2.5	2.5	2.5	10.1	37.2
Salalah	2.5	2.5	2.5	2.5	2.5	5.1	27.8	25.3	2.5	12.6	2.5	7.6	95.9
Riyan (Mukala)	7.5	3.4	14.4	13.2	5.2	0.1	4.2	3.9	0.8	1.0	3.0	3.5	60.2
Khormaksar (Aden)	6.9	2.9	5.2	2.3	1.2	0.1	3.0	2.5	3.8	0.8	3.4	5.6	37.6
Dhala	14.3	7.8	10.1	16.6	42.3	13.4	105.9	112.7	41.0	6.3	6.2	0.0	376.6
Mukayras	1.0	8.6	6.9	32.7	13.3	2.7	46.6	94.7	30.3	0.6	0.0	0.0	240.1
Iraq													
Mosul	67.2	63.4	69.3	50.8	25.3	0.7	0.1	0.0	0.7	9.9	36.1	65.3	388.8
Kirkuk	60.2	63.2	75.3	50.7	20.7	0.3	0.0	0.0	0.1	4.1	40.2	57.8	372.6
Khanaqin	62.0	43.7	66.4	37.0	18.2	0.3	0.2	0.1	0.0	4.3	30.6	47.8	310.6
Rutbah	13.1	13.7	15.4	16.0	14.5	0.1	0.0	0.0	0.6	5.6	12.6	16.7	108.3
Habbaniya	27.9	12.0	10.5	19.9	6.2	0.0	0.0	0.0	0.2	2.3	20.0	13.1	112.1
Baghdad	24.8	24.0	23.1	21.5	7.3	0.1	0.0	0.0	0.3	3.7	17.4	22.7	144.9
Kut-al-Hai	24.7	20.6	19.2	16.6	6.8	0.0	0.0	0.0	0.0	3.3	19.4	23.2	133.8
Diwaniya	21.6	15.0	16.9	18.0	7.9	0.0	0.0	0.0	0.0	3.9	15.4	20.1	118.8
Nasiriyah	19.2	13.4	15.7	16.9	7.1	0.0	0.0	0.0	0.0	2.2	15.7	19.4	109.6
Basrah	24.2	14.3	20.3	20.9	7.8	0.0	0.0	0.0	0.0	0.8	22.5	29.3	140.1
Kuwait													
Kuwait	18.2	10.4	5.5	10.9	4.8	0.0	0.0	0.0	0.0	0.5	26.4	20.3	97.0
Iran													
Tabriz	48.1	33.3	63.8	44.2	37.9	14.8	2.3	0.1	22.2	62.5	10.7	10.4	350.3
Mashhad	20.5	24.6	36.5	45.6	24.4	0.0	6.0	2.7	1.1	25.1	9.0	3.7	199.2
Tehran	46.8	20.8	25.6	16.3	12.0	0.6	0.3	0.2	3.5	14.8	11.6	12.1	164.6
Abadan	28.2	38.5	10.7	3.2	5.0	0.0	0.0	0.0	0.8	7.5	37.5	3.8	135.2
Bushehr	77.4	30.9	1.5	8.8	0.3	0.0	0.0	0.0	0.0	2.8	52.9	22.1	196.7
Shiraz	143.6	50.2	3.5	26.0	1.3	0.0	0.3	0.0	0.0	1.1	46.3	23.9	296.2
Kerman	39.4	30.9	5.7	16.6	1.3	0.9	0.3	0.0	0.0	0.0	0.9	4.6	100.6
Esfahan	19.4	13.8	8.8	5.4	1.6	0.0	0.0	0.0	0.0	94.0	100.3	57.7	301.0

TABLE VI

DURATION OF SUNSHINE (%)

	Jan.	Feb.	Mar.	Apr.	May	June	July	Aug.	Sep.	Oct.	Nov.	Dec.
Turkey												
Samsun	29	30	28	33	44	60	66	66	52	46	40	29
Trabzon	34	35	32	32	41	51	47	51	42	50	40	35
Florya (Ist.)	29	34	39	48	60	72	80	80	66	56	40	30
Goztepe (Ist.)	28	32	38	48	60	71	78	79	66	53	38	29
Erzurum	36	45	44	50	59	71	80	82	77	67	53	38
Sivas	28	32	38	48	58	76	87	86	76	62	47	29
Ankara	35	38	46	53	64	74	85	87	78	67	53	33
Izmir	42	50	53	64	70	79	86	92	83	68	45	44
Antalya	49	56	60	66	73	83	89	90	85	76	67	52
Adana	51	54	55	61	70	79	83	87	81	74	64	48
Diyarbakir	41	48	48	56	69	86	88	49	58	71	58	42
Cyprus												
Famagusta	38	46	50	64	76	84	89	88	81	71	57	39
Nicosia	54	59	64	51	76	85	89	90	87	78	64	55
Syria												
Aleppo	41	51	63	71	82	93	96	97	93	81	69	44
Damascus	56	65	73	78	87	93	98	98	94	84	79	61
Kameshli	48	55	57	62	72	88	91	91	91	78	70	52
El Haseke	45	57	58	61	72	86	87	90	87	75	67	51
Deir ez Zor	59	68	69	69	77	90	93	94	93	85	79	64
Palmyra	57	71	71	75	80	95	98	96	91	85	76	63
Lebanon												
Beirut	42	47	51	62	72	81	82	81	78	70	64	45
Riyaq	48	55	59	67	77	89	90	92	91	76	64	45
Jordan												
Deir Alla	63.2	67.3	62.5	72.1	86.4	92.6	95.5	95.3	87.7	85.0	74.3	65.3
Aqaba Airport	72.8	80.2	77.1	75.2	91.9	88.0	96.0	95.0	94.6	88.6	85.4	77.3
Amman Airport	64.7	66.1	72.8	78.1	85.3	93.6	94.9	96.2	93.4	87.5	76.2	65.0
H.4	61.9	75.2	71.7	74.4	87.5	98.3	93.9	90.5	94.1	85.2	74.7	70.5
Mafraq	58.8	67.0	70.7	70.9	83.8	87.1	91.4	91.7	85.8	83.0	79.0	63.0
Ma'an Airport	69.6	82.2	80.0	75.0	91.6	90.4	94.7	91.0	94.1	86.4	81.6	76.0
West Bank and Gaza Strip												
Gaza	63	62	66	69	80	82	84	84	84	79	75	65
Jerusalem	61.8	64.8	65.2	72.2	72.1	84.9	93.5	93.9	86.8	84.7	75.2	65.0
Jericho	59.7	61.3	67.2	75.4	85.9	93.8	97.8	97.6	91.8	84.1	80.2	67.1

TABLE VII

MEAN SEA LEVEL PRESSURE (mbar)

	Jan.	Feb.	Mar.	Apr.	May	June	July	Aug.	Sep.	Oct.	Nov.	Dec.
Turkey												
Samsun	1019.9	1015.5	1016.9	1013.8	1014.3	1013.8	1012.3	1013.3	1017.0	1019.2	1017.8	1017.6
Trabzon	1019.9	1017.1	1017.1	1014.6	1014.4	1013.4	1011.7	1012.7	1016.4	1019.6	1018.8	1017.9
Florya (Ist.)	1019.1	1016.5	1016.1	1013.6	1012.9	1013.0	1012.0	1012.7	1016.1	1018.5	1017.3	1015.7
Goztepe (Ist.)	1019.9	1017.1	1017.1	1014.1	1013.7	1013.7	1012.8	1013.7	1017.0	1019.4	1018.2	1016.7
Sivas	1024.3	1021.1	1018.4	1013.3	1012.5	1011.1	1008.5	1009.7	1014.2	1019.8	1022.4	1023.4
Ankara	1020.3	1017.7	1015.4	1011.7	1011.7	1010.8	1008.3	1009.1	1013.4	1018.1	1020.0	1018.8
Izmir	1018.1	1015.6	1015.3	1012.9	1012.6	1011.7	1009.7	1010.4	1014.1	1017.6	1018.5	1015.8
Antalya	1016.8	1014.2	1013.7	1011.3	1011.2	1008.8	1005.4	1006.3	1010.6	1015.1	1017.1	1016.0
Adana	1018.2	1015.7	1014.3	1011.7	1011.7	1009.1	1005.5	1006.3	1010.5	1015.3	1017.7	1017.7
Erzurum	1027.3	1024.2	1020.6	1014.0	1012.6	1009.5	1005.5	1006.2	1011.3	1009.0	1023.1	1026.9
Diyarbakir	1021.5	1017.9	1015.1	1010.7	1009.4	1004.8	999.2	1001.2	1007.5	1015.3	1019.9	1021.0
Cyprus												
Nicosia	1016.5	1014.5	1014.2	1012.4	1011.2	1008.4	1004.9	1005.9	1010.1	1014.3	1016.4	1017.4
Limassol	1014.5	1012.8	1012.7	1011.3	1010.1	1007.6	1004.5	1005.3	1009.1	1013.0	1014.3	1017.1
Syria												
Aleppo	1018.4	1017.4	1013.9	1011.8	1010.1	1006.7	1002.6	1003.5	1008.7	1014.4	1018.9	1019.1
Kameshli	1019.8	1018.5	1015.7	1012.0	1009.7	1004.2	999.6	1000.7	1007.0	1014.3	1019.0	1020.0
El Haseke	1017.7	1016.3	1012.4	1009.9	1008.3	1003.2	998.3	999.9	1006.2	1013.0	1017.4	1018.0
Deir ez Zor	1019.1	1017.8	1014.0	1011.3	1009.6	1005.4	1000.7	1002.0	1007.8	1014.3	1018.8	1020.0
Palmyra	1018.9	1017.3	1013.8	1011.1	1009.9	1006.4	1002.4	1003.5	1008.5	1014.1	1018.2	1018.7
Abu Kamal	1018.7	1016.9	1013.7	1010.6	1008.8	1004.7	1000.0	1001.5	1007.3	1013.6	1018.2	1018.5
Lebanon												
Tripoli—El Mina	1016.5	1016.2	1014.7	1013.8	1012.5	1010.3	1006.8	1007.5	1011.3	1015.1	1016.8	1017.3
Beirut	1016.9	1016.5	1014.2	1012.9	1012.4	1010.0	1008.8	1009.3	1011.0	1014.6	1016.9	1017.1
Jordan												
Aqaba Airport	1017.1	1015.9	1014.1	1011.2	1010.0	1007.6	1004.8	1005.3	1008.6	1013.1	1015.4	1016.9
Amman Airport	1019.2	1017.8	1015.9	1013.3	1011.8	1009.3	1006.1	1006.4	1010.7	1014.9	1017.7	1019.0
H.4	1019.4	1017.5	1014.9	1011.4	1010.0	1006.5	1002.9	1003.4	1008.4	1014.0	1018.6	1019.0
Mafraq	1019.0	1017.6	1015.4	1012.5	1011.4	1008.6	1004.3	1005.6	1010.2	1014.4	1017.6	1018.8
Ma'an Airport	1018.4	1016.4	1014.2	1011.4	1010.0	1008.0	1004.6	1005.0	1008.6	1013.8	1016.9	1018.0
West Bank and Gaza Strip												
Gaza	1017.8	1016.5	1015.0	1012.8	1012.1	1010.6	1007.5	1007.6	1011.2	1014.8	1016.8	1017.4
Jerusalem	1018.2	1017.3	1015.4	1013.0	1012.5	1010.3	1007.1	1007.5	1011.2	1014.8	1017.0	1017.7
Jericho	1018.6	1017.5	1015.8	1013.0	1012.4	1010.0	1006.8	1007.2	1011.0	1013.0	1017.2	1018.2
Arabian Peninsula												
Jiddah	1014.1	1013.4	1011.6	1009.8	1007.5	1005.0	1004.1	1003.8	1006.2	1010.3	1012.6	1013.9
Riyadh	1021.4	1018.2	1013.4	1010.8	1007.7	1003.7	1000.1	1004.4	1007.3	1014.1	1018.0	1019.7
Medina	1013.8	1012.9	1008.8	1006.6	1003.6	1001.4	998.9	999.9	1002.8	1014.3	1015.8	1014.0
Iraq												
Mosul	1020.2	1018.0	1015.2	1012.4	1009.3	1003.7	998.7	1000.8	1007.3	1014.6	1018.8	1020.9
Kirkuk	1019.3	1017.7	1014.6	1011.9	1008.4	1002.9	997.8	999.7	1006.2	1013.4	1017.9	1019.7
Khanaqin	1019.0	1017.2	1014.4	1011.4	1007.8	1002.0	996.9	998.7	1005.2	1012.3	1017.1	1019.9
Rutbah	1019.5	1017.5	1015.0	1012.3	1010.1	1006.3	1002.4	1003.6	1008.6	1014.6	1018.4	1020.2
Habbaniya	1019.0	1017.2	1014.2	1011.3	1008.5	1003.6	999.1	1000.9	1006.7	1013.4	1017.7	1019.8
Baghdad	1019.1	1017.5	1014.5	1011.6	1008.3	1003.2	998.5	1000.2	1006.5	1013.2	1017.6	1019.8
Kut-al-Hai	1019.1	1017.5	1014.3	1010.9	1007.6	1001.9	997.6	999.1	1005.2	1012.4	1017.2	1019.3
Diwaniya	1019.0	1017.2	1014.0	1010.9	1007.9	1002.7	998.3	999.8	1005.7	1012.6	1017.2	1019.4
Nasiriyah	1018.7	1017.0	1013.9	1010.7	1007.4	1002.1	997.8	999.4	1005.3	1012.4	1017.0	1019.1
Basrah	1018.8	1017.2	1014.1	1011.1	1007.3	1001.5	997.4	999.1	1005.1	1012.5	1017.1	1019.2
Kuwait												
Kuwait	1019	1017	1014	1011	1007	1001	997	999	1005	1012	1017	1019

TABLE VIII

MEAN NUMBER OF DAYS WITH SNOW

	Jan.	Feb.	Mar.	Apr.	May	June	July	Aug.	Sep.	Oct.	Nov.	Dec.	Year
Turkey													
Samsun	1.9	3.2	1.0	0.0	0.0	0.0	0.0	0.0	0.0	0.0	0.1	0.4	6.7
Trabzon	0.9	1.9	0.2	0.1	0.0	0.0	0.0	0.0	0.0	0.0	0.0	0.3	3.4
Florya (Ist.)	3.0	3.4	1.6	0.0	0.0	0.0	0.0	0.0	0.0	0.0	0.2	0.6	8.8
Goztepe (Ist.)	2.8	3.2	1.6	0.0	0.0	0.0	0.0	0.0	0.0	0.0	0.2	0.5	8.3
Sivas	8.9	10.3	5.4	0.9	0.0	0.0	0.0	0.0	0.0	0.1	1.1	4.7	31.4
Ankara	4.5	5.6	1.6	0.3	0.0	0.0	0.0	0.0	0.0	0.0	0.7	2.1	14.8
Izmir	0.0	0.1	0.0	0.0	0.0	0.0	0.0	0.0	0.0	0.0	0.0	0.0	0.1
Antalya	0.0	0.0	0.0	0.0	0.0	0.0	0.0	0.0	0.0	0.0	0.0	0.0	0.0
Adana	0.1	0.0	0.0	0.0	0.0	0.0	0.0	0.0	0.0	0.0	0.0	0.0	0.1
Erzurum	11.2	11.9	11.0	3.7	0.2	0.0	0.0	0.0	0.0	0.4	3.6	9.9	51.9
Diyarbakir	2.5	2.0	0.9	0.0	0.0	0.0	0.0	0.0	0.0	0.0	0.0	0.7	6.1
Cyprus													
Nicosia	0.4	0.6	0.1	0.0	0.0	0.0	0.0	0.0	0.0	0.0	0.0	0.1	1.2
Syria													
Aleppo	1.7	1.4	0.1	0.0	0.0	0.0	0.0	0.0	0.0	0.0	0.0	0.3	3.5
Damascus	1.5	1.4	0.2	0.0	0.0	0.0	0.0	0.0	0.0	0.0	0.0	0.5	3.6
Kameshli	1.6	1.4	0.4	0.0	0.0	0.0	0.0	0.0	0.0	0.0	0.1	0.2	3.7
El Haseke	1.0	0.6	0.2	0.0	0.0	0.0	0.0	0.0	0.0	0.0	0.0	0.2	2.0
Deir ez Zor	0.2	0.1	0.0	0.0	0.0	0.0	0.0	0.0	0.0	0.0	0.0	0.0	0.3
Palmyra	0.3	0.3	0.1	0.0	0.0	0.0	0.0	0.0	0.0	0.0	0.0	0.1	0.8
Abu Kamal	0.1	0.1	0.1	0.0	0.0	0.0	0.0	0.0	0.0	0.0	0.0	0.0	0.3
Lebanon													
Tripoli—El Mina	0.0	0.0	0.0	0.0	0.0	0.0	0.0	0.0	0.0	0.0	0.0	0.0	0.0
Beirut	0.0	0.0	0.0	0.0	0.0	0.0	0.0	0.0	0.0	0.0	0.0	0.0	0.0
Riyaq	2.0	2.0	1.0	1.0	0.0	0.0	0.0	0.0	0.0	0.0	0.0	1.0	7.0
Jordan													
Aqaba Airport	0.0	0.0	0.0	0.0	0.0	0.0	0.0	0.0	0.0	0.0	0.0	0.0	0.0
Amman Airport	0.51	0.95	0.16	0.0	0.0	0.0	0.0	0.0	0.0	0.0	0.0	0.21	1.83
H.4	0.0	0.0	0.0	0.0	0.0	0.0	0.0	0.0	0.0	0.0	0.0	0.20	0.20
Mafraq	0.11	0.11	0.10	0.0	0.0	0.0	0.0	0.0	0.0	0.0	0.0	0.0	0.32
Ma'an Airport	0.0	0.33	0.0	0.0	0.0	0.0	0.0	0.0	0.0	0.0	0.0	0.14	0.47
West Bank and Gaza Strip													
Jerusalem	0.23	0.85	0.14	0.0	0.0	0.0	0.0	0.0	0.0	0.0	0.0	0.21	1.43
Jericho	0.0	0.0	0.0	0.0	0.0	0.0	0.0	0.0	0.0	0.0	0.0	0.0	0.0
Iraq													
Mosul	0.5	0.4	0.0	0.0	0.0	0.0	0.0	0.0	0.0	0.0	0.0	0.0	0.9
Kirkuk	0.2	0.5	0.2	0.0	0.0	0.0	0.0	0.0	0.0	0.0	0.0	0.0	0.9
Khanaqin	0.0	0.2	0.0	0.0	0.0	0.0	0.0	0.0	0.0	0.0	0.0	0.0	0.2
Rutbah	0.2	0.2	0.0	0.0	0.0	0.0	0.0	0.0	0.0	0.0	0.0	0.1	0.5
Habbaniya	0.1	0.1	0.0	0.0	0.0	0.0	0.0	0.0	0.0	0.0	0.0	0.0	0.2
Baghdad	0.1	0.1	0.0	0.0	0.0	0.0	0.0	0.0	0.0	0.0	0.0	0.0	0.2
Kut-al-Hai	0.1	0.0	0.0	0.0	0.0	0.0	0.0	0.0	0.0	0.0	0.0	0.0	0.1
Diwaniya	0.1	0.0	0.0	0.0	0.0	0.0	0.0	0.0	0.0	0.0	0.0	0.1	0.2
Nasiriyah	0.0	0.0	0.1	0.0	0.0	0.0	0.0	0.0	0.0	0.0	0.0	0.0	0.1
Basrah	0.0	0.0	0.0	0.0	0.0	0.0	0.0	0.0	0.0	0.0	0.0	0.0	0.0

TABLE VIII *(continued)*

	Jan.	Feb.	Mar.	Apr.	May	June	July	Aug.	Sep.	Oct.	Nov.	Dec.	Year
Kuwait													
Kuwait	0.0	0.0	0.0	0.0	0.0	0.0	0.0	0.0	0.0	0.0	0.0	0.0	0.0
Iran													
Tabriz	10.0	9.0	4.0	1.0	0.0	0.0	0.0	0.0	0.0	1.0	0.6	3.3	28.9
Mashhad	3.3	4.0	1.6	0.6	0.0	0.0	0.0	0.0	0.0	0.0	0.0	1.6	11.1
Tehran	3.6	3.0	1.0	0.3	0.0	0.0	0.0	0.0	0.0	0.0	0.0	0.6	8.5
Abadan	0.0	0.0	0.0	0.0	0.0	0.0	0.0	0.0	0.0	0.0	0.0	0.0	0.0
Bushehr	0.0	0.0	0.0	0.0	0.0	0.0	0.0	0.0	0.0	0.0	0.0	0.0	0.0
Shiraz	4.0	0.0	0.0	0.0	0.0	0.0	0.0	0.0	0.0	0.0	0.0	0.3	4.3
Kerman	0.6	0.6	0.3	0.0	0.0	0.0	0.0	0.0	0.0	0.0	0.0	0.0	1.5
Esfahan	3.0	2.0	0.6	0.6	0.3	0.0	0.0	0.0	0.0	0.0	0.0	1.3	7.8

TABLE IX

DAYS WITH GALES (Surface wind speed ⩾ 34 knots)

	Jan.	Feb.	Mar.	Apr.	May	June	July	Aug.	Sep.	Oct.	Nov.	Dec.	Year
Turkey													
Samsun	0.8	0.4	0.4	0.5	0	0.2	0	0	0	0.2	0.6	0.5	3.4
Trabzon	0.5	0.5	0.3	0.1	0	0.1	0	0	0.1	0.2	0.2	0.2	2.2
Florya (Ist.)	1.8	1.3	1.7	0.6	0.3	0.4	0.4	0.9	0.7	0.4	0.6	1.2	10.4
Goztepe (Ist.)	0.2	0.2	0.2	0.1	0	0.1	0.1	0.1	0.0	0.0	0.3	0.1	1.5
Sivas	0.5	0.6	0.6	1.0	0.6	0.3	0.3	0.3	0.3	0.2	0.2	0.6	5.2
Ankara	0.3	0.9	1.3	1.7	0.9	0.6	0.5	0.2	0.1	0.2	0.6	0.5	8.0
Izmir	4.1	4.5	4.4	2.8	1.0	0.9	0.6	0.6	0.5	0.9	2.6	4.1	26.9
Antalya	1.4	0.9	0.4	0.3	0	0.1	0	0.1	0.1	0.1	0.4	1.3	4.9
Erzurum	0.2	0.4	0.5	0.5	0.2	0.0	0	0.2	0	0	0.2	0.0	2.3
Diyarbakir	0.3	0.7	0.4	0.7	0.9	0.6	0.5	0.1	0.3	0.3	0.1	0.3	5.1
Adana	0.8	1.7	1.3	0.8	0.3	0.4	0.3	0	0.3	0.4	0.4	0.7	7.3
Cyprus													
Kyrenia	0.1	0.1	0.0	0.0	0.0	0.0	0.0	0.1	0.0	0.0	0.1	0.2	0.6
Nicosia	0.2	0.2	0.2	0.2	—	—	—	—	—	0.1	0.2	0.6	1.7
Limassol	0.4	0.1	0.0	0.4	0.1	0	0	0	0	0	0	0.5	1.5
Syria													
Aleppo	0.5	0.4	0.2	0.5	0.2	0.1	0.3	0.0	0.0	0.0	0.0	0.1	2.3
Damascus	2.3	1.8	2.0	2.5	2.0	1.3	1.5	0.5	0.4	0.1	0.5	0.7	15.6
Kameshli	0.6	0.6	1.1	0.7	0.8	0.7	0.0	0.0	0.3	0.3	0.2	0.7	6.0
El Haseke	0.5	0.6	0.8	0.9	0.7	0.3	0.0	0.1	0.1	0.1	0.2	0.6	4.8
Deir ez Zor	0.3	0.5	0.2	0.4	0.8	0.1	0.5	0.3	0.1	0.5	0.0	0.2	3.8
Palmyra	0.4	0.5	0.7	1.1	0.8	0.6	0.3	0.4	0.0	0.2	0.4	0.3	5.7
Abu Kamal	0.2	0.3	0.4	0.5	1.0	0.2	0.4	0.0	0.0	0.4	0.0	0.2	3.6
Lebanon													
Tripoli—El													
Mina	8	7	9	6	4	6	5	3	2	2	3	6	61
Beirut	10	8	9	6	3	1	1	0	1	3	4	8	54
Riyaq	3	4	4	2	2	2	2	1	1	1	1	2	25
Jordan													
Amman Air-													
port	1.88	1.75	1.27	0.56	0.34	0.14	0.02	0.07	0.04	0.07	0.36	0.48	6.98
Mafraq	0.50	0.57	0.40	0.67	0.53	0.0	0.0	0.07	0.0	0.08	0.08	0.23	3.13
West Bank and Gaza Strip													
Gaza	0.4	1.0	0.0	0.0	0.2	0.0	0.0	0.0	0.0	0.2	0.2	2.4	4.4
Jerusalem	1.25	1.94	2.00	0.50	0.19	0.13	0.13	0.0	0.06	0.31	0.50	1.19	8.19
Kuwait													
Kuwait	1	1	3	2	2	3	4	1	0	0	1	0	18

245

TABLE X

DAYS OF FOG (vis. <1,000 m)

	Jan.	Feb.	Mar.	Apr.	May	June	July	Aug.	Sep.	Oct.	Nov.	Dec.	Year
Turkey													
Samsun	0.7	0.8	2.1	2.6	2.5	0.2	0.0	0.0	0.0	0.2	0.0	0.5	10.0
Trabzon	0.4	0.8	2.1	3.1	2.5	0.5	0.0	0.0	0.0	0.0	0.2	0.2	9.8
Florya (Istanbul)	1.2	1.4	1.5	1.2	1.3	0.5	0.6	0.6	1.2	1.4	1.3	1.2	13.2
Goztepe (Istanbul)	0.7	0.8	0.7	0.9	1.0	0.3	0.4	0.6	1.2	1.1	0.9	0.8	9.6
Sivas	4.5	3.2	1.9	0.8	0.5	0.1	0.1	0.2	0.4	1.8	4.1	4.7	22.1
Ankara	5.6	3.3	3.0	2.0	0.5	0.3	0.2	0.1	0.4	1.4	4.0	5.2	25.9
Izmir	0.2	0.1	0	0.1	0	0	0	0.0	0	0	0.1	0.1	0.6
Antalya	0.2	0.0	0.3	0.4	0.2	0.1	0.4	0.1	0.3	0.3	0.1	0.3	2.6
Adana	0.2	0.5	0.8	1.2	0.6	0.4	0.3	0.3	0.2	0.0	0.2	0.1	4.8
Erzurum	3.7	2.6	2.1	0.6	0.0	0.1	0	0	0.1	0.2	1.3	3.4	14.1
Diyarbakir	3.4	2.1	0.8	0.2	0.4	0	0	0	0	0.5	1.1	3.4	11.9
Cyprus													
Kyrenia	0	0	0	0	0	0	0.1	0	0	0	0	0	0.1
Famagusta	0.1	0.1	0.2	0.2	0	0	0.1	0	0	0.1	0	0	0.4
Nicosia	2.3	1.8	2.1	2.4	2.0	2.3	2.0	1.8	1.2	0.6	1.1	2.5	22.1
Limassol	0	0	0	0	0	0	0	0	0	0	0	0.1	0.1
Syria													
Aleppo	5.9	1.9	1.1	0.1	0.0	0.0	0.0	0.0	0.0	0.4	1.5	7.1	1.0
Damascus	2.5	1.7	0.6	0.6	0.5	0.4	0.1	1.2	0.7	0.2	0.0	3.2	11.7
Kameshli	1.7	1.3	1.1	0.2	0.3	0.0	0.0	0.0	0.0	0.0	0.6	2.2	7.4
El Haseke	2.6	1.5	1.0	0.7	0.0	0.0	0.1	0.0	0.0	0.3	0.6	3.6	10.4
Deir ez Zor	3.7	0.8	0.1	0.1	0.0	0.0	0.0	0.0	0.0	0.1	0.8	4.5	10.1
Palmyra	2.3	0.9	0.0	0.0	0.0	0.0	0.0	0.0	0.1	0.3	0.2	2.6	6.4
Abu Kamal	3.5	1.2	0.0	0.2	0.0	0.0	0.0	0.0	0.0	0.3	0.8	2.8	8.8
Lebanon													
Tripoli—El Mina	0	0	0	0	0	0	0	0	0	0	0	0	0.0
Beirut	0	0	0	0	0	0	0	0	0	0	0	0	0.0
Riyaq	3	2	1	1	1	0	0	1	1	1	1	3	15.0
Jordan													
Aqaba Airport	0.0	0.0	0.0	0.0	0.0	0.0	0.0	0.0	0.0	0.0	0.14	0.43	0.57
Amman Airport	1.72	1.02	0.58	0.16	0.12	0.0	0.02	0.02	0.0	0.12	0.59	1.16	5.51
H.4	1.60	0.60	0.0	0.0	0.20	0.0	0.0	0.0	0.0	0.20	0.60	0.60	3.80
Mafraq	5.44	3.0	1.20	1.20	0.60	0.20	1.20	1.91	2.10	1.10	1.40	2.40	21.75
Ma'an Airport	0.33	0.0	0.16	0.0	0.0	0.0	0.0	0.0	0.0	0.0	0.29	0.43	1.21
West Bank and Gaza Strip													
Gaza	0.2	1.0	1.2	1.6	1.0	1.6	1.6	1.0	2.2	1.6	1.6	0.4	13.0
Jerusalem	2.6	1.93	2.07	1.53	0.87	0.80	1.73	2.20	1.86	0.64	1.64	2.00	19.87
Jericho	0.20	0.0	0.0	0.0	0.0	0.0	0.0	0.0	0.0	0.0	0.0	0.40	0.60

TABLE XI

DAYS WITH THUNDERSTORMS

	Jan.	Feb.	Mar.	Apr.	May	June	July	Aug.	Sep.	Oct.	Nov.	Dec.	Year
Turkey													
Samsun	0.1	0.1	0.4	1.4	2.8	2.7	1.5	1.3	1.5	1.1	0.4	0.1	13.4
Trabzon	0.3	0.1	0.1	0.8	2.9	3.6	2.1	1.9	2.4	1.5	0.4	0.1	16.3
Florya (Ist.)	0.4	0.3	0.6	0.7	1.3	2.3	1.2	1.1	0.9	1.0	0.8	0.4	11.0
Goztepe (Ist.)	0.2	0.3	0.4	0.8	1.6	2.1	1.8	1.1	1.3	1.1	0.8	0.4	11.9
Sivas	0.0	0.1	0.4	2.0	6.7	4.7	1.1	1.2	1.2	1.0	0.3	0.1	18.8
Ankara	0.0	0.1	0.6	2.1	6.0	5.4	2.4	1.4	1.3	1.0	0.3	0.1	20.7
Izmir	2.8	2.8	2.2	1.7	2.7	1.4	0.4	0.3	1.9	1.9	3.1	2.5	23.2
Antalya	3.5	3.2	1.9	1.9	3.8	2.9	0.8	0.8	1.6	3.4	3.0	3.8	30.4
Adana	1.7	2.1	2.4	3.4	4.6	2.7	1.0	0.7	2.1	3.3	1.9	1.5	27.4
Erzurum	0.0	0.0	0.0	0.7	4.1	5.2	3.4	2.1	0.6	0.7	0.0	0.0	17.8
Diyarbakir	0.0	0.6	1.3	4.3	4.9	2.5	0.9	0.4	0.9	1.7	0.9	0.2	18.0
Cyprus													
Kyrenia	1	1	1	1	2	1	0.5	0	1	2	3	2	15
Famagusta	1	1	1	1	1	0	0.5	0	1	1	1	1	9
Nicosia	2.6	2.1	2.7	2.0	3.8	1.3	0.4	0.4	1.4	4.1	3.1	3.6	27.5
Limassol	1	1	0.5	1	0.5	0.5	0	0	0	0.5	1	2	7
Syria													
Aleppo	0.5	0.3	0.7	2.8	3.6	1.0	0.0	0.1	0.3	2.2	0.5	0.5	12.3
Damascus	0.5	0.3	0.4	0.9	1.6	0.0	0.0	0.0	0.3	1.7	1.4	0.7	7.8
Kameshli	0.6	0.7	2.3	3.8	4.4	1.2	0.2	0.0	0.7	2.1	1.2	0.5	17.7
El Haseke	0.2	0.4	1.2	3.2	2.5	0.5	0.0	0.0	0.5	1.7	0.6	0.1	10.9
Deir ez Zor	0.4	0.2	0.6	2.5	2.7	0.2	0.0	0.0	0.5	1.5	0.4	0.8	9.8
Palmyra	0.4	0.6	0.8	2.3	2.5	0.3	0.0	0.0	0.3	2.1	0.7	0.3	10.3
Abu Kamal	0.7	0.5	1.1	2.9	2.8	0.2	0.0	0.0	0.0	1.8	0.8	0.6	11.4
Lebanon													
Tripoli—El													
Mina	4	3	3	2	1	0	0	0	1	3	3	4	24
Beirut	5	4	4	2	1	0	0	0	1	2	3	5	27
Riyaq	1	1	1	1	1	0	0	0	0	1	1	1	8
Jordan													
Aqaba Air-port	0.57	0.59	0.14	0.43	1.0	0.14	0.0	0.0	0.06	0.86	0.38	0.29	4.10
Amman Air-port	1.10	0.40	0.70	0.80	0.70	0.03	0.0	0.03	0.22	0.70	0.80	0.40	5.88
H.4	0.40	0.60	0.40	1.20	1.00	0.40	0.0	0.0	0.0	1.40	0.80	0.40	6.60
Mafraq	0.77	0.11	0.60	1.30	0.30	0.0	0.0	0.0	0.11	0.78	0.67	0.56	5.20
Ma'an Air-port	0.17	0.17	0.33	1.0	0.67	0.0	0.0	0.07	0.57	0.43	0.72	0.40	4.46
West Bank and Gaza Strip													
Gaza	2.2	1.0	0.4	1.0	1.0	0.2	0.0	0.0	0.0	1.2	1.6	3.0	11.6
Jerusalem	1.31	1.46	1.57	1.00	0.50	0.07	0.0	0.0	0.0	0.93	1.79	1.43	10.06
Jericho	1.20	1.00	1.00	1.60	1.20	0.20	0.0	0.0	0.0	1.67	0.83	1.33	10.03
Arabian Peninsula													
Jiddah	0.25	0	0.08	0.42	0.17	0	0	0	0	0	0.58	0.33	
Riyadh	0.50	0.25	0.58	0.50	0.91	0	0	0	0	0	0.08	0.7	
Dhahran	0.44	0.44	0.56	1.44	0.22	0	0	0	0	0.11	0.67	0	
Riyan (Mukala)	0	0	0.30	0.00	0.70	0.20	0.70	1.10	0.30	0	0	0	
Khormaksar (Aden)	0	0	0.13	0.09	0.43	0.30	2.17	1.78	1.17	0.13	0	0	

TABLE XII

SURFACE WIND ANALYSIS (%)

		N	NE	E	SE	S	SW	W	NW	Calm
Turkey										
Samsun	Jan.	3	3	1	1.5	17.5	33.5	5.5	14	21
	Apr.	13	14	1	1	4.5	7.5	1.5	13.5	44
	July	15	15	0.5	0	8.5	6.0	4.0	15.0	35
	Oct.	11	9.5	1	0	15.5	17.0	5.0	11.0	30
Trabzon	Jan.	1.0	2.5	8.5	18.0	18.5	18.0	10.5	9.0	20
	Apr.	6.5	12.0	17.0	5.5	10.0	8.5	11.0	9.0	19
	July	7.5	12.0	10.0	3.5	9.0	8.5	8.5	12.0	27
	Oct.	2.0	8.0	9.5	7.0	18.5	17.0	17.0	9.5	20
Florya (Ist.)	Jan.	14.0	29.0	6.0	1.5	4.5	19.0	2.5	3.5	20
	Apr.	13.5	35.5	3.0	1.0	3.0	18.0	2.0	2.0	23
	July	8.5	54.5	7.5	0.0	0.0	9.0	1.0	1.5	18
	Oct.	9.5	38.0	4.5	1.0	1.5	13.5	3.0	1.0	29
Goztepe (Ist.)	Jan.	9.0	34.5	5.0	8.0	5.5	15.5	6.0	2.5	14
	Apr.	9.0	40.5	4.0	2.0	3.0	14.0	8.0	1.5	18
	July	7.0	62.0	4.0	1.0	1.0	4.5	4.5	1.0	15
	Oct.	4.5	51.0	5.5	2.0	1.5	9.0	4.5	1.0	21
Erzurum	Jan.	2.5	4.5	8.5	12.5	8.5	16.5	13.0	7.0	27
	Apr.	2.5	5.5	11.5	12.0	12.0	23.5	13.0	6.5	14
	July	8.5	20.0	19.5	7.0	6.5	13.5	8.5	9.5	7
	Oct.	4.0	6.0	9.5	13.5	13.0	19.0	18.0	6.0	11
Sivas	Jan.	10.5	13.0	5.5	9.5	15.0	15.5	2.5	12.5	14
	Apr.	9.0	8.5	4.0	5.5	6.5	11.5	6.0	11.0	38
	July	19.0	10.0	2.0	4.0	2.0	5.5	3.5	15.0	39
	Oct.	7.5	3.0	2.0	3.0	3.5	8.0	7.0	10.0	56
Cyprus										
Marphou Bay	Jan.	6.3	23.0	22.0	7.0	2.0	5.3	21.6	20.0	3
	Apr.	11.0	17.0	7.3	1.0	0.3	2.3	22.0	38.3	0
	July	12.0	9.0	0.9	0.0	0.0	0.0	17.0	61.0	0
	Oct.	13.0	22.7	5.0	0.6	0.0	0.7	21.0	37.0	0
Famagusta	Jan.	3.5	21.0	23.0	13.5	5.0	8.0	17.0	9.0	0
	Apr.	2.6	12.0	32.0	4.0	2.0	12.0	25.0	10.0	0
	July	0.9	16.4	45.0	17.0	3.5	5.5	8.0	4.0	0
	Oct.	3.5	13.5	9.5	6.0	2.0	12.5	39.0	9.0	5
Limassol	Jan.	7.0	13.0	12.0	5.0	4.0	10.0	18.0	24.0	7
	Apr.	2.0	8.0	15.0	13.0	8.0	18.0	20.0	6.0	10
	July	0.4	3.0	14.0	12.0	13.0	26.0	22.0	0.9	9
	Oct.	4.0	7.0	7.0	7.0	9.0	15.0	18.0	18.0	15
Syria										
Aleppo	Jan.	11.8	18.8	22.0	2.7	3.7	5.4	9.8	4.1	21.7
	Apr.	6.1	7.7	8.8	1.7	3.2	11.6	36.3	8.7	15.9
	July	1.2	0.2	0.3	0.1	0.9	14.7	71.0	9.5	2.1
	Oct.	8.8	6.9	11.3	2.6	2.1	6.8	25.2	10.5	25.8
Damascus	Jan.	4.5	6.5	7.0	2.0	4.2	10.4	15.2	8.0	42.2
	Apr.	6.1	5.5	6.4	3.2	6.2	9.6	19.5	18.6	24.9
	July	0.7	2.7	7.6	4.8	8.2	8.6	17.3	26.0	20.7
	Oct.	7.4	8.4	11.1	4.4	5.1	5.4	9.1	10.4	38.7
Kameshli	Jan.	24.7	18.4	23.8	4.6	2.2	2.2	3.7	5.8	14.6
	Apr.	17.6	15.6	18.0	5.2	5.1	6.9	11.1	7.3	13.2
	July	20.5	5.3	4.4	2.4	5.0	10.8	21.6	14.1	15.9
	Oct.	27.1	12.2	10.0	2.9	3.9	5.1	9.0	11.9	17.9

TABLE XII *(continued)*

		N	NE	E	SE	S	SW	W	NW	Calm
Syria										
El Haseke	Jan.	6.0	8.1	11.9	10.7	8.8	3.3	13.6	10.9	26.7
	Apr.	8.9	9.0	8.3	6.4	5.2	3.5	23.5	12.8	22.4
	July	10.2	1.5	0.7	0.7	3.5	5.9	42.7	19.6	15.2
	Oct.	9.4	5.1	4.6	2.7	3.4	2.5	21.3	14.4	36.6
Deir ez Zor	Jan.	7.7	4.3	9.6	8.7	7.0	7.1	11.1	10.8	33.7
	Apr.	12.5	4.8	6.5	4.9	5.5	9.0	18.7	19.2	18.9
	July	9.7	0.8	0.4	0.2	0.8	5.7	40.2	38.7	3.5
	Oct.	12.3	3.2	4.7	3.8	3.3	4.3	16.0	17.8	34.5
Abu Kamal	Jan.	5.5	4.4	12.6	13.9	7.0	10.4	21.7	10.8	13.7
	Apr.	8.9	5.5	9.6	8.2	5.4	11.0	28.1	15.9	7.4
	July	9.0	1.1	0.6	0.2	0.4	3.8	50.3	30.2	4.4
	Oct.	8.9	4.8	8.1	6.4	5.0	7.5	23.8	15.9	19.6
Palmyra	Jan.	7.9	14.2	10.2	1.8	3.5	7.9	25.6	7.0	21.8
	Apr.	7.1	7.0	6.2	2.3	3.3	6.6	41.1	19.0	7.4
	July	2.3	0.5	0.3	0.2	0.9	5.7	67.2	21.6	1.3
	Oct.	7.1	6.4	4.7	1.6	3.0	8.1	41.1	14.4	13.6
Lebanon										
Tripoli—El Mina	Jan.	6.1	13.4	7.0	16.4	16.4	21.3	4.8	3.0	11.6
	Apr.	3.9	7.9	1.3	2.6	10.7	38.1	11.6	4.5	19.4
	July	2.0	3.5	2.2	3.3	4.4	50.5	16.7	2.2	14.2
	Oct.	2.9	4.8	7.7	16.7	10.7	14.4	11.3	8.1	23.4
Beirut	Jan.	4.5	9.4	23.0	11.3	18.4	11.3	5.2	3.7	13.2
	Apr.	5.8	11.9	8.4	5.6	12.7	21.9	9.3	6.8	17.6
	July	1.3	3.0	3.0	6.4	18.0	32.0	12.2	3.2	20.9
	Oct.	9.5	16.5	21.0	5.4	5.7	7.3	6.2	7.4	20.8
Cedres	Jan.	5.9	7.6	9.4	4.8	13.1	8.4	7.8	19.6	23.4
	Apr.	10.1	10.4	5.6	4.3	13.4	9.0	13.9	18.3	15.0
	July	10.7	13.5	9.4	5.0	5.4	2.2	17.5	18.2	16.2
	Oct.	6.2	13.2	12.0	6.8	10.2	18.6	14.9	14.0	14.0
Dahr el Baidar	Jan.	2.0	1.1	20.3	0.0	0.3	0.8	26.3	24.3	21.7
	Apr.	2.2	1.0	14.4	0.8	0.1	0.3	29.0	34.6	17.7
	July	5.4	1.4	2.7	0.2	0.0	0.6	31.8	51.0	6.9
	Oct.	4.8	1.6	18.4	3.7	0.0	0.5	10.3	25.6	35.1
Riyaq	Jan.	9.8	17.1	1.0	0.7	10.2	28.4	10.1	7.2	25.0
	Apr.	6.4	11.4	1.6	0.6	12.3	35.2	14.9	6.6	24.0
	July	3.8	10.3	3.2	0.8	12.6	39.2	14.6	3.8	29.4
	Oct.	10.1	21.2	2.6	0.3	7.8	26.4	13.4	7.6	28.3
Ksara	Jan.	1.8	3.9	7.7	1.8	5.2	16.2	26.0	8.1	29.2
	Apr.	2.1	3.8	4.1	1.8	5.6	18.4	29.8	11.2	22.8
	July	3.0	2.8	3.2	1.4	3.4	11.2	23.2	23.0	23.6
	Oct.	2.7	5.1	6.4	2.3	4.6	10.7	19.7	11.4	37.2
Marjayoun	Jan.	7.2	13.8	9.2	8.1	10.5	7.7	15.2	10.0	18.2
	Apr.	6.3	10.2	6.0	2.6	3.0	4.8	31.0	14.9	21.2
	July	1.0	0.4	0.0	0.0	0.3	3.2	59.5	19.8	15.8
	Oct.	10.1	12.7	4.2	0.9	4.6	3.4	28.6	14.8	20.6

TABLE XII *(continued)*

		N	NE	E	SE	S	SW	W	NW	Calm
Jordan										
Aqaba Airport	Jan.	63.2	7.8	1.9	0.3	11.7	2.2	3.9	3.2	5.8
	Apr.	43.3	3.7	2.7	0.6	36.3	5.0	3.7	4.0	0.7
	July	69.3	5.5	1.3	0.0	19.0	3.0	0.6	1.3	0
	Oct.	70.0	4.8	0.0	0.6	13.8	2.6	1.1	2.6	4.5
Amman Airport	Jan.	3.4	4.5	11.9	9.1	7.3	19.3	28.3	7.8	7.5
	Apr.	3.5	1.3	3.0	4.0	5.4	20.6	50.5	10.7	1.1
	July	3.4	0.5	0	0	0.9	9.0	60.9	25.3	0
	Oct.	8.9	4.3	6.1	5.7	7.7	12.2	32.9	18.1	4.1
H.4	Jan.	9.7	2.4	3.6	7.7	23.7	12.9	25.4	8.9	5.6
	Apr.	9.2	1.4	3.4	0.6	14.3	11.2	47.7	9.5	2.7
	July	10.3	0.6	1.7	0.3	9.3	13.0	42.7	20.0	2.0
	Oct.	7.1	1.3	2.7	2.7	18.3	15.6	36.4	12.3	3.4
Mafraq	Jan.	2.6	1.3	3.8	28.7	13.9	11.6	18.1	13.6	6.5
	Apr.	2.0	0.0	3.1	7.0	9.7	14.3	38.7	22.6	2.7
	July	1.6	0.0	0.0	0.0	0.0	4.4	49.9	43.9	0.0
	Oct.	4.4	0.8	1.6	8.9	9.3	6.8	30.2	22.6	15.3
Ma'an Airport	Jan.	6.1	2.3	9.7	11.3	13.2	7.4	34.8	8.1	7.1
	Apr.	2.0	2.6	4.0	8.0	9.0	18.7	41.0	14.0	1.3
	July	10.6	5.5	8.0	9.2	3.5	4.5	35.5	16.4	12.9
	Oct.	6.1	4.2	14.8	10.3	11.3	10.6	22.9	6.1	13.5
West Bank and Gaza Strip										
Gaza	Jan.	1.75	4.62	5.21	37.37	11.78	17.07	5.29	10.58	6.33
	Apr.	7.65	14.24	4.42	19.95	5.18	8.25	13.88	21.34	5.19
	July	7.94	3.79	1.24	21.11	6.85	7.28	13.78	33.00	5.01
	Oct.	13.75	13.21	8.32	28.19	2.98	3.93	4.43	19.53	5.85
Jerusalem	Jan.	1.4	4.9	23.9	8.9	2.5	5.7	36.4	14.5	2.0
	Apr.	5.3	1.6	6.8	7.9	3.5	2.1	39.2	31.4	2.3
	July	4.8	0.0	0.0	0.0	0.0	0.3	46.2	48.2	0.5
	Oct.	12.7	1.1	8.1	5.1	2.5	0.7	31.4	36.5	1.9
Jericho	Jan.	37.4	2.0	5.1	7.8	29.6	2.4	4.8	2.7	8.2
	Apr.	18.8	1.0	3.1	9.3	27.4	2.8	23.0	1.4	13.2
	July	3.0	1.0	4.3	24.5	48.0	2.7	4.6	1.3	10.6
	Oct.	27.9	7.7	5.7	6.6	17.6	2.3	6.6	4.7	21.2
Dead Sea North	Jan.	31.0	4.8	3.0	10.7	16.2	8.5	7.8	18.0	0.0
	Apr.	31.2	7.7	2.8	4.8	17.4	5.7	15.8	14.6	0.0
	July	39.0	7.1	1.3	2.5	26.2	1.9	4.2	18.0	0.0
	Oct.	40.8	10.1	2.1	6.8	18.5	3.1	4.3	14.3	0.0
Israel										
Eilat	Jan.	72.0	2.3	0.9	0.3	6.7	6.1	5.8	6.1	0.0
	Apr.	50.2	29.2	3.2	1.6	4.7	3.4	1.1	7.2	0.0
	July	72.0	3.9	0.3	0.3	8.2	5.5	0.0	9.8	0.0
	Oct.	69.0	18.2	0.9	0.0	2.0	7.2	1.2	1.7	0.0
Arabian Peninsula										
Jiddah	Jan.	36	7					7	36	
	Apr.	43							50	
	July	21							79	
	Oct.					7		7	43	

TABLE XII (continued)

		N	NE	E	SE	S	SW	W	NW	Calm
Arabian Peninsula										
Medina	Jan.		21	14	14			21		
	Apr.						14	57	7	7
	July						7	72	21	
	Oct.	7		21	21			37	7	7
Taif	Jan.					10	45	35	10	
	Apr.					10	10	80		
	July							80	20	
	Oct.			10			10	70	10	
Riyadh	Jan.	7	7	7	43	7			14	7
	Apr.	14	14		36	7	36		29	
	July	36							64	
	Oct.	14			36				21.5	7
Dhahran	Jan.								70	
	Apr.	30	30	20	10					
	July	60	10	10					20	
	Oct.		10	20					40	20
Iraq										
Mosul	Jan.	8.4	6.8	15.4	12.0	7.7	4.2	0.9	11.9	23.7
	Apr.	17.8	9.1	13.1	7.7	8.4	5.5	14.4	12.9	11.1
	July	16.5	8.3	7.5	6.8	8.6	8.2	18.8	15.6	9.7
	Oct.	19.9	7.3	8.0	5.4	9.4	6.2	13.3	14.0	16.5
Kirkuk	Jan.	9.2	3.4	8.1	15.1	20.3	8.5	11.3	15.0	9.1
	Apr.	8.8	6.2	6.0	10.5	17.3	14.4	15.6	15.6	5.6
	July	9.2	3.8	1.2	3.4	10.0	19.8	25.5	23.3	3.8
	Oct.	8.4	2.6	3.1	6.5	15.7	12.0	22.7	19.3	10.6
Khanaqin	Jan.	7.8	1.5	8.1	4.2	19.7	5.3	26.4	9.1	17.9
	Apr.	11.2	1.3	8.1	4.2	16.4	6.5	32.8	9.6	9.9
	July	10.4	0.7	2.4	0.6	5.5	3.8	54.0	17.3	5.3
	Oct.	9.1	0.8	4.5	2.1	10.4	3.9	46.1	14.5	8.6
Rutbah	Jan.	7.5	4.0	8.3	10.6	11.6	10.8	21.6	15.2	10.4
	Apr.	8.6	3.4	5.7	8.3	11.1	14.0	28.0	15.8	5.1
	July	11.5	1.2	1.5	0.8	3.9	10.2	40.6	28.2	2.1
	Oct.	10.5	6.1	10.5	9.3	10.0	9.8	20.7	14.4	8.7
Habbaniya	Jan.	14.2	5.0	4.8	9.2	13.5	7.6	12.0	25.8	7.9
	Apr.	13.0	6.8	9.0	5.7	10.9	9.5	14.0	27.5	3.6
	July	9.1	2.6	0.7	0.1	3.1	4.4	18.3	59.8	1.9
	Oct.	18.6	7.2	4.6	4.1	10.4	6.1	9.8	33.8	5.4
Baghdad	Jan.	18.2	3.4	6.7	11.8	10.4	7.0	9.8	29.6	3.1
	Apr.	23.2	6.7	5.7	12.2	10.7	4.8	12.8	22.6	1.3
	July	18.1	1.7	0.3	1.3	2.9	1.7	23.6	48.7	1.7
	Oct.	30.0	5.9	3.1	6.6	8.3	2.9	9.7	28.7	4.8
Kut-al-Hai	Jan.	7.3	1.8	14.8	9.5	5.5	3.7	25.2	26.1	6.1
	Apr.	10.8	5.5	15.0	11.4	4.7	2.3	23.4	23.7	3.2
	July	3.7	0.0	1.9	1.8	1.6	2.1	49.2	37.9	1.8
	Oct.	8.9	2.6	6.5	8.6	4.7	1.9	24.8	35.5	6.5

TABLE XII *(continued)*

		N	NE	E	SE	S	SW	W	NW	Calm
Iraq										
Diwaniya	Jan.	14.4	3.1	8.2	12.9	4.3	4.9	24.8	18.7	8.5
	Apr.	21.2	4.0	6.7	9.4	6.4	3.5	23.2	18.6	7.0
	July	21.1	0.3	0.3	0.5	0.9	2.6	43.0	28.8	2.5
	Oct.	28.2	2.2	4.1	5.1	5.7	5.2	19.7	24.1	5.7
Nasiriyah	Jan.	12.1	2.6	10.7	8.7	6.3	3.5	17.6	34.3	4.2
	Apr.	20.8	3.3	20.3	10.2	2.1	0.6	12.5	27.6	2.6
	July	10.3	0.3	3.4	1.5	0.2	0.1	18.6	63.8	1.8
	Oct.	27.6	2.6	10.3	6.1	3.4	1.4	9.1	34.8	4.7
Basrah	Jan.	8.1	2.7	7.2	12.1	5.3	1.7	25.4	30.0	7.4
	Apr.	15.1	4.1	8.1	13.6	4.3	0.7	13.0	36.8	4.3
	July	10.4	0.4	1.7	2.1	0.0	0.1	28.4	55.1	1.8
	Oct.	11.4	3.8	6.9	5.3	2.8	1.4	22.7	37.7	8.0
Kuwait										
Kuwait Airport	Jan.	10.7	3.3	6.7	12.6	10.4	3.6	8.9	27.3	16.5
	Apr.	15.0	5.7	13.7	16.7	12.5	4.4	6.1	16.6	9.3
	July	14.7	3.0	6.1	5.4	5.0	2.7	12.3	39.4	11.4
	Oct.	13.2	5.2	10.0	9.6	13.2	5.5	8.7	19.2	15.4
Iran										
Tabriz	Jan.	4.3	6.8	24.0	4.6	1.2	6.1	11.8	2.9	38.3
	Apr.	2.5	5.9	19.3	7.4	1.5	11.4	31.5	6.7	13.8
	July	0.7	5.1	64.1	11.5	2.5	3.9	8.7	0.7	2.8
	Oct.	2.2	3.5	24.1	5.3	1.5	6.1	20.7	4.4	32.2
Mashhad	Jan.	8.7	2.5	3.6	7.9	15.5	2.5	3.9	15.5	39.9
	Apr.	7.8	2.9	11.9	17.7	14.5	3.3	4.9	12.9	24.1
	July	8.3	27.2	19.4	10.0	7.6	0.3	0.4	6.4	20.4
	Oct.	4.4	5.7	9.4	24.7	17.6	5.3	1.8	7.5	23.6
Tehran	Jan.	16.9	5.3	3.6	4.6	7.6	2.8	13.9	13.7	31.6
	Apr.	8.8	6.3	6.6	6.3	7.1	9.7	32.9	6.7	15.6
	July	2.7	4.3	2.1	25.0	21.7	6.2	15.7	6.7	15.6
	Oct.	13.6	5.0	2.2	7.1	7.6	5.7	21.6	8.9	28.3
Esfahan	Jan.	2.8	2.8	5.1	4.7	6.4	5.8	13.9	4.0	54.5
	Apr.	4.9	4.8	6.7	1.8	5.9	14.0	30.1	10.0	21.8
	July	6.1	18.6	19.7	7.8	1.7	1.7	10.7	10.1	23.6
	Oct.	5.1	6.4	10.7	4.3	5.1	8.9	11.2	5.7	42.6
Abadan	Jan.	7.5	4.3	3.9	10.4	10.0	1.8	12.2	25.5	24.4
	Apr.	3.4	3.3	5.2	15.1	11.9	4.8	15.2	27.4	13.7
	July	0.7	0.1	1.4	5.7	2.8	5.1	33.3	40.1	10.8
	Oct.	12.9	6.4	5.1	6.1	10.4	4.6	13.3	20.7	20.5
Kerman	Jan.	9.6	9.3	1.3	5.1	10.1	11.1	17.1	6.9	29.5
	Apr.	11.0	5.7	2.8	4.3	3.8	17.1	31.9	11.0	12.4
	July	25.0	26.6	6.4	1.6	0.2	7.2	8.1	19.3	5.6
	Oct.	15.2	13.8	5.1	3.6	4.2	12.4	24.5	8.7	12.5
Bushehr	Jan.	16.4	7.4	5.3	9.2	5.6	7.1	7.8	28.9	12.3
	Apr.	5.2	6.6	4.5	2.9	8.6	16.2	30.4	19.3	6.3
	July	1.1	2.1	2.8	0.2	6.4	27.5	33.4	15.1	11.4
	Oct.	4.6	4.7	3.5	2.2	6.1	12.9	23.7	27.2	15.1
Shiraz	Jan.	8.9	2.6	4.3	5.1	7.1	2.1	10.1	20.4	39.4
	Apr.	11.2	1.1	6.3	4.0	7.1	5.5	31.2	21.1	12.5
	July	7.8	4.3	9.4	6.0	7.1	5.6	18.6	22.9	18.3
	Oct.	7.8	0.4	6.4	5.6	5.1	1.6	21.8	28.7	22.6

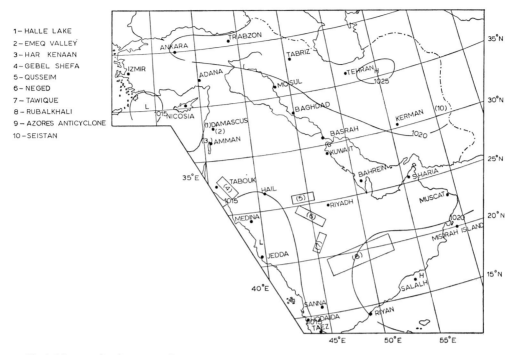

Fig.1. Mean sealevel pressure, January.

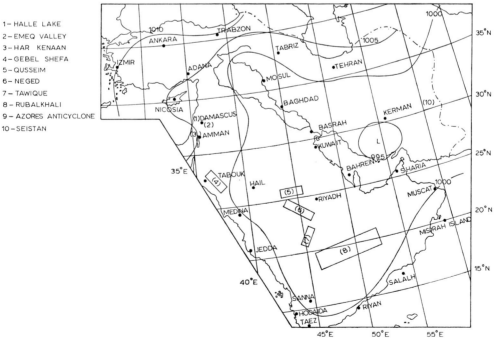

Fig.2. Mean sealevel pressure, July.

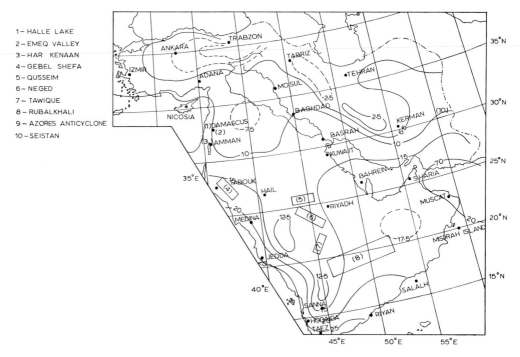

Fig.3. Daily mean temperature, January, (°C).

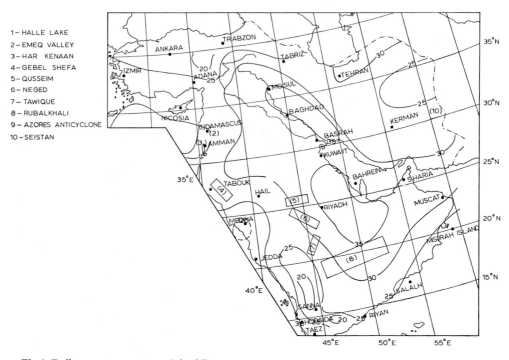

Fig.4. Daily mean temperature, July (°C).

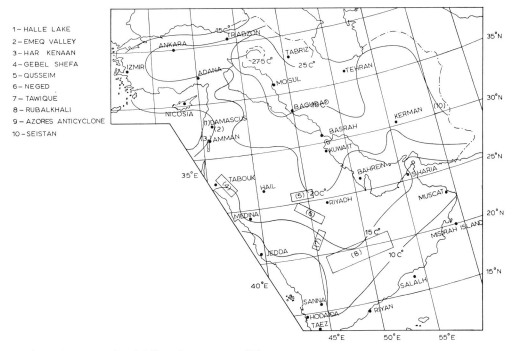

Fig.5. Mean annual variability of temperature (°C).

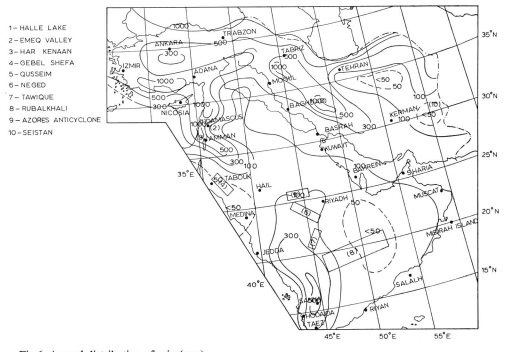

Fig.6. Annual distribution of rain (mm).

Chapter 4

Tropical Cyclones of the Indian Seas

K. N. RAO

Introduction

Tropical cyclones of the Bay of Bengal and the Arabian Sea form an important feature of weather over India and neighbouring countries. As is well known, the passage of these cyclonic disturbances over land is associated with very heavy rain and strong winds, particularly in coastal areas, resulting in considerable loss of life and property. An Indian estimate for the five-year period 1961–1965 shows that more than 500 lives and several thousands of cattle were lost, and the damage to crops, houses, public utilities, etc., cost approximately 500 million rupees.

As an example of heavy rain it may be of interest to mention that in association with the passage of a depression from the Bay of Bengal towards the end of June, 1941, both the Bay of Bengal and Arabian Sea branches of the monsoon had strengthened and the rainfall was concentrated and exceptionally heavy in the coastal strip to the north of Bombay and in Gujarat. Dharampur in Surat District (20°33′N 73°11′E) registered a rainfall of 97.5 cm during the 24 h ending at 08h local time on July 2, 1941, which is probably a rainfall record for a day for low-level stations in the world. Several stations in this area recorded more than 25 cm of rain in 24 h and a few recorded more than 50 cm. Systematic accounts of cyclonic disturbances in the Bay of Bengal and the Arabian Sea for more than the last 70 years are available in published form. From 1891, the accounts of such disturbances have appeared in the *India Weather Review*. In 1923, accounts of storms and depressions were included in a separate part of the *India Weather Review* entitled "Part C–Storms and Depressions". At the end of the text of Part C, two plates show the tracks of cyclonic storms and depressions, respectively.

In this chapter the main climatological features of the depressions and storms forming in the Bay of Bengal and Arabian Sea north of 5°N are described. The data used for the present account are generally available for the period 1890–1969. For the period 1890–1960 data and charts contained in the excellent publication *Tracks of Storms and Depressions in the Bay of Bengal and Arabian Sea* (ANONYMOUS, 1964) have been used. This publication gives a historical account and also contains a list of important references on the subject, which may be consulted for more details. For 1961–1969, individual yearly accounts have been used. After the definitions of cyclonic disturbances, their frequency and variability are discussed in the next two sections. The areas of formation are then described followed by an account of the frequency and the passage of cyclonic disturbances over the different parts of the coast. The chapter concludes with a discussion of the tracks of storms and depressions and their dissipation.

257

Definitions

The specifications of the tropical cyclonic disturbances in general use are: (*a*) depression when winds reach up to 33 knots (Beaufort 7); (*b*) moderate storm when winds are from 34 to 47 knots (Beaufort 8–9); (*c*) severe cyclonic storm when winds are from 48 to 63 knots.

These specifications conform to those of the World Meteorological Organization (W.M.O.) in their publication No. 9, T.P. 4, Vol. D, Part D, *Tropical Disturbances*. There is also a provision in the W.M.O. specifications for hurricanes, though such an expression is not in current usage in India. It is the practice to describe a storm as "Severe storm" when it reaches a stage where the wind speed exceeds 63 knots with a core of hurricane winds.

Bay of Bengal

Table I gives the monthly and annual number of depressions and storms that formed in the Bay of Bengal during the period 1890–1969. The monthly frequency of depressions and storms is summarised in Table II. The percentage of depressions which intensified into storms is also included in this table. Decade-wise annual frequency is shown in Table III. Statistical parameters, such as variability and skewness and kurtosis and W_1, W_2, for testing the normality of the monthly, seasonal and annual frequency distributions are detailed in Table IV.

TABLE I

MONTHLY AND ANNUAL NUMBER OF DEPRESSIONS AND STORMS IN THE BAY OF BENGAL, 1890–1969

Years	Jan.	Feb.	Mar.	Apr.	May	June	July	Aug.	Sep.	Oct.	Nov.	Dec.	Annual
1890	0	0	0	0	1	1	2	3	2	2	0	0	11
1891	0	0	0	0	1	0	0	2	5	1	2	0	11
1892	0	0	0	0	0	1	2	1	2	3	1	0	10
1893	0	0	0	1	1	1	1	3	3	3	1	0	14
1894	0	0	0	1	0	3	3	1	3	3	0	0	14
1895	0	0	0	1	0	2	2	3	3	1	0	1	13
1896	0	0	0	0	0	2	2	4	1	0	2	1	12
1897	0	0	0	0	1	1	1	1	2	2	0	2	10
1898	0	0	0	0	1	1	2	2	2	1	1	0	10
1899	0	0	0	1	0	2	0	3	2	3	1	0	12
1900	0	0	0	0	1	2	3	3	1	2	0	0	12
1901	0	0	0	0	1	2	2	2	1	2	2	1	13
1902	0	0	0	0	1	1	2	2	3	2	0	0	11
1903	0	0	0	0	1	1	3	2	2	2	3	1	15
1904	0	0	0	0	1	3	1	2	3	1	2	0	13
1905	0	0	0	0	1	1	4	1	4	1	0	1	13
1906	1	0	0	0	0	2	2	1	2	1	1	2	12
1907	0	0	1	0	2	2	0	3	1	2	2	1	14
1908	0	0	0	0	1	2	1	4	4	2	2	1	17
1909	0	0	0	0	2	1	3	0	3	2	1	1	13
1910	0	0	0	1	0	1	1	1	2	1	2	0	9
1911	0	0	0	1	0	2	0	3	2	2	0	0	10
1912	0	0	0	0	0	0	1	3	2	3	1	1	11
1913	0	0	0	0	0	2	3	2	2	3	2	1	15

TABLE I(*continued*)

Years	Jan.	Feb.	Mar.	Apr.	May	June	July	Aug.	Sep.	Oct.	Nov.	Dec.	Annual
1914	0	0	0	0	2	2	2	1	2	1	1	1	12
1915	0	0	0	0	0	1	0	2	3	2	3	1	12
1916	0	0	0	0	1	2	0	1	1	4	3	0	12
1917	0	0	0	0	1	3	1	2	2	2	1	0	12
1918	1	0	0	0	1	1	2	1	2	2	3	1	14
1919	0	0	0	0	0	2	2	4	1	1	3	1	14
1920	1	0	0	1	0	1	2	2	2	2	1	0	12
1921	1	0	0	1	1	2	2	1	3	3	1	1	16
1922	0	0	0	1	0	2	3	2	4	1	3	1	17
1923	1	0	0	1	1	0	1	3	2	1	1	3	14
1924	0	0	1	1	1	1	2	4	1	3	2	0	16
1925	1	0	1	1	2	1	3	2	1	2	3	1	18
1926	0	0	0	0	1	0	2	2	2	2	1	1	11
1927	0	0	0	0	3	1	4	4	4	2	1	1	20
1928	0	0	1	0	1	1	2	2	3	2	1	1	14
1929	1	0	0	0	1	2	3	4	0	2	2	0	15
1930	0	0	0	0	2	2	3	0	2	1	3	1	14
1931	0	0	0	0	1	0	1	4	2	4	0	2	14
1932	0	0	0	0	1	1	3	1	2	2	2	0	12
1933	0	0	0	0	1	1	3	2	3	1	1	1	13
1934	1	0	0	1	0	2	2	3	2	1	2	0	14
1935	0	0	0	1	0	1	2	1	2	2	2	1	12
1936	0	0	0	1	1	2	2	2	3	1	1	2	15
1937	0	0	0	2	0	1	3	1	4	2	2	1	16
1938	0	0	1	0	1	0	1	1	1	1	2	0	8
1939	1	0	0	1	0	0	2	3	3	2	2	1	15
1940	0	0	0	0	1	2	1	4	2	1	2	1	14
1941	0	0	0	0	1	4	2	2	2	1	2	1	15
1942	0	0	0	1	0	1	3	2	2	2	1	0	12
1943	0	0	0	0	2	0	3	1	1	3	1	0	11
1944	0	1	0	0	1	1	4	2	1	3	2	1	16
1945	1	0	0	1	0	0	3	1	3	1	2	2	14
1946	0	0	0	0	1	0	3	4	2	2	3	2	17
1947	1	1	0	1	0	2	2	2	2	3	1	2	17
1948	1	0	0	1	1	1	2	2	2	3	0	0	13
1949	0	0	0	1	2	1	1	1	3	2	0	0	11
1950	0	0	0	1	1	2	2	1	3	1	1	1	13
1951	0	0	0	0	0	2	3	3	1	1	1	1	12
1952	0	0	0	0	2	1	3	2	0	4	2	1	15
1953	1	0	0	1	0	2	0	2	2	1	2	0	11
1954	0	0	0	0	0	0	2	1	3	2	0	2	10
1955	0	0	0	0	2	2	0	1	3	3	2	0	13
1956	0	1	0	1	1	2	1	3	1	1	1	0	12
1957	0	0	0	0	0	0	0	1	2	0	0	1	4
1958	0	0	0	0	2	0	2	1	2	3	2	0	12
1959	0	0	0	0	0	2	3	1	1	1	1	1	10
1960	0	0	0	0	1	1	0	3	1	2	3	0	11
1961	1	1	0	0	2	2	1	0	4	1	0	0	12
1962	0	0	0	0	2	1	1	0	3	2	1	1	11
1963	1	0	0	0	1	2	1	1	2	3	0	1	12
1964	0	0	0	0	1	0	1	4	2	2	2	1	13
1965	0	0	0	0	2	0	2	2	1	2	1	2	12
1966	0	0	0	1	1	3	2	0	2	1	4	2	16
1967	1	0	0	0	1	1	2	1	3	3	1	1	14
1968	0	0	0	0	1	1	2	3	2	1	2	1	13
1969	0	0	0	0	1	1	1	2	3	2	1	2	13

TABLE II

FREQUENCY OF DEPRESSIONS AND STORMS IN THE BAY OF BENGAL, 1890–1969

Number of depressions and storms	Jan.	Feb.	Mar.	Apr.	May	June	July	Aug.	Sep.	Oct.	Nov.	Dec.	Annual
(1) 0	64	76	75	54	26	15	10	5	2	2	15	30	
1	16	4	5	25	40	31	18	24	16	27	28	38	
2				1	13	29	30	26	36	32	26	11	
3					1	4	19	15	19	15	10	1	
4						1	3	10	6	4	1		
5									1				
(2) Total	16	4	5	27	69	105	147	161	174	152	114	63	1,038
(3) Average	.20	.05	.06	.34	.86	1.3	1.8	2.0	2.2	1.9	1.4	.79	13.0
(4) Percentage of annual	1.5	.4	.5	2.6	6.7	10	14	15	17	15	11	6.1	
(5) Median	0	0	0	0	1	1	2	2	2	2	1	1	13
(6) Mode	0	0	0	0	1	1	2	2	2	2	1	1	12
(7) Number which intensified into storms	5	1	4	19	39	35	40	26	34	61	71	33	370
(8) Percentage of annual	1.3	.3	.1	.5	10.5	9.5	10.8	7.0	9.2	16.5	19.1	9.5	
(9) Percentage which intensified into storms (7)/(2) × 100	31	25	80	70	57	33	27	16	19	40	62	56	35

TABLE III

DEPRESSIONS AND STORMS IN THE BAY OF BENGAL, ANNUAL FREQUENCY OF 1890–1969

Number of depressions and storms	1890–1899	1900–1909	1910–1919	1920–1929	1930–1939	1940–1949	1950–1959	1960–1969	1890–1969
4							1		1
5									
6									
7									
8					1				1
9			1						1
10	3		1				2		6
11	2	1	1	1		2	1	2	10
12	2	2	4	1	2	1	3	3	18
13	1	4			1	1	2	3	12
14	2	1	2	2	3	2			12
15		1	1	1	2	1	1		7
16				2	1	1		2	6
17		1		1		2			4
18				1					1
19									
20				1					1
Average	11.7	13.3	12.1	15.3	12.3	14.0	11.2	13.4	13.0
Median									13.0
Mode									12.0

TABLE IV

VARIABILITY OF STORMS AND DEPRESSIONS IN THE BAY OF BENGAL, 1890–1969

| | Average | S.D. | C.V. | Skewness | Kurtosis | W_1 | W_2 |
				β_1	β_2		
Monthly							
Jan.	0.20	0.40	200	2.25	3.25	5.69	0.65
Feb.	0.05	0.22	436	17.05	18.05	15.65	30.31
Mar.	0.06	0.24	387	13.07	14.07	13.70	22.32
Apr.	0.34	0.50	148	0.97	2.66	3.74	—0.53
May	0.86	0.72	83	0.17	2.65	1.56	—0.55
June	1.30	0.87	67	0.06	2.90	0.93	—0.04
July	1.80	1.04	57	0.02	2.37	0.51	—1.12
Aug.	2.00	1.12	56	0.07	2.24	1.02	—1.37
Sep.	2.20	0.96	44	0.08	3.11	1.10	0.36
Oct.	1.90	0.88	47	0.11	2.63	1.25	—0.59
Nov.	1.50	0.99	68	0.03	2.32	0.67	—1.21
Dec.	0.79	0.72	91	0.29	2.77	2.06	—0.31
Seasonal							
Jan.–Feb.	0.25	0.5	194	3.18	5.33	6.75	4.82
Mar.–May	1.3	0.5	64	0.43	3.83	2.48	1.82
June–Sep.	7.3	1.7	24	0.05	3.74	0.85	1.62
Oct.–Dec.	4.1	1.4	35	0.00	2.46	0.21	—0.94
May–July	4.0	1.5	38	0.016	2.99	0.48	0.13
Annual	13.0	2.4	19	0.03	4.70	0.63	3.55

Let $X_1, X_2, \ldots X_n$ be n values.

$$\text{Average} = \bar{X} = \sum_{r=1}^{n} X_r / n$$

$$\text{Standard deviation (S.D.)} = \left\{ \sum_{r=1}^{n} (X_r - \bar{X})^2 / n - 1 \right\}^{\frac{1}{2}}$$

$$\text{Variance} = \sum_{r=1}^{n} (X_r - X)^2 / n - 1$$

$$\text{Skewness} = \beta_1 = \mu_3 / \mu_2^{3/a}$$

$$\text{Kurtosis} = \beta_2 = \mu_4 / \mu_2^{a} - 3$$

$$\mu_K = k\text{th moment}$$

$$= \sum_{r=1}^{n} (X_r - \bar{X})^k / n$$

$$\text{S.E. } (\beta_1) = (6/n)^{\frac{1}{2}}; \text{ S.E. } (\beta_2) = (24/n)^{\frac{1}{2}} \text{ approximately}$$

$$\text{S.E.} = \text{Standard error}$$

$$W_1 = \beta_1 / \text{S.E.}(\beta_1)$$

$$W_2 = \beta_2 / \text{S.E.}(\beta_2)$$

W_1 and W_2 are the ratios for testing normality and should be greater than 2 for significance at 5% level and greater than 2.6 for significance at 1% level.

Averages

The average number of depressions and storms, which formed in the Bay of Bengal annually, is 13. During January to March, cyclonic disturbances are very rare; only 25 have been reported during a period of 80 years. September with 2.2 has the highest monthly average; July, August and October have averages of nearly 2 and June and November about 1.3 each; May and December each average only about 0.8. The mean (average) median and mode are very close to each other.

Frequency and variability

Monthly

About 50% of the yearly total number of Bay disturbances develop during August–October. 82% of the total develop during the months June–November. September, with five disturbances, had the highest number formed in any month. September and October are the only two months during the 80-year period which have, except in two years, recorded one or more disturbances. The standard deviation is nearly 1 for almost all the months from June to November. The coefficient of variation (C.V.) has a wide range during these months, suggesting large variations from year to year. The lowest CV is 44% for September, and October has nearly the same value. W_1 and W_2 should be greater than 2 for significance at the 5% level and greater than 2.6 for the 1% level. The monthly frequency distributions for May–December are not skew. Statistically the frequency distributions for these months are not significantly different from normal.

Seasonal

The average number of cyclonic disturbances for the monsoon season June–September is 7.3, and they account for 56% of the annual number. The S.D. is 1.7 and the C.V. 24%. The extreme variations are 13 in 1927 and 3 in 1957. Compared to the post-monsoon season October–December which has an average of 4.1 and a higher CV of 34%, (nearly 50% higher) the southwest monsoon season is less variable and more dependable. The statistical nature of the frequency distributions for both seasons is normal.

Annual

Considering the entire year, the number of cyclonic disturbances varied between 4 and 20 though the frequency in these extreme ranges is almost negligibly small. The modal value is 12. The mean and median are equal to 13. Frequencies of 12 and 13 represent 45% of the total. The probability of the annual frequency, which is in the range 11–14, is 75% or three times in four years. The decade-frequencies also show that there has been no large variations in the means of the different decades. The S.D. is only 2.4 and the C.V. 19% which is not high.

It is noticeable when examining the variations from year to year that 50% of the annual variations are small and $< \pm 1$. 90% of the yearly variations lie in the range -4 to $+4$. The extreme variations are -8 and $+9$.

Storms

Table V gives the monthly and annual frequencies of storms. The average number of Bay storms in a year is 4.6 with the highest monthly average of 0.9 for November and 0.8 for October. Two to three storms occur during the non-monsoon months of May, October, November and December. The entire monsoon period may have only one to two storms in a year. The average for the southwest monsoon period is 1.7 and is the same as for October and November together. The highest number of storms in a year was 9 in 1893

TABLE V

STORM FREQUENCY IN THE BAY OF BENGAL 1890–1969

Number	Jan.	Feb.	Mar.	Apr.	May	June	July	Aug.	Sep.	Oct.	Nov.	Dec.	Annual
0	75	79	76	61	44	49	49	56	51	33	35	46	
1	5	1	4	19	33	27	24	22	25	34	25	33	252
2					3	4	5	2	3	12	15	1	45
3							2		1	1	4		8
4											1		1
Total	5	1	4	19	39	35	40	26	34	61	71	35	370
Average	.06	.01	.05	.24	.49	.44	.50	.33	.43	.76	.89	.44	4.63

Annual

Number	1	2	3	4	5	6	7	8	9
Frequency (years)	3	7	11	16	20	10	9	3	1

Av. = 4.6; S.D. = 1.8; C.V. = 38%; median = 5; mode = 5.

while the most frequent number of storms is 5. In recent years (1961–1969) the highest number was 7 in 1966. Once in two years, on an average, there will be 5 or more storms. During the 80-year period, 135 storms were listed for the southwest monsoon season of June–September; the monthly averages in these months were almost uniform. Almost an equal number of storms (132) were reported during October and November. The general impression, that the southwest monsoon period is one of few storms, is not therefore supported. There are, however, as is well known, certain major differences in the structure of the monsoon period storms which form mainly at the head of the Bay of Bengal, and storms which form in the non-monsoon months at low latitudes.

The percentage of disturbances which develop into cyclonic storms varies widely from month to month (Table II). It is as low as 16 and 19 % in August and September, respectively. May, November and December have the highest percentage of disturbances developing into storms. It is of interest to mention that during the last decade almost all the disturbances developed into storms in the months of May and November. Annually, 35 % of the Bay disturbances intensify into storms, i.e., one out of every three; of these storms 35 % further intensify into severe cyclonic storms. The percentages for the different months are:

J.	F.	M.	A.	M.	J.	J.	A.	S.	O.	N.	D.	Annual
40	100	50	42	66	12	19	4	30	39	44	40	35(%)

Arabian Sea

Table VI gives the monthly and annual numbers of depressions and storms which formed in the Arabian Sea from 1930 to 1969 (40 years) and Table VII their monthly and annual frequencies. The frequency of depressions and storms in the Arabian Sea is less than a sixth of those occurring in the Bay of Bengal. The annual average is only 1.9 with the S.D. 1.2 and the C.V. 63%. The C.V. is here more than three times the value for the Bay of Bengal. The percentage which develops into storms is about 50. The principal months when storms form are May, June, October and November. None of the months has, however, an average number higher than 0.4. There have been six years during the 40-year period when no depressions or storms have been reported from the Arabian Sea. The variability is high and the frequency low.

TABLE VI

MONTHLY AND ANNUAL NUMBER OF DEPRESSIONS AND STORMS IN THE ARABIAN SEA, 1930–1969

Year	Jan.	Feb.	Mar.	Apr.	May	June	July	Aug.	Sep.	Oct.	Nov.	Dec.	Annual
1930						2				1			3
1931													0
1932					2					1			3
1933					1		1						2
1934						1			1				2
1935	1					1							2
1936						1					1		2
1937						1				1			2
1938					1						1		2
1939										1	1		2
1940										1	1		2
1941					1					1			2
1942			1										1
1943											1		1
1944						1		1					2
1945											1		1
1946						1							1
1947			1						1				2
1948						2				2	1		5
1949					1								1
1950							1		1				2
1951			1			1					1		3
1952												1	1
1953													0
1954			1			1	1						3
1955													0
1956						1				1			2
1957					1				1	1			3
1958													0
1959					1	1	1			1			4
1960					1		1				2		4
1961					1	1	2						4
1962					1		1						2
1963					1					1	2		4
1964						1		1			1		3
1965												1	1
1966									1	1			2
1967													0
1968													0
1969						1							1

Areas of formation

The areas where depressions originally form and also their frequencies are briefly discussed. The data considered are for the 70-year period 1891–1960.

Bay of Bengal

January–March. The number of depressions during these months is practically negligible — 21 in a period of 70 years. Most of them were formed between 5° and 10°N.
April. The depressions originated from higher latitudes 7.5° to 15°N.

TABLE VII

MONTHLY AND ANNUAL FREQUENCIES OF STORMS AND DEPRESSIONS IN THE ARABIAN SEA, 1930–1969

Number of depressions and storms	Jan.	Feb.	Mar.	Apr.	May	June	July	Aug.	Sep.	Oct.	Nov.	Dec.
0	39	40	40	36	29	25	32	38	37	27	28	38
1	1	—	—	4	10	13	7	2	3	12	10	2
2	—	—	—	—	1	2	1	—	—	1	2	—
Total	1	0	0	4	12	17	9	2	3	14	14	2

	1930–1939	1940–1949	1950–1959	1960–1969	1930–1969
0	1	–	3	2	6
1	–	5	1	2	8
2	7	4	2	2	15
3	2	–	3	1	6
4	–	–	1	3	4
5		1			1
Average	2.0	1.8	1.8	2.1	1.9
Median					2
Mode					2
S.D.					1.2
C.V.					63%

May. This is a pre-monsoon month with an average of slightly less than one. No depressions in this month were observed during a third of the 70-year period examined (1891–1960). 66% of all the disturbances were formed in the latitudinal belt 10°–15°N but the longitudinal positions varied widely between 80° and 100°E. 74% were formed in the 10°- longitudinal belt 82.5°–92.5°E.

June–September. More than 80% of all the disturbances were formed in June, July and August in the 5°-belt of latitudes 17.5°–22.5°N. The percentage for September was only 60, suggesting a wider area of formation. The average for the season for the latitudes mentioned above is nearly 75% which shows that this area is a preferred position for the formation of depressions during the southwest monsoon season. Again, considering the longitudinal position, the concentration of 60% of the depressions in the 2.5° interval 87.5°–90°E is significant. Combined, the 5° square Bay area between longitudes 85°–90°E and latitudes 17.5°–22.5°N contained two-thirds of the total number of depressions which formed in the Bay of Bengal. The standard deviation of the positions of formation is about 2° (both for latitude and longitude) and the standard error of the mean position is less than 0.2°.

October. 50% of the depressions and storms in this month were formed between latitudes 10° and 15°N. The percentage rises to 80 if the latitude interval considered is 7.5°–17.5°N. 75% were reported in the longitudinal belt 82.5°–92.5°E.

November. A point of interest is that about 45% of the total number for this month were formed in the 2.5°-interval 7.5°–10°N. 75% were reported between 7.5° and 12.5°N and 90% between 7.5° and 15°N. The range of latitudinal variation is less than the range of variation in October. Though nearly 50% of the depressions were formed in the longitudinal belt 82.5°–87.5°E, the areas of formation extended widely, ranging from 77.5° to 100°E.

TABLE VIII

AREAS OF FORMATION OF DEPRESSIONS AND STORMS IN THE BAY OF BENGAL, 1891–1960

Long. °E	Lat. °N: 12.5°–15°	–17.5°	–20°	–22.5°	–25°	Total	Mean Lat. °N
June–Sep.:							
80.0°–82.5°					3	3	23.75
–85.0°	3	6	1		6	16	18.75
–87.5°	6	29	68	21	13	137	18.86
–90.0°	4	13	114	153	22	306	20.15
–92.5°	3	9	21	18	5	56	19.33
–95.0°	1	7	1	2		11	17.16
–97.5°	3	2				5	14.75
Total	20	66	205	194	49	534	

Long. °E	Lat. °N: 2.5°–5.0°	–7.5°	–10.0°	–12.5°	–15°	–17.5°	–20°	–22.5°	Total
Oct.–Dec.:									
77.5–80.0		3	3						6
–82.5		3	8	8	3				22
–85.0		2	16	19	10	3			52
–87.5	3	8	20	16	10	7	4		68
–90.0	3	2	18	9	3	3	8	5	53
–92.5	2	6	12	9	8	3	2	1	43
–95.0	1		6	6	10	1	1	1	26
–97.5		2	1	2	8				13
–100.0			2	2	1				5
Total	9	26	86	71	53		15	7	288

December. Positions are much lower, as many as 15% of the depressions were formed in latitudes less than 5°N. 80% were in the interval latitude 5°–12.5°N.

Table VIII gives the areas where depressions and storms were formed for the two principal seasons, June–September and October–December. The frequencies are grouped at intervals of 2.5°.

Figs.4–6 show the areas where depressions and storms were formed during the southwest monsoon season June–September, October–December and annual.

Arabian Sea

In the 70-year period 1891–1960, 17, 24, 21 and 15 depressions and storms have been listed as having formed in the Arabian Sea in the months May, June, October and November, respectively. The areas of formation have been examined. The latitudinal and longitudinal positions of formation range widely from less than 7.5° to 23°N and from 55° to 77.5°E, respectively.

The main latitudinal interval is 10°–12.5°N in May, October and November and 17.5°–20°N in June. 80% of the depressions and storms were formed between 10° and 12.5°N in May and June together. The longitudinal interval 67.5°–72.5°E accounted for 70% of the total. In October and November 50% were formed between 10° and 12.5°N. The longitudinal positions varied widely between 62.5° and 75°E.

Frequency of Bay depressions and storms crossing the coasts of India, Burma, Bangladesh and Sri Lanka

Tables X and XI give the frequency of depressions and storms which formed in the Bay of Bengal and crossed the coasts of India and Pakistan, Burma and Sri Lanka during the

TABLE IX

AREAS OF FORMATION OF DEPRESSIONS AND STORMS IN THE ARABIAN SEA, 1890–1969

Long. °E	Lat. °N: 5°–7.5°	7.5°–10°	10°–12.5°	12.5°–15°	15°–17.5°	17.5°–20°	20°–22.5°	22.5°–25°	Total
May–June:									
55.0°–57.5°		1							1
–60.0°		1							1
–62.5°		1							1
–65.0°						2			2
–67.5°				1		2	1		4
–70.0°	1		3	5	6				15
–72.5°	3	3	2	3	3				14
–75.0°		2							2
–77.5°								1	1
Total	4	8	6	8	13	1		1	41

Long. °E	Lat. °N: 5°–7.5°	7.5°–10°	10°–12.5°	12.5°–15°	15°–17.5°	17.5°–20°	20°–22.5°	22.5°–25°	Total
Oct.–Nov.:									
57.5°–60.0°		1							1
–62.5°		1							1
–65.0°	1	1	1						3
–67.5°		2		2		1			5
–70.0°		3		1					4
–72.5°	1	3	2	2	1				9
–75.0°	3	8	1						12
–77.5°	1								1
Total	1	5	19	4	5	1	1		36

TABLE X

FREQUENCY OF STORMS AND DEPRESSIONS CROSSING COASTS IN THE BAY OF BENGAL, 1890–1969

	Jan.	Feb.	Mar.	Apr.	May	June	July	Aug.	Sep.	Oct.	Nov.	Dec.	Total
(I) Storms and depressions													
1. Frequency	16	4	5	27	69	105	147	161	174	152	114	63	1,038
2. Number which crossed coast	9	1	5	19	54	77	120	136	132	125	73	24	772
3. (2) as percentage of (1)	56	25	100	70	78	73	81	85	76	81	64	43	74
(II) Storms													
1. Frequency	5	1	4	19	39	35	40	26	34	61	71	35	370
2. No. of storms which crossed coast	5	1	3	14	37	32	41	27	31	63	60	20	334
3. (2) as percentage of (1)	100	100	75	74	95	91	100	100	91	100	85	57	90
(III) Storms crossing as percentage of storms and depressions II(2)/I(2)	56	100	60	74	70	42	34	20	25	51	82	80	43

TABLE XI

FREQUENCY OF DEPRESSIONS AND STORMS CROSSING DIFFERENT PARTS OF COASTS OF INDIA, BURMA, BANGLA-
DESH AND SRI LANKA, 1890–1969

Coastal belt	Lat. (°N)	J.	F.	M.	A.	M.	J.	J.	A.	S.	O.	N.	D.	Total	Aver.	(%)
India																
TRV–PBN*	0829–0916	1										2	2	5	.06	0.6
PBN–NGP	0916–1106	2	1	1	1	1						8	5	19	.2	2
MGP–MDS	1106–1300	1		2	1	5					9	17	2	37	.5	5
MDS–MPT	1300–1611	1			2	4	1			5	20	12		45	.6	6
MPT–VZG	1611–1743	1				2	1			12	16	7		39	.5	5
VZG–GPL	1743–1916					1	5	3	9	20	11	2		51	.6	7
GPL–CAL	1916–2239					4	49	99	102	79	21	5		359	4.5	47
Bangladesh																
CAL–CHT	2239–2239				11	15	17	24	15	24	6	2		114	1.4	15
CHT–AKY	2239–2008			2	9	5	1	1		12	4	4		33	.4	4
Burma																
AKY–RGN	2008–1646				10	18	1			1	11	8	2	49	.6	6
RGN–TVY	1646–1405			3							1			4	.05	4
Sri Lanka		3		2							1	1	10	17	.2	2
Total		9	1	5	19	54	77	120	136	132	125	73	27	772		
Average		.1		.06	.2	.7	1.0	1.5	1.7	1.7	1.6	.9	.3	9.7		
Percentage of annual		1		1	2	7	10	16	18	17	16	9	4			

* Abbreviations explained in legend of Fig. 1.

period 1890–1969. In Table IX the frequency of the number of depressions and storms is given which crossed the different parts of the coasts.

(*1*) The average number of depressions and storms, which crossed the coasts, is 9.7 in a year, which is 75% of the total annual number of the Bay depressions and storms.

(*2*) 43% of all the disturbances (depressions and storms) which formed in the Bay crossed the coasts.

(*3*) The average number of disturbances which crossed in each of the months July–October is nearly identical — about 1.6.

About half the number of Bay depressions and storms crossed the coast between Gopalpur and Calcutta (19°–22°39′N). Two-thirds of all the Bay depressions and storms which crossed the coasts were during the southwest monsoon period (June–September). Of the 256 disturbances which crossed during July and August, 85% were between Gopalpur and Calcutta.

Burma

The coastal belt considered is from Akyab to Rangoon and from Rangoon to Tavoy. The annual average for this area is only 0.65 or 2 crossings in three years. This figure corresponds to only 7% of the number of disturbances which crossed the coast from Trivandrum to Tavoy.

The number which crossed between Rangoon and Tavoy was a negligible 4 in 80 years. April, May, October and November are the principal months when Bay depressions and storms strike the coast of Burma. May has the highest monthly average. The Burma coast accounts for a third of the crossings in this month. Though more than 50% of all

Bay disturbances cross the Burma coast in April, the average is only a third of the May average.

Bangladesh

(*1*) Calcutta to Chittagong is a very important coastal belt and has crossings amounting to more than 15% of the annual total. The percentages for the individual months are:

May	June	July	Aug.	Sep.	Oct.	Nov.
20	20	14	18	11	9	8%

(*2*) Chittagong to Akyab adds another 4% to the total of crossings mentioned above.

(*3*) The average number of crossings in a year between Calcutta and Akyab is 1.8.

Sri Lanka

December is the principal month for crossings; in that month 60% of the total number of disturbances which affected Sri Lanka crossed its coasts. The annual average is a very small 0.2.

Tracks

Tracks of cyclonic storms in the Bay of Bengal and the Arabian Sea are shown in Figs.1–3 for the representative months of May (pre-monsoon), July (monsoon) and October (post-monsoon). The frequency of storms and depressions in different parts of the Bay of Bengal, in 2.5°-squares, is shown for the southwest monsoon season June–September and post-monsoon season October–December in Figs.4 and 5, respectively. To supplement, the frequency of dissipation of depressions and storms in the two seasons June–September and October–December and annual is also shown in Figs.6–8, respectively.

Bay of Bengal

The tracks of storms and depressions vary widely in the different months and seasons. In the pre-monsoon months the Bay storms mostly recurve and 70% of them strike the coastal belt between Calcutta and Rangoon. With the onset of the monsoon in June, the areas of formation are over the northern portion of the Bay. Striking mostly between Gopalpur (19°16′N) and Calcutta, they move in a northwesterly direction. In July and August, the features are similar, most of them move in a west-northwest to northwesterly direction. In September the direction of movement is northwest, though occasionally it might be north or in other directions.

In his report on the Orissa rivers, MAHALANOBIS (1941) has given tables for the mean latitude position of the storm track corresponding to longitudes at 1°-intervals from 74°E to 90°E. The study covers storm tracks during 1891 to 1928 and the mean latitudinal positions are given for the months July to September. These are reproduced in Tables XII and XIII. The mean tracks for each of the months July, August and September and for

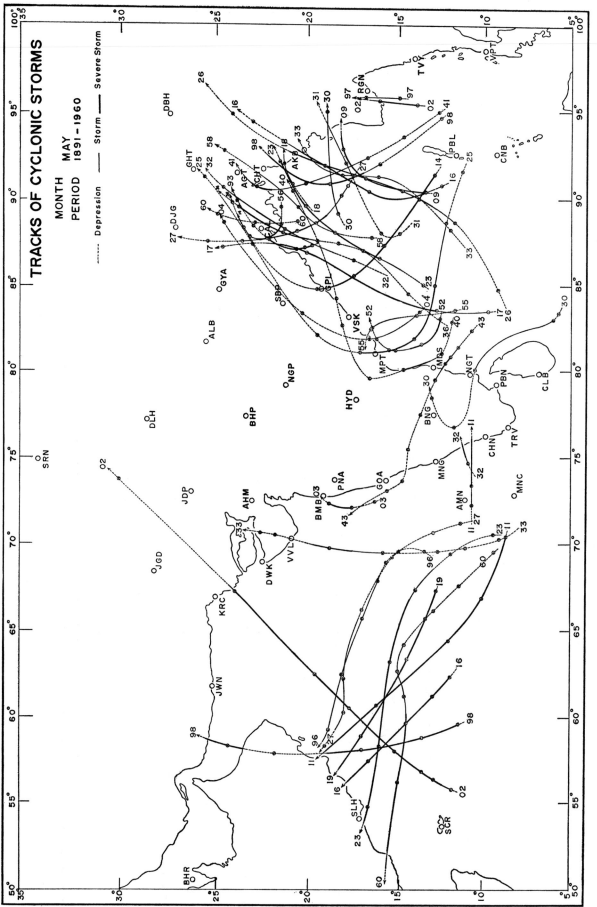

Fig. 1. Tracks of cyclonic storms, May 1891–1960. *AGT* = Agartala; *AHM* = Ahmadabad; *AKB* = Akyab; *ALB* = Allahabad; *AMN* = Aminidevi; *BHP* = Bhopal; *BHR* = Bahrain; *BMB* = Bombay; *BNG* = Bangalore; *CAL* = Calcutta; *CHN* = Cochin; *CHT* = Chittagong; *CLB* = Colombo; *CNB* = Carnicobar; *DBH* = Dibrugarh; *DJG* = Darjeeling; *DLH* = Delhi; *DWK* = Dwaraka; *GHT* = Gauhati; *GOA* = Goa; *GPL* = Gopalpur; *GYA* = Gaya; *HYD* = Hyderabad; *JCD* = Jacobabad; *JDP* = Jodhpur; *JWN* = Jiwani; *KRC* = Karachi; *MDS* = Madras; *MNC* = Minicoy; *MNG* = Mangalore; *MPT* = Machilipattnam; *NGP* = Nagpur; *NGT* = Nagappattinam; *PBL* = Port Blair; *PBN* = Pamban; *PNA* = Poona (Pune); *RGN* = Rangoon; *SBP* = Sambalpur; *SCR* = Socotra; *SLH* = Salalah; *SRN* = Srinagar; *TRV* = Trivandrum; *TVY* = Tavoy; *VPT* = Victoria Point; *VSK* = Vishakhapatnam;

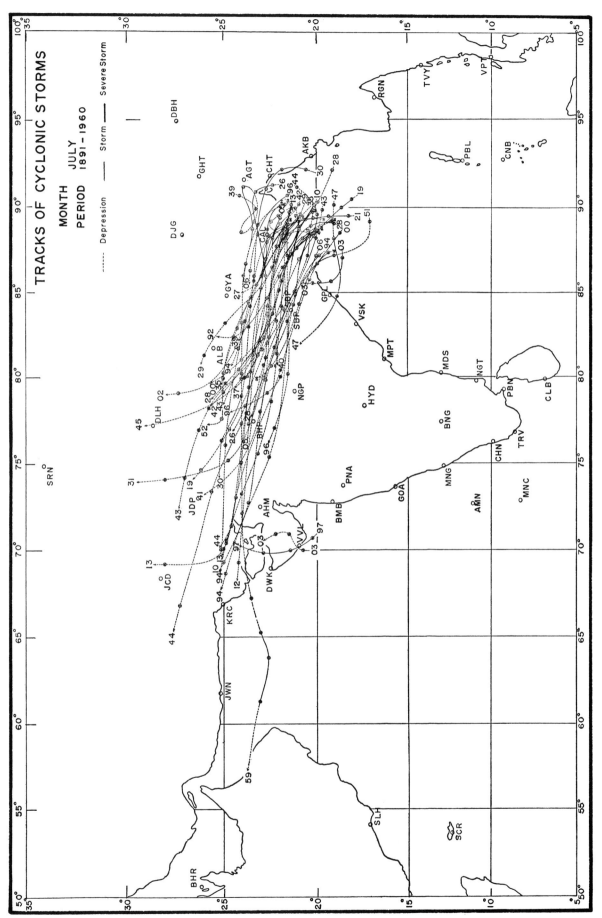

Fig. 2. Tracks of cyclonic storms, July 1891–1960. Legend Fig. 1.

Tropical cyclones of the Indian Seas

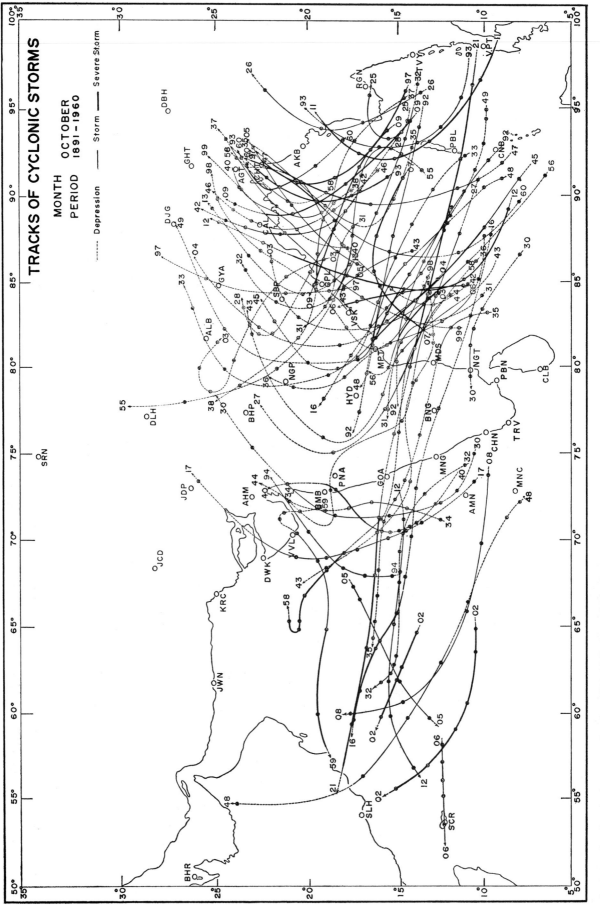

Fig. 3. Tracks of cyclonic storms, October 1891–1960. Legend Fig. 1.

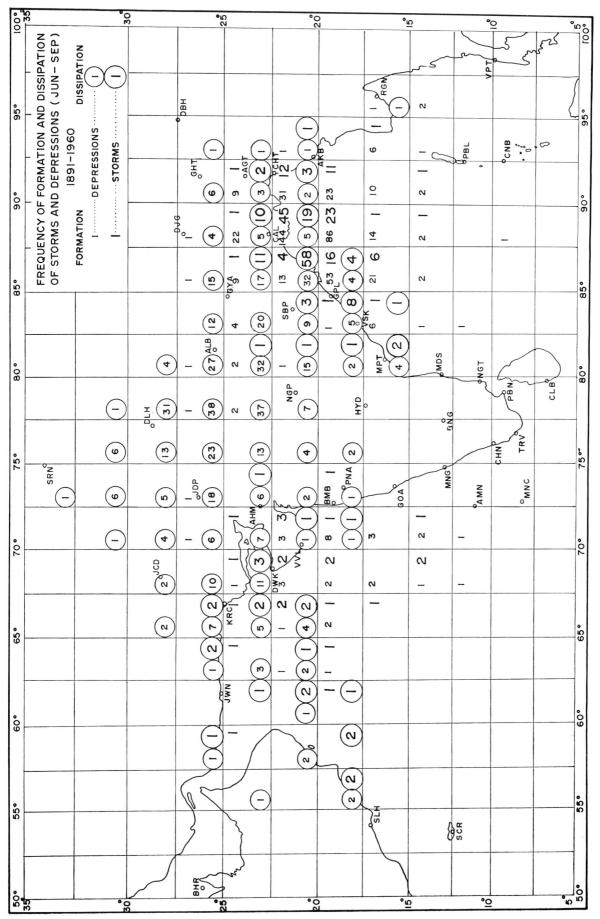

Fig. 4. Frequency of formation and dissipation of storms and depressions, June–September 1891–1960. Legend Fig. 1.

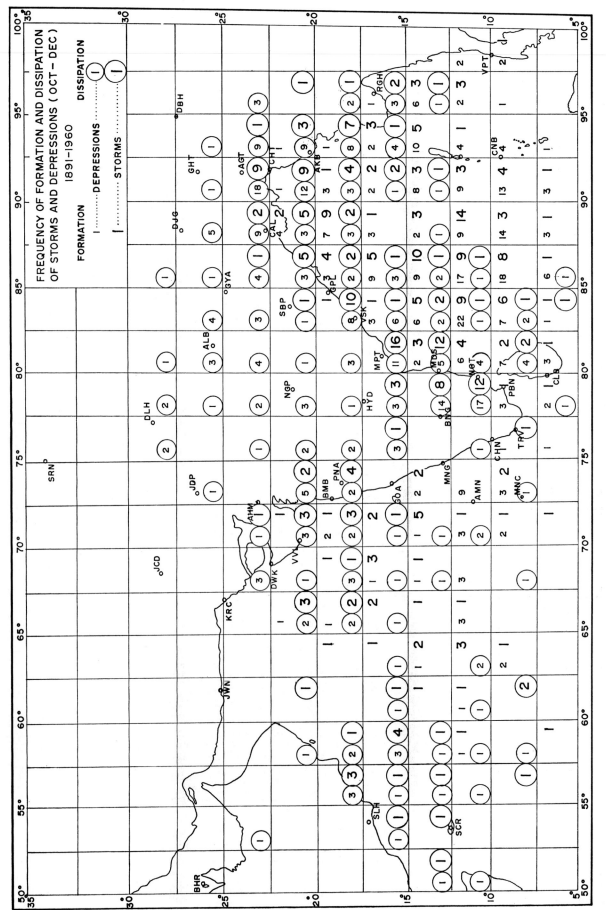

Fig. 5. Frequency of formation and dissipation of storms and depressions, October–December 1891–1960. Legend Fig. 1.

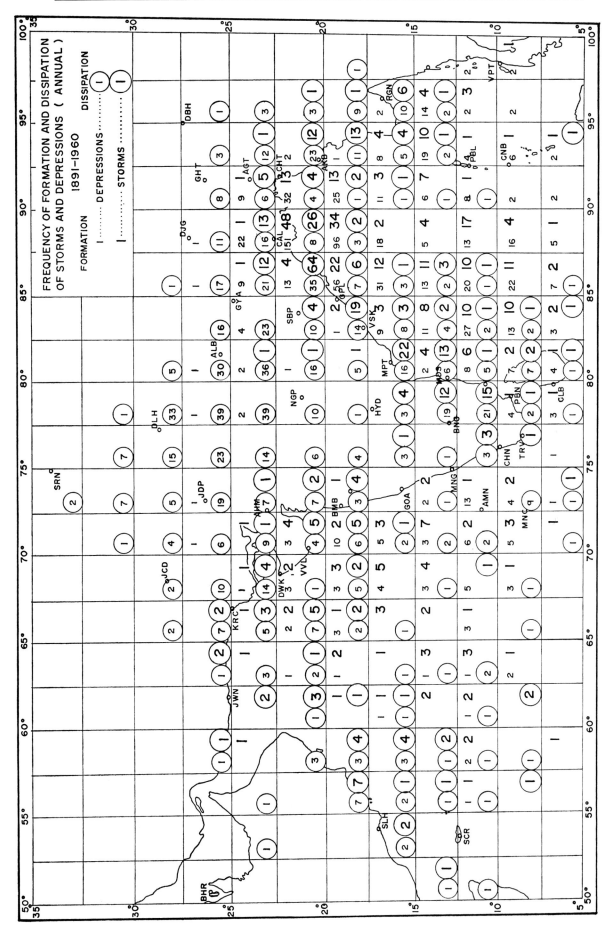

Fig. 6. Annual frequency of formation and dissipation of storms and depressions, 1891–1960. Legend Fig. 1.

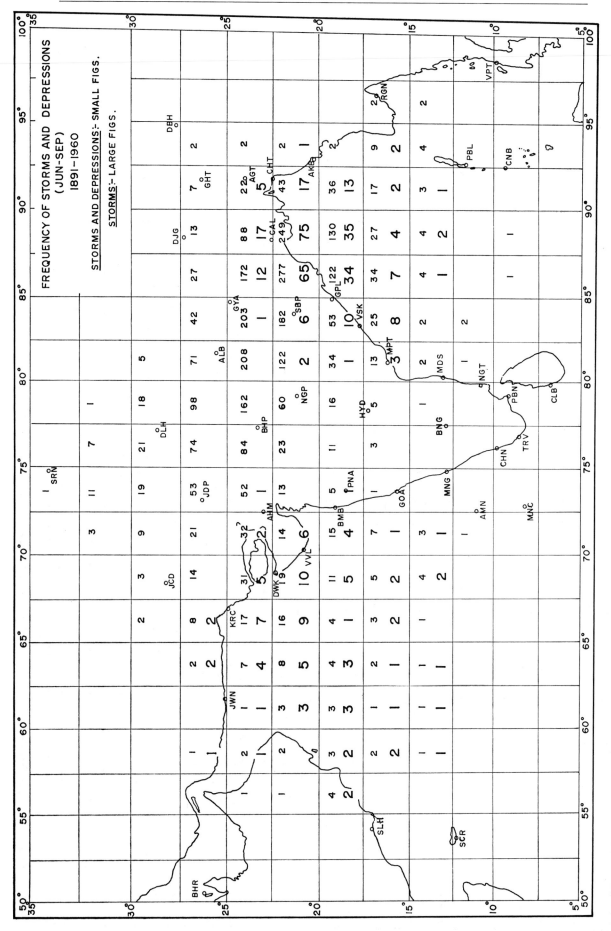

Fig. 7. Frequency of formation and dissipation of storms and depressions, June–September 1891–1960. Legend Fig. 1.

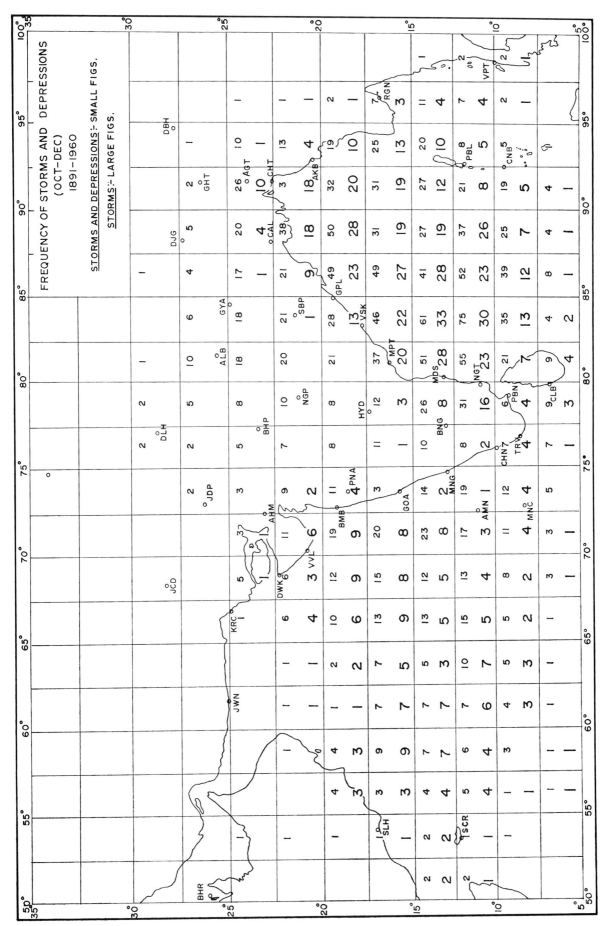

Fig. 8. Frequency of formation and dissipation of storms and depressions, October–December 1891–1960. Legend Fig. 1.

TABLE XII

STORM TRACKS 1891–1928: MEAN LATITUDE WITH PROBABLE ERROR AND STANDARD DEVIATION FOR JULY, AUGUST AND SEPTEMBER

Long. (°E)	July			August			September		
	no. of storms	mean ± P.E.	S.D.	no. of storms	mean ± P.E.	S.D.	no. of storms	mean ± P.E.	S.D.
90°	—	—	—	8	20.86 ± .31	1.29	20	18.43 ± .31	2.04
89°	—	—	—	19	21.36 ± .20	1.29	22	19.81 ± .23	1.62
88°	34	21.47 ± .15	1.30	30	21.26 ± .17	1.42	26	20.44 ± .19	1.43
87°	40	21.51 ± .13	1.23	34	21.11 ± .17	1.46	29	20.61 ± .23	1.87
86°	41	22.04 ± .15	1.45	39	21.30 ± .21	1.90	33	21.13 ± .27	2.32
85°	38	22.15 ± .13	1.20	37	21.97 ± .19	1.71	30	21.74 ± .30	2.47
84°	36	22.35 ± .13	1.18	35	22.31 ± .17	1.49	28	21.69 ± .30	2.34
83°	35	22.58 ± .13	1.11	34	22.79 ± .13	1.15	26	21.87 ± .27	2.07
82°	34	22.84 ± .14	1.19	32	23.31 ± .14	1.20	25	22.24 ± .28	2.10
81°	33	23.12 ± .15	1.29	29	23.62 ± .14	1.13	24	22.77 ± .25	1.81
80°	31	23.68 ± .21	1.75	25	24.07 ± .18	1.36	22	23.02 ± .28	1.96
79°	28	23.89 ± .21	1.68	21	24.14 ± .18	1.19	17	23.44 ± .22	1.35
78°	22	23.90 ± .22	1.55	19	24.40 ± .32	2.08	10	23.87 ± .30	1.42
77°	15	23.89 ± .27	1.55	12	23.83 ± .41	2.13	9	24.11 ± .42	1.89
76°	11	24.01 ± .32	1.55	11	23.94 ± .46	2.27	6	24.38 ± .47	1.70
75°	11	24.36 ± .35	1.74	8	24.06 ± .64	2.70			
74°	8	23.40 ± .68	2.84	4	24.18 ± .21	.62			

TABLE XIII

MEAN STORM TRACKS FOR COMBINED PERIOD JULY–SEPTEMBER, 1891–1928.

Long. (°E)	Number of storms	Mean lat. with P.E.	S.D.
90°	28	19.13 ± .28	1.46
89°	41	20.52 ± .18	1.13
88°	90	19.79 ± .21	1.95
87°	103	21.12 ± .10	.96
86°	113	21.52 ± .12	1.30
85°	105	21.97 ± .12	1.23
84°	99	22.15 ± .12	1.15
83°	95	22.46 ± .10	1.00
82°	91	22.84 ± .11	1.06
81°	86	23.19 ± .12	1.08
90°	78	23.62 ± .11	.93
79°	66	23.85 ± .12	1.00
78°	51	23.08 ± .23	2.01
77°	36	24.00 ± .21	1.25
76°	28	24.06 ± .24	1.28
75°	19	24.24 ± .34	1.48
74°	12	23.66 ± .45	1.51

the combined period are shown in Fig.9. It will be observed that the standard deviations of the mean positions are generally 1° to 2°; these are higher in September. The probable error of these tracks of the mean positions is as small as 0.1° to 0.2° for longitudes 79° to 89°E. In October–December they have struck almost all parts of the coast; recurvature is their principal feature.

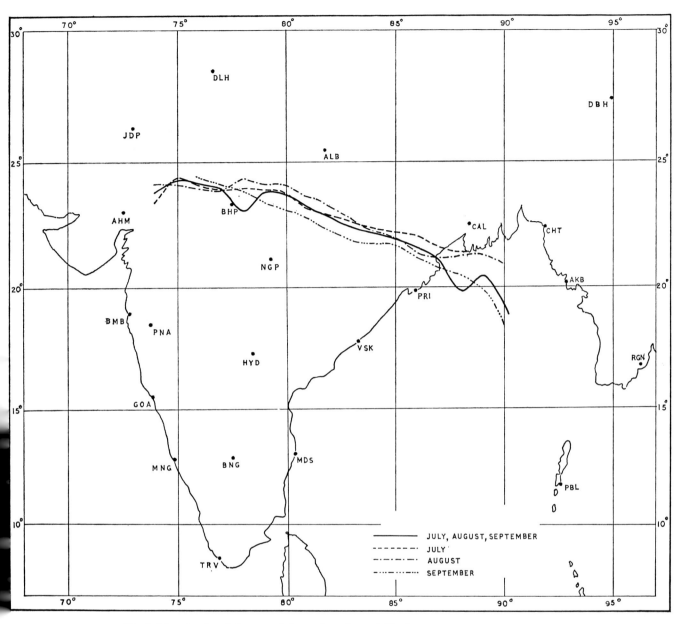

Fig. 9. Mean tracks of storms and depressions. Legend Fig. 1.

Arabian Sea

The number of storms is considerably smaller, and those which cross the coast and move over land are even fewer. During a period of 70 years, in April and May only four storms crossed the west coast. In the monsoon season, the frequency is practically nil. The main months during which storms, which rise in the Arabian Sea, strike the coast and move over land are October and November. These storms strike mostly the north Bombay and Gujarat coasts. Storms or depressions which form in the Arabian Sea seldom strike south of Bombay.

Dissipation

The duration of a depression over sea/land areas varies widely according to month and season. In the monsoon season when the disturbances form at the head of the Bay of Bengal, the average duration over sea is only 1.5 days. Individually they have varied up to 4 days over sea. The period they remain over sea is thus very small; nearly half the cases considered during 1961–1969 have a duration of 1 day or less. Over land they last longer, the average duration being 3 days. The duration is longer in August and September. In the months October to December, the duration over the sea is longer, the monthly averages being 3, 5 and 5 days, respectively. Of the 41 cyclonic disturbances during 1961–1969, 63% remained over sea for 4 or more days:

No. of days:	2	3	4	5	6	7	8	9	13
Frequency:	10	6	12	4	5	1	1	1	1

After crossing, they decrease in intensity and dissipate rapidly. As may be seen from Figs.4, 5 and 8, most of them dissipate by the time they reach longitude 77.5°E or latitude 27.5°N.

Acknowledgements

I wish to express my grateful thanks to Dr. A. S. Ramanathan (Meteorologist) and Mr. A. Tiruvengadathan (Assistant Meteorologist) of the Office of the Deputy Director General (Forecasting), Poona, for considerable help in preparing and checking the material of this chapter.

References

ANONYMOUS, 1964. *Tracks of Storms and Depressions in the Bay of Bengal and Arabian Sea, 1877–1960.* India Meteorological Department.
India Weather Review, Annual Summary, 1945; Part C, 1941. India Meteorological Department.
MAHALANOBIS, P. C., 1941. *Report on Orissa Rivers*. Government of Orissa, India.

Appendix—Climatic tables

TABLE XIV

CLIMATIC TABLE FOR PESHAWAR
Latitude 34°01′N, longitude 71°35′E, elevation 1,164 m

Month	Mean sta. press. (mbar)	Temperature (°C)				Mean vapour press. (mbar)	Precipitation (mm)	
		mean daily	mean daily range	extremes			mean	24-h max.
				max.	min.			
Jan.	977.1	10.9	12.5	25.0	−3.3	6.9	36.6	77.0
Feb.	974.9	12.8	12.3	30.0	−2.2	8.8	38.9	55.1
Mar.	972.1	17.6	12.5	37.2	0.6	10.5	62.0	55.9
Apr.	968.5	22.7	13.8	42.2	5.0	12.9	44.7	61.5
May	963.4	28.7	14.8	47.8	10.6	12.7	19.6	97.8
June	957.7	32.8	15.5	48.9	17.2	16.5	7.9	67.3
July	956.7	32.9	12.4	50.0	18.9	24.1	32.0	69.9
Aug.	959.1	31.4	10.7	47.8	18.9	24.9	51.6	150.9
Sep.	964.3	28.6	12.9	43.3	14.4	18.9	20.6	51.3
Oct.	970.9	23.4	15.2	38.3	6.1	11.9	4.8	31.0
Nov.	975.3	17.2	15.5	32.8	0.0	12.3	7.9	50.5
Dec.	977.3	12.1	14.4	28.3	−2.2	7.2	17.0	40.9
Annual	968.1	22.6	13.6	50.0	−3.3	13.5	344.4	150.9

Month	Mean evap. (mm)	Number of days with			Mean cloudiness (oktas)	Mean sunshine (h)	Wind	
		precip.	thunder-storms	fog			prevalent direct.	mean speed (m/sec)
Jan.		6	0.2	0.4	5.5		180	0.6
Feb.		7	1.5	0.3	6.5		180	0.8
Mar.		8	5	0.1	5.5		180	0.9
Apr.		7	6	0.1	5.0		270	1.0
May		4	7	0	3.5		310	1.1
June		2	6	0	2.5		360	1.1
July		4	9	0	3.7		360	1.0
Aug.		3	8	0	3.9		360	0.9
Sep.		2	5	0	2.1		360	0.8
Oct.		1.8	4	0	2.1		180	0.5
Nov.		1.2	0.8	0	2.5		180	0.5
Dec.		2	0.3	0.2	4.7		180	0.5
Annual		48	53	1.1	4.0		270	0.8

TABLE XV

Latitude 34°05′N, longitude 74°50′E, elevation 1,586 m

Month	Mean sta. press. (mbar)	Temperature (°C)				Mean vapour press. (mbar)	Precipitation (mm)	
		mean daily	mean daily range	extremes			mean	24-h max.
				max.	min.			
Jan.	844.9	1.1	6.7	17.2	−14.4	5.4	72.8	147.8
Feb.	843.2	3.5	8.7	20.6	−20.0	6.0	72.3	66.3
Mar.	842.3	8.5	9.9	25.6	−5.6	7.9	104.1	70.1
Apr.	841.7	13.3	11.9	31.1	0	10.0	78.1	65.3
May	839.1	17.9	13.4	35.6	2.8	12.3	63.4	52.8
June	835.3	21.7	14.6	37.8	7.2	15.0	35.6	65.8
July	832.9	24.6	12.4	38.3	10.6	19.5	61.0	79.8
Aug.	834.2	23.9	12.0	36.7	10.0	18.7	62.8	67.3
Sep.	838.2	20.5	15.6	35.0	4.4	14.1	31.8	102.4
Oct.	843.1	14.1	16.9	33.9	−1.7	9.7	28.7	59.9
Nov.	845.0	7.7	15.6	23.9	−7.8	6.8	17.5	64.3
Dec.	845.3	3.5	10.6	18.3	−11.7	5.8	35.9	64.5
Annual	840.4	13.3	12.3	38.3	−20.0	10.9	564.0	147.8

Month	Mean evap. (mm)	Number of days with			Mean cloudi-ness (oktas)	Mean sun-shine (h)	Wind	
		precip.	thunder-storms	fog			prevalent direct.	mean speed (m/sec)
Jan.	3.4	11	0.4	1.1	6.6	2.0	110	1.0
Feb.		8	0.2	0.7	5.9	3.7	330	1.2
Mar.		14	3	0.1	5.4	4.0	320	1.6
Apr.		12	7	0	4.6	6.0	130	1.5
May		10	10	0	4.0	7.6	130	1.2
June		7	7	0	3.1	7.5	120	1.1
July		9	7	0	3.8	8.0	140	1.1
Aug.		10	6	0	4.2	7.4	140	1.0
Sep.		6	4	0	2.7	7.9	140	1.0
Oct.		5	2	0	2.3	7.4	140	0.9
Nov.		4	0	0.4	2.9	7.0	140	0.9
Dec.		7	0.2	1.1	5.1	4.0	130	0.9
Annual		103	47	3	4.2	6.0	150	1.1

TABLE XVI

CLIMATIC TABLE FOR LEH
Latitude 34°09′N, longitude 77°34′E, elevation 3,514 m

Month	Mean sta. press. (mbar)	Temperature (°C)				Mean vapour press. (mbar)	Precipitation (mm)	
		mean daily	mean daily range	extremes			mean	24-h max.
				max.	min.			
Jan.	664.5	−8.4	11.8	8.3	−28.3	1.8	11.8	24.4
Feb.	663.7	−5.5	12.6	12.8	−25.6	2.1	8.6	16.8
Mar.	664.3	0.0	12.7	19.4	−19.4	3.0	11.9	16.0
Apr.	666.0	5.6	13.6	23.9	−12.8	3.6	6.5	22.1
May	655.8	9.9	14.3	28.9	−4.4	4.1	6.5	22.3
June	664.1	13.9	14.4	33.9	−1.1	5.3	4.3	19.6
July	663.1	17.3	14.5	33.3	0.6	8.1	15.7	23.6
Aug.	663.4	16.9	14.6	32.2	2.8	8.4	19.5	51.3
Sep.	665.5	13.1	15.5	30.6	−4.4	6.1	12.2	25.9
Oct.	667.7	6.7	15.1	25.6	−7.8	3.5	7.1	39.1
Nov.	667.3	0.6	14.4	20.0	−13.9	2.5	2.9	16.2
Dec.	666.0	−4.7	12.7	12.8	−25.6	1.9	8.0	15.2
Annual	665.1	5.5	13.8	33.9	−28.3	4.2	115.0	51.3

Month	Mean evap. (mm)	Number of days with			Mean cloudiness (oktas)	Mean sunshine (h)	Wind	
		precip.	thunder-storms	fog			prevalent direct.	mean speed (m/sec)
Jan.		6	0	0.1	4.9		050	0.9
Feb.		4	0	0	4.7		050	1.1
Mar.		4	0	0.3	4.8		230	1.5
Apr.		3	0.1	0	4.3		180	1.8
May		3	0.5	0	4.1		190	1.9
June		2	1.8	0	3.3		210	1.8
July		4	0.9	0	3.7		200	1.4
Aug.		5	0.8	0	4.1		220	1.3
Sep.		3	0.1	0	2.9		170	1.3
Oct.		1.6	0	0	2.3		130	1.4
Nov.		0.9	0	0	2.9		050	1.4
Dec.		5	0	0.1	4.3		050	1.0
Annual		41	4	0.5	3.8		140	1.4

TABLE XVII

CLIMATIC TABLE FOR KABUL
Latitude 34°30′N, longitude 69°13′E, elevation 1,985 m

Month	Mean sta. press. (mbar)	Temperature (°C)				Mean vapour press.* (mbar)	Precipitation (mm)	
		mean daily	mean daily range	extremes			mean	24-h max.
				max.	min.			
Jan.		−2.1	10.0	14.4	−21.1	2.8	32.77	49.78
Feb.		0.6	10.1	23.3	−20.6	3.2	37.85	25.15
Mar.		6.2	10.7	20.0	−14.4	4.6	91.19	44.70
Apr.		12.2	12.6	27.8	−2.8	7.2	83.82	28.70
May		18.2	15.1	35.0	1.1	8.9	21.59	17.02
June		21.9	17.2	37.8	5.6	9.2	4.32	15.24
July		24.6	17.4	38.3	10.6	10.7	1.78	7.37
Aug.		23.7	18.2	37.2	8.3	10.5	2.29	6.35
Sep.		19.9	18.4	36.1	2.2	7.5	1.02	7.62
Oct.		14.1	17.7	31.7	−1.1	4.5	10.16	6.60
Nov.		8.2	16.0	25.0	−20.6	3.4	15.24	15.24
Dec.		2.8	11.6	18.9	−20.6	3.3	14.48	26.42
Annual		12.5	14.6	38.3	−21.1	6.3	316.48	49.78

Month	Mean evap. (mm)	Number of days with			Mean cloudi-ness* (oktas)	Mean sun-shine (h)	Wind	
		precip.	thunder-storms	fog			prevalent direct.	mean speed (m/sec)
Jan.		8	0	0.5	4.6		230	1.6
Feb.		9	0.1	0.2	5.4		230	1.6
Mar.		10	0.7	0	4.4		180	1.3
Apr.		9	6	0.1	3.7		180	1.0
May		6	8	0.5	2.3		180	1.0
June		1.5	3	0	0.7		180	1.1
July		0.5	1.3	0	0.7		180	1.0
Aug.		0.7	2	0	0.9		180	0.8
Sep.		0.2	2	0.1	0.5		180	0.8
Oct.		1.5	1.8	0	1.1		230	0.9
Nov.		1.8	0.3	0.1	2.5		230	0.9
Dec.		5	0	0.1	4.8		230	1.1
Annual		53	25	1.6	2.6		200	1.1

* Only 08h30 data available.

TABLE XVIII

CLIMATIC TABLE FOR KABUL (AIRPORT)
Latitude 34°33′N, longitude 69°12′E, elevation 1,803 m

Month	Mean sta. press. (mbar)	Temperature (°C)				Relat. humid. (%)	Precipitation (mm)	
		mean daily	mean daily range	extremes			mean	24-h max.
				max.	min.			
Jan.	825.5	−0.3	15.5	15.0	−11.3	72	13.1	9.0
Feb.	822.3	4.2	11.2	17.0	−5.4	76	83.4	18.0
Mar.	820.2	6.1	13.1	21.3	−5.8	72	79.7	25.7
Apr.	820.4	11.0	12.7	24.0	−1.0	76	70.8	16.0
May	820.0	17.0	17.6	30.0	3.0	51	11.9	3.2
June	816.9	23.1	19.7	35.0	8.8	29	0	0
July	814.5	24.6	16.9	35.7	10.2	34	3.5	2.5
Aug.	815.2	23.9	17.0	35.2	10.5	35	6.8	6.8
Sept.	819.2	18.7	18.9	32.2	5.4	41	3.0	3.0
Oct.	823.0	12.1	16.6	26.8	−2.7	55	4.1	2.1
Nov.	824.2	3.1	19.8	20.6	−9.3	47	0	0
Dec.	822.5	−0.7	17.2	14.7	−11.5	47	4.8	2.8
Annual	820.3	11.9	16.3	35.7	−11.5	53	281.1	25.7

Month	Mean evap. (mm)	Number of days with			Mean cloudi-ness (oktas)	Mean sun-shine (h)
		precip.	thunder-storms	fog		
Jan.		2	0		3	
Feb.		14	0		5	
Mar.		16	1		5	
Apr.		13	7		5	238.3
May		9	8		4	321.9
June		0	0		1	367.8
July		4	2		2	—
Aug.		1	3		1	356.5
Sep.		1	1		1	291.5
Oct.		4	0		2	277.3
Nov.		0	0		tr	280.0
Dec.		3	0		2	224.9
Annual		67	22		2.6	2358.2

TABLE XIX

CLIMATIC TABLE FOR KANDAHAR (AIRPORT)
Latitude 31°30′N, longitude 65°31′E, elevation 1,004 m

Month	Mean sta. press. (mbar)	Temperature (°C)				Relat. humid. (%)	Precipitation (mm)	
		mean daily	mean daily range	extremes			mean	24-h max.
				max.	min.			
Jan.	903.0	7.4	17.0	25.0	−2.3	49	3.1	3.0
Feb.	898.8	11.6	12.3	26.2	1.7	67	76.5	21.4
Mar.	896.3	14.0	15.8	32.0	−0.6	46	11.6	8.4
Apr.	895.4	18.5	14.8	33.3	5.9	52	27.4	12.8
May	893.2	25.5	19.2	38.4	9.4	26	0	0
June	887.2	31.2	19.7	43.7	8.5	22	0	0
July	885.2	31.5	17.2	42.6	16.5	25	0	0
Aug.	887.2	28.6	19.2	42.2	16.0	21	0	0
Sep.	893.2	23.2	20.1	40.1	9.5	19	0	0
Oct.	898.8	17.2	17.6	37.2	3.0	33	1.8	1.8
N(v.	903.0	6.2	19.9	25.0	−8.3	32	0	0
Dec.	900.7	5.1	19.1	22.7	−10.5	32	tr	tr
Annual	895.2	18.3	17.7	43.7	−10.5	35	120.4	21.4

Month	Mean evap. (mm)	Number of days with			Mean cloudiness (oktas)	Mean sunshine (h)
		precip.	thunderstorms	fog		
Jan.		2	0		2	294.0
Feb.		10	0		4	—
Mar.		4	0		3	241.5
Apr.		11	1		3	252.0
May		0	0		1	362.8
June		0	0		1	366.5
July		0	0		1	361.0
Aug.		0	0		0	370.2
Sep.		0	0		0	314.5
Oct.		1	0		2	307.7
Nov.		0	0		0	301.9
Dec.		1	0		2	270.0
Annual		29	1		1.6	3442.1

TABLE XX

CLIMATIC TABLE FOR LAHORE

Latitude 31°35′N, longitude 74°20′E, elevation 234 m

Month	Mean sta. press. (mbar)	Temperature (°C)				Mean vapour press. (mbar)	Precipitation (mm)	
		mean daily	mean daily range	extremes			mean	24-h max.
				max.	min.			
Jan.	991.9	12.3	15.5	27.8	−2.2	8.2	26.4	74.7
Feb.	989.5	14.6	15.4	32.2	−1.1	9.9	24.6	61.7
Mar.	986.2	19.9	16.3	41.1	3.9	10.3	20.1	59.2
Apr.	982.1	26.1	17.4	44.4	7.8	11.5	14.5	41.7
May	977.3	31.1	17.5	47.8	15.0	13.9	15.0	76.2
June	972.7	33.6	15.0	48.3	18.3	21.9	41.7	125.5
July	972.5	32.2	10.9	47.8	20.6	28.9	138.4	210.1
Aug.	974.7	31.1	10.2	44.4	19.4	28.5	130.8	128.5
Sep.	979.5	29.6	13.5	43.3	17.2	23.6	55.9	167.9
Oct.	985.5	24.9	19.0	41.1	7.8	15.3	6.1	51.8
Nov.	990.0	18.4	19.8	34.4	2.2	11.3	2.5	16.8
Dec.	992.3	13.6	17.6	30.6	−1.7	9.2	11.9	55.4
Annual	982.9	23.9	15.7	48.3	−2.2	16.0	487.9	210.1

Month	Mean evap. (mm)	Number of days with			Mean cloudi- ness (oktas)	Mean sun- shine (h)	Wind	
		precip.	thunder- storms	fog			prevalent direct.	mean speed (m/sec)
Jan.		4	0.7	3	4.0		310	0.5
Feb.		5	2	1.6	4.5		310	0.8
Mar.		4	3	0.1	3.7		310	1.0
Apr.		3	4	0	2.9		310	1.0
May		3	5	0	1.7		050	1.0
June		6	7	0	2.1		130	1.1
July		9	8	0	3.7		130	1.1
Aug.		9	9	0	3.9		130	0.9
Sep.		3	5	0.1	1.5		130	0.8
Oct.		0.9	1.1	0	0.7		130	0.5
Nov.		0.4	0.4	0.7	1.5		310	0.4
Dec.		1.8	0.5	2	3.5		310	0.4
Annual		49	46	7	2.8		210	0.8

TABLE XXI

CLIMATIC TABLE FOR AMRITSAR
Latitutde 31°38′N, longitude 74°52′E, elevation 234 m

Month	Mean sta. press. (mbar)	Temperature (°C)				Mean vapour press. (mbar)	Precipitation (mm)	
		mean daily	mean daily range	extremes			mean	24-h max.
				max.	min.			
Jan.	989.3	11.5	14.1	25.0	−1.7	9.9	38.0	35.5
Feb.	985.8	14.5	16.1	32.2	−0.6	10.5	10.6	25.9
Mar.	983.9	19.5	16.0	35.6	3.9	12.8	25.6	39.4
Apr.	980.4	25.2	18.0	43.3	6.9	12.3	9.5	15.2
May	974.9	30.1	17.5	46.1	12.9	13.3	10.7	24.6
June	970.4	32.8	15.2	46.7	15.6	19.3	32.1	39.6
July	971.0	32.7	9.7	45.6	20.6	29.7	169.4	66.0
Aug.	972.7	29.7	8.9	40.0	19.3	31.1	168.2	79.8
Sep.	977.3	28.9	11.1	40.6	17.2	26.3	105.8	122.4
Oct.	983.1	24.3	15.3	38.3	8.3	18.7	54.3	190.4
Nov.	987.9	17.7	17.7	32.2	−0.6	12.1	9.6	31.8
Dec.	989.7	13.2	16.4	27.7	−2.8	10.0	15.3	21.2
Annual	980.6	28.2	14.6	46.7	−2.8	17.2	649.1	190.4

Month	Mean evap. (mm)	Number of days with			Mean cloudi-ness (oktas)	Mean sun-shine (h)	Wind	
		precip.	thunder-storms	fog			prevalent direct.	mean speed (m/sec)
Jan.	5	1.8	2	3.3	6.8	360	1.9
Feb.	4	2	0.7	2.7	7.5	350	2.1
Mar.	5	4	0	3.3	7.4	020	2.6
Apr.	3	5	0.2	2.4	9.8	030	2.8
May	3	4	0.2	1.8	10.4	070	3.4
June	3	6	0	1.3	8.4	110	3.4
July	10	10	0	3.5	7.3	110	3.2
Aug.	11	9	0	3.8	7.7	110	2.5
Sep.	5	5	0.2	2.1	8.7	140	1.9
Oct.	1.9	1.7	0.2	0.8	9.3	080	1.8
Nov.	0.9	0.6	0.4	1.5	9.1	360	1.4
Dec.	1.7	0.6	1.9	2.7	7.2	330	1.4
Annual	53	50	6	2.4	8.3	170	2.3

TALBE XXII

CLIMATIC TABLE FOR QUETTA
Latitude 30°10′N, longitude 67°01′E, elevation 1,830 m

Month	Mean sta. press. (mbar)	Temperature (°C)				Mean vapour press. (mbar)	Precipitation (mm)	
		mean daily	mean daily range	extremes			mean	24-h max.
				max.	min.			
Jan.	834.1	3.8	12.5	25.6	−16.1	5.0	49.3	55.9
Feb.	832.9	5.7	12.7	26.7	−13.3	6.0	50.3	41.4
Mar.	832.7	10.5	14.1	28.9	−9.4	6.5	44.2	46.2
Apr.	831.9	15.5	15.6	32.8	−2.8	9.2	24.9	35.6
May	830.5	19.9	17.7	36.7	1.1	12.5	9.9	24.9
June	827.1	23.9	18.3	39.4	6.1	14.9	4.3	55.1
July	824.7	26.4	16.1	39.4	8.3	17.2	11.7	37.6
Aug.	826.5	24.9	17.0	39.4	7.2	14.1	8.4	34.8
Sep.	831.0	19.9	20.3	36.1	0.6	9.8	1.0	13.5
Oct.	834.9	14.1	20.5	32.8	−5.6	6.4	7.1	24.1
Nov.	836.3	9.3	18.5	27.2	−11.1	4.5	7.1	26.2
Dec.	835.5	5.6	15.0	24.4	−19.4	4.9	25.7	42.7
Annual	831.5	14.9	16.5	39.4	−19.4	9.3	239.8	55.9

Month	Mean evap. (mm)	Number of days with			Mean cloudi-ness (oktas)	Mean sun-shine (h)	Wind	
		precip.	thunder-storms	fog			prevalent direct.	mean speed (m/sec)
Jan.		9	0.6	1.0	4.6		130	1.1
Feb.		9	0.7	0	5.1		130	1.3
Mar.		8	4	0	4.5		130	1.4
Apr.		5	4	0	3.4		130	1.4
May		1.7	1.1	0	2.0		130	1.2
June		1.6	1.7	0	1.5		130	1.1
July		2	5	0	1.9		130	1.2
Aug.		0.3	0.9	0	1.3		180	1.0
Sep.		0.1	0.1	0	0.5		180	0.8
Oct.		0.3	0.1	0	0.9		180	0.8
Nov.		0.7	0.3	0	1.3		130	0.8
Dec.		6	0.6	0	3.5		130	0.9
Annual		44	19	1.0	2.5		140	1.1

TABLE XXIII

CLIMATIC TABLE FOR MULTAN
Latitude 30°12′N, Longitude 71°31′E, elevation 138 m

Month	Mean sta. press. (mbar)	Temperature (°C)				Mean vapour press. (mbar)	Precipitation (mm)	
		mean daily	mean daily range	extremes			mean	24-h max.
				max.	min.			
Jan.	1002.5	12.8	14.4	28.3	−1.7	8.5	9.4	29.2
Feb.	999.9	15.4	14.5	33.9	−0.6	10.5	9.7	27.9
Mar.	996.3	21.4	14.8	42.2	4.1	11.6	10.2	35.6
Apr.	991.9	27.4	15.5	46.1	8.9	14.0	6.9	26.7
May	986.9	32.8	15.2	50.0	15.0	17.7	8.4	38.6
June	981.6	35.4	12.1	49.4	15.0	25.5	14.0	44.5
July	981.1	34.4	9.4	48.0	21.7	30.7	51.1	174.5
Aug.	983.5	32.7	8.9	45.0	21.1	30.3	46.2	128.8
Sep.	988.5	31.2	11.7	44.4	18.3	24.7	13.7	108.5
Oct.	995.1	26.2	17.1	42.2	10.0	15.7	2.0	29.2
Nov.	1000.1	19.4	17.8	36.1	5.0	10.9	1.5	22.1
Dec.	1002.9	14.2	15.8	29.4	0.0	9.5	6.1	22.4
Annual	992.5	25.3	14.0	50.0	−1.7	17.3	179.1	174.5

Month	Mean evap. (mm)	Number of days with			Mean cloudi-ness (oktas)	Mean sun-shine (h)	Wind	
		precip.	thunder-storms	fog			prevalent direct.	mean speed (m/sec)
Jan.		2	0.1	1.9	3.1		360	0.5
Feb.		4	0.4	0.8	3.7		360	0.7
Mar.		3	0.8	0.2	2.8		360	0.8
Apr.		1.7	1.2	0	2.3		360	0.9
May		0.7	1.4	0	1.0		360	0.9
June		1.8	2	0	1.3		230	1.2
July		3	1.7	0	2.2		230	1.1
Aug.		1.6	1.0	0	2.1		230	1.0
Sep.		0.6	0.8	0	0.7		230	0.9
Oct.		0.2	0.2	0	0.5		130	0.5
Nov.		0	0	0.1	1.1		360	0.5
Dec.		1.1	0	0.9	2.5		360	0.5
Annual		20	10	4	1.9		300	0.8

TABLE XXIV

CLIMATIC TABLE FOR DARJEELING
Latitude 27°03′N, longitude 88°16′E, elevation 2,127 m

Month	Mean sta. press. (mbar)	Temperature (°C)				Mean vapour press. (mbar)	Precipitation (mm)	
		mean daily	mean daily range	extremes			mean	24-h max.
				max.	min.			
Jan.	790.9	6.1	6.3	18.9	−3.9	7.4	21.7	38.1
Feb.	789.7	7.7	6.8	17.2	−5.0	8.1	26.7	42.9
Mar.	790.1	11.3	7.1	23.3	−0.6	9.9	52.4	72.9
Apr.	790.1	14.4	7.2	26.7	1.1	11.4	109.0	135.1
May	788.5	15.7	5.7	23.9	5.6	15.3	187.1	232.9
June	786.5	17.0	4.6	26.7	8.3	18.5	522.3	454.1
July	786.7	17.6	4.4	25.0	3.9	19.2	712.9	200.9
Aug.	787.7	17.6	4.4	26.7	8.3	18.7	572.5	237.5
Sep.	789.9	17.3	5.3	26.7	10.0	17.9	418.5	492.8
Oct.	792.5	15.1	7.1	23.3	4.4	14.1	116.1	334.5
Nov.	792.7	11.3	7.9	22.2	−0.6	9.9	14.2	219.7
Dec.	791.9	8.1	7.5	20.0	−1.7	7.8	5.0	31.2
Annual	789.7	13.3	6.2	26.7	−5.0	13.2	2758.4	492.8

Month	Mean evap. (mm)	Number of days with			Mean cloudi-ness (oktas)	Mean sun-shine (h)	Wind	
		precip.	thunder-storms	fog			prevalent direct.	mean speed (m/sec)
Jan.		2	0.3	11	4.3		090	0.6
Feb.		6	0.4	8	4.9		100	1.0
Mar.		6	2	7	4.5		090	1.1
Apr.		1	4	9	4.8		120	1.4
May		19	4	15	6.1		190	1.1
June		22	1.8	21	7.1		110	1.0
July		29	1.7	25	7.5		110	0.8
Aug.		27	1.5	23	7.1		110	0.9
Sep.		23	1.4	19	6.8		100	0.7
Oct.		9	0.2	9	4.9		090	0.6
Nov.		5	0	7	3.9		090	0.5
Dec.	104		0	7	3.7		090	0.5
Annual	160		17	161	5.5		110	0.8

TABLE XXV

CLIMATIC TABLE FOR AGRA
Latitude 27°10′N, longitude 78°02′E, elevation 169 m

Month	Mean sta. press. (mbar)	Temperature (°C) mean daily	mean daily range	extremes max.	min.	Mean vapour press. (mbar)	Precipitation (mm) mean	24-h max.
Jan.	996.9	14.8	14.8	31.1	−2.2	9.6	16.2	49.5
Feb.	994.2	18.0	15.4	35.6	−1.7	9.5	8.8	51.8
Mar.	991.2	23.8	16.2	42.8	5.6	10.4	10.9	41.3
Apr.	987.3	29.7	16.1	45.0	11.7	11.1	5.3	32.3
May	982.1	34.5	14.6	47.2	16.7	15.4	10.0	32.3
June	978.3	35.0	11.0	48.3	19.4	22.9	60.0	97.3
July	978.3	30.9	7.8	45.6	21.1	31.1	210.2	152.7
Aug.	980.4	29.3	7.0	42.2	20.8	31.3	263.2	149.9
Sep.	984.5	28.9	8.6	40.6	17.2	27.7	151.5	286.0
Oct.	990.8	26.2	14.2	41.1	9.4	17.1	23.5	169.7
Nov.	994.9	20.6	17.2	36.1	2.8	10.9	2.1	45.7
Dec.	997.1	16.1	15.9	30.0	−0.6	9.5	3.7	26.7
Annual	988.0	25.7	13.3	48.3	−2.2	17.2	765.4	286.0

Month	Mean evap. (mm)	Number of days with precip.	thunder-storms	fog	Mean cloudiness (oktas)	Mean sunshine (h)	Wind prevalent direct.	mean speed (m/sec)
Jan.	2.8	2	0.3	1.1	2.2		330	1.0
Feb.	4.3	1.7	0.8	0.3	1.8		290	1.2
Mar.	7.2	1.9	1.9	0.1	1.7		300	1.4
Apr.	10.8	0.9	1.3	0	1.5		250	1.4
May	13.7	2	3	0	1.3		280	1.6
June	13.8	4	3	0.2	2.9		280	1.9
July	8.9	15	2	0.8	5.3		060	1.6
Aug.	5.9	16	1.5	0.5	5.6		200	1.3
Sep.	4.6	8	1.3	0.6	3.3		040	1.3
Oct.	4.5	2	0.4	0.1	0.9		300	0.9
Nov.	3.8	0.4	0.1	0.1	0.7		270	0.7
Dec.	2.9	1.1	0.6	0.8	1.4		290	0.3
Annual	6.9	55	16	5	2.4		25	1.3

TABLE XXVI

CLIMATIC TABLE FOR KATMANDU
Latitude 27°42′N, longitude 85°12′E, elevation 1,324 m

Month	Mean sta. press. (mbar)	Temperature (°C)				Mean vapour press. (mbar)	Precipitation (mm)	
		mean daily	mean daily range	extremes			mean	24-h max.
				max.	min.			
Jan.	870.8	9.9	15.9	25.0	−2.8	8.7	23.5	36.6
Feb.	869.0	12.3	16.7	28.3	−1.7	9.8	20.2	27.4
Mar.	868.4	16.6	17.2	33.3	1.1	11.7	33.3	47.2
Apr.	866.9	20.2	17.4	37.2	4.4	13.6	57.5	43.4
May	864.3	23.1	14.5	37.5	9.4	17.7	106.9	55.6
June	861.3	24.3	10.4	37.8	13.2	22.7	207.0	108.5
July	861.1	24.3	8.4	32.8	16.1	24.5	378.6	167.4
Aug.	862.6	24.2	8.6	33.3	16.1	24.5	356.8	101.6
Sep.	865.3	23.3	9.5	33.3	13.3	23.0	159.6	71.4
Oct.	869.3	19.9	12.8	33.3	5.6	17.7	43.4	47.5
Nov.	870.9	15.1	15.3	29.4	0.6	12.5	5.5	28.5
Dec.	871.3	11.1	16.1	28.3	−2.8	9.7	1.7	15.2
Annual	866.7	18.7	13.6	37.8	−2.8	16.3	1394.0	167.4

Month	Mean evap. (mm)	Number of days with			Mean cloudi-ness (oktas)	Mean sun-shine (h)	Wind	
		precip.	thunder-storms	fog			prevalent direct.	mean speed (m/sec)
Jan.		3	0.6	17	1.3		100	0.3
Feb.		3	1.1	12	1.1		110	0.4
Mrt.		4	4	6	1.1		090	0.5
Apr.		7	7	1.2	1.7		090	0.5
May		13	11	0.9	2.1		100	0.5
June		20	9	0	3.0		100	0.4
July		25	8	0.1	3.9		110	0.4
Aug.		26	8	0.5	3.8		120	0.3
Sep.		17	9	3	3.5		090	0.3
Oct.		6	3	16	1.9		100	0.3
Nov.		0.8	0.4	18	1.3		100	0.3
Dec.		0.9	0.1	17	1.1		110	0.3
Annual		126	61	92	2.1		100	0.4

TABLE XXVII

CLIMATIC TABLE FOR JODHPUR
Latitude 26°18′N, longitude 73°01′E, elevation 224 m

Month	Mean sta. press. (mbar)	Temperature (°C)				Mean vapour press. (mbar)	Precipitation (mm)	
		mean daily	mean daily range	extremes			mean	24-h max.
				max.	min.			
Jan.	989.8	17.1	15.1	32.8	−2.2	7.1	7.3	40.1
Feb.	987.5	19.9	15.9	38.3	−0.6	7.3	5.1	22.6
Mar.	984.9	25.2	16.2	41.8	5.0	8.1	1.9	20.6
Apr.	981.6	30.3	15.9	48.0	9.4	9.8	2.2	26.7
May	977.3	34.5	14.3	48.9	17.2	15.5	6.4	38.1
June	973.5	34.3	11.6	47.8	19.4	23.1	30.9	152.9
July	972.7	31.3	8.9	45.6	19.4	27.9	121.8	194.1
Aug.	974.9	29.2	8.0	42.9	20.6	28.0	145.5	184.4
Sep.	979.0	29.4	10.6	42.8	17.8	24.2	47.4	215.9
Oct.	984.7	27.7	16.1	42.2	10.0	12.7	6.8	142.0
Nov.	988.4	22.7	17.5	37.2	5.6	8.1	3.3	26.9
Dec.	990.0	18.7	16.0	33.3	0.6	7.8	1.5	22.9
Annual	982.1	26.7	13.8	48.9	−2.2	15.0	380.1	215.9

Month	Mean evap. (mm)	Number of days with			Mean cloudi-ness (oktas)	Mean sun-shine (h)	Wind	
		precip.	thunder-storms	fog			prevalent direct.	mean speed (m/sec)
Jan.	3.4	0.9	0.3	0.5	2.1	8.9	040	2.5
Feb.	5.0	1.2	1.1	0	2.1	9.5	050	2.5
Mar.	7.9	1.1	1.2	0	2.1	9.0	050	2.7
Apr.	11.3	0.6	1.3	0	1.7	10.2	280	2.8
May	14.4	1.3	2	0	1.1	10.5	230	4.2
June	13.1	3	5	0	2.3	9.4	240	5.1
July	9.6	9	6	0	5.5	6.9	240	4.6
Aug.	7.1	11	5	0.1	5.9	6.3	230	3.6
Sep.	7.0	5	4	0	3.5	8.7	250	3.0
Oct.	6.4	0.6	0.9	0	1.1	10.3	030	1.8
Nov.	4.6	0.6	0.5	0	1.0	9.9	050	1.6
Dec.	3.4	0.3	0.1	0	1.8	9.4	050	2.0
Annual	7.8	35	27	0.6	2.5	9.1	140	3.0

TABLE XXVIII

CLIMATIC TABLE FOR SIBSAGAR
Latitude 26°59′N, longitude 94°38′E, elevation 97 m

Month	Mean sta. press. (mbar)	Temperature (°C)				Mean vapour press. (mbar)	Precipitation (mm)	
		mean daily	mean daily range	extremes			mean	24-h max.
				max.	min.			
Jan.	1004.9	16.3	12.8	28.9	3.3	15.6	29.7	43.7
Feb.	1002.4	18.3	11.1	30.5	2.8	16.9	46.5	42.9
Mar.	999.9	21.7	11.2	35.6	7.2	19.4	94.6	113.0
Apr.	997.5	24.0	9.4	37.2	12.8	23.1	218.2	148.6
May	994.2	26.1	7.3	42.8	16.7	27.6	361.1	188.2
June	990.1	28.1	6.7	37.8	19.4	31.9	390.8	189.2
July	989.7	28.9	7.0	38.9	20.6	31.7	476.3	218.9
Aug.	990.9	28.7	6.7	37.8	18.3	32.5	400.2	154.4
Sep.	994.4	28.1	6.8	36.4	19.4	31.8	301.8	182.4
Oct.	999.7	25.7	7.9	35.6	16.0	28.1	135.5	99.6
Nov.	1003.0	21.4	10.8	32.2	9.4	21.7	30.1	53.9
Dec.	1004.8	17.6	12.4	28.9	4.4	16.9	19.5	36.3
Annual	997.6	23.7	9.1	42.8	2.8	24.8	2504.3	218.9

Month	Mean evap.* (mm)	Number of days with			Mean cloudi-ness (oktas)	Mean sun-shine (h)	Wind	
		precip.	thunder-storms	fog			prevalent direct.	mean speed (m/sec)
Jan.	1.9	5	1.2	20	4.7		050	0.7
Feb.	2.4	9	3	11	5.1		060	1.1
Mar.	3.8	11	7	3	4.7		040	1.5
Apr.	4.9	17	11	0.7	5.4		050	1.7
May	4.8	22	11	0	6.2		040	1.7
June	5.2	24	12	0	6.8		030	1.5
July	4.8	26	12	0	6.8		180	1.5
Aug.	5.1	25	18	0	6.7		110	1.4
Sep.	4.5	19	14	0	6.4		150	1.3
Oct.	3.5	15	4	3	5.0		030	0.9
Nov.	2.5	5	0.4	11	4.3		050	0.7
Dec.	1.9	4	0.8	21	4.3		110	0.6
Annual	3.8	182	94	70	5.5		070	1.3

* Data for Mohanbari.

TABLE XXIX

CLIMATIC TABLE FOR ALLAHABAD
Latitude 25°27′N, longitude 81°44′E, elevation 98 m

Month	Mean sta. press. (mbar)	Temperature (°C)				Mean vapour press. (mbar)	Precipitation (mm)	
		mean daily	mean daily range	extremes			mean	24-h max.
				max.	min.			
Jan.	1004.7	16.4	14.6	31.1	2.2	12.1	20.2	70.9
Feb.	1002.0	19.1	15.1	36.1	1.1	11.6	22.2	51.3
Mar.	998.6	25.1	16.3	41.7	7.2	10.7	14.3	34.5
Apr.	994.6	30.7	16.3	45.0	12.8	10.7	4.8	26.4
May	989.5	34.3	14.7	47.2	17.2	15.3	8.2	35.6
June	986.1	34.3	10.9	47.8	19.4	24.1	101.7	176.0
July	986.4	30.1	7.0	45.6	22.2	32.4	274.8	209.3
Aug.	988.3	29.1	6.1	40.0	21.1	32.7	333.1	335.3
Sep.	992.2	29.0	7.6	39.4	18.3	30.9	195.1	266.2
Oct.	998.7	26.5	12.2	40.6	11.7	21.9	39.7	163.3
Nov.	1002.9	21.1	15.9	35.6	5.6	14.1	6.9	96.0
Dec.	1004.9	17.1	15.5	31.1	2.2	12.3	6.3	54.6
Annual	995.7	26.1	12.6	47.8	1.1	19.1	1027.3	335.3

Month	Mean evap. (mm)	Number of days with			Mean cloudiness (oktas)	Mean sunshine (h)	Wind	
		precip.	thunderstorms	fog			prevalent direct.	mean speed (m/sec)
Jan.	2.3	3	1.8	1.8	2.4	8.1	230	1.6
Feb.	3.5	3	2	0.6	2.0	9.3	230	1.5
Mar.	5.8	2	3	0.1	1.8	8.9	230	1.8
Apr.	8.6	1.1	2	0.1	1.6	9.9	220	2.0
May	10.1	1.5	3	0	1.5	10.0	070	2.4
June	10.0	8	9	0	4.1	7.2	260	2.6
July	5.3	19	13	0	6.7	5.2	150	2.3
Aug.	4.3	21	12	0	6.8	4.9	120	2.0
Sep.	4.1	13	9	0	5.0	6.9	360	1.8
Oct.	3.7	5	2	0	2.1	8.9	260	1.2
Nov.	3.1	1.0	0	0	1.1	9.3	250	0.9
Dec.	2.2	1.2	0.6	1.7	1.5	8.9	250	1.0
Annual	5.3	79	57	4	3.1	8.1	220	1.8

TABLE XXX

CLIMATIC TABLE FOR SHILLONG
Latitude 25°34′N, longitude 91°53E, elevation 1,500 m

Month	Mean sta. press. (mbar)	Temperature (°C)				Mean vapour press. (mbar)	Precipitation (mm)	
		mean daily	mean daily range	extremes			mean	24-h max.
				max.	min.			
Jan.	851.3	9.5	11.9	21.1	−2.8	9.5	15.2	52.3
Feb.	850.0	11.7	10.7	24.4	−2.8	9.3	28.5	41.1
Mar.	849.5	16.0	11.0	20.9	−0.6	9.9	59.4	189.5
Apr.	848.4	18.9	9.7	30.0	6.7	12.9	136.4	117.9
May	846.2	19.6	8.2	30.7	5.6	17.4	325.4	169.7
June	843.5	20.5	6.3	28.3	11.7	20.5	544.6	415.3
July	843.5	21.1	6.0	28.3	15.0	21.1	394.9	205.7
Aug.	844.7	20.9	6.3	29.4	14.4	20.9	334.6	118.1
Sep.	847.3	20.1	7.0	27.8	11.7	21.3	314.9	256.2
Oct.	850.9	17.3	8.9	27.2	5.6	17.2	220.2	295.2
Nov.	851.7	13.3	11.2	25.6	1.1	12.8	34.9	96.0
Dec.	851.7	10.5	11.9	22.8	−1.7	10.3	6.3	41.1
Annual	848.2	16.7	9.1	30.7	−2.8	15.2	2415.3	415.3

Month	Mean evap. (mm)	Number of days with			Mean cloudiness (oktas)	Mean sunshine (h)	Wind	
		precip.	thunder-storms	fog			prevalent direct.	mean speed (m/sec)
Jan.	1.7	3	0.2	1.5	3.7		170	0.7
Feb.	2.4	5	1.2	0.4	3.7		220	1.1
Mar.	3.8	7	4	0.2	3.1		220	1.7
Apr.	4.5	13	10	0.3	4.4		220	2.2
May	4.1	22	14	0.9	5.6		220	1.9
June	2.7	25	10	1.2	7.1		210	1.2
July	2.8	27	5	1.2	7.5		190	1.0
Aug.	2.6	23	7	1.5	7.0		180	0.9
Sep.	2.6	22	10	1.7	6.5		160	0.7
Oct.	2.1	17	6	1.0	5.1		150	0.6
Nov.	1.8	5	0.5	0.6	4.1		130	0.6
Dec.	1.6	2	0.1	0.6	3.7		140	0.6
Annual	2.7	171	68	11	5.1		230	1.1

TABLE XXXI

CLIMATIC TABLE FOR KARACHI (MANORA)
Latitude 24°48′N, longitude 66°59′E, elevation 4 m

Month	Mean sta. press. (mbar)	Temperature (°C)				Mean vapour press. (mbar)	Precipitation (mm)	
		mean daily	mean daily range	extremes			mean	24-h max.
				max.	min.			
Jan.	1016.0	19.2	10.1	30.0	6.1	11.5	11.7	41.7
Feb.	1013.9	20.5	8.8	32.2	6.1	15.3	11.2	29.2
Mar.	1010.9	23.8	7.6	37.2	12.2	19.1	7.4	53.3
Apr.	1007.4	26.6	6.3	40.0	17.2	26.1	3.8	104.4
May	1003.9	28.8	5.3	42.8	21.7	31.7	1.5	30.7
June	998.6	30.2	4.5	40.6	22.8	33.8	18.3	182.1
July	997.5	29.3	4.1	36.7	21.7	32.7	81.3	199.6
Aug.	1000.1	27.8	4.1	35.0	21.1	30.0	39.6	137.4
Sep.	1004.7	27.3	5.0	38.3	18.3	29.1	13.2	206.0
Oct.	1009.9	26.9	7.5	40.0	17.2	26.1	0.5	13.2
Nov.	1013.9	24.4	10.2	36.7	11.1	17.9	2.0	22.1
Dec.	1016.1	20.8	10.3	30.6	8.3	16.3	5.1	46.5
Annual	1007.7	25.5	6.9	42.8	6.1	23.9	195.6	206.0

Month	Mean evap. (mm)	Number of days with			Mean cloudi-ness (oktas)	Mean sun-shine (h)	Wind	
		precip.	thunder-storms	fog			prevalent direct.	mean speed (m/sec)
Jan.		2	0.4	0.1	3.3		050	2.8
Feb.		3	0.9	0.6	2.7		050	3.0
Mar.		0.7	0.4	0	2.7		270	3.7
Apr.		0.7	0.1	0.1	2.1		270	4.3
May		0.1	0	0.3	2.8		270	5.2
June		1.0	1.1	0	5.1		270	5.7
July		6	1.9	0	7.1		270	5.7
Aug.		3	0.7	0	7.3		270	5.3
Sep.		0.7	0	0.2	4.3		270	4.5
Oct.		0.2	0	0.7	1.1		270	2.9
Nov.		0.4	0	1.8	1.5		360	2.4
Dec.		1.3	0.2	0.3	2.5		050	2.3
Annual		19	6	3	3.4		220	4.0

TABLE XXXII

CLIMATIC TABLE FOR BOGRA
Latitude 24°51′N, longitude 89°23′E, elevation 21 m

Month	Mean sta. press. (mbar)	Temperature (°C)				Mean vapour press. (mbar)	Precipitation (mm)	
		mean daily	mean daily range	extremes			mean	24-h max.
				max.	min.			
Jan.	1013.2	17.7	13.1	30.0	4.4	13.6	10.9	75.2
Feb.	1010.9	19.8	14.1	34.4	3.9	14.3	20.1	105.4
Mar.	1007.3	24.9	15.4	41.1	8.3	15.1	29.0	71.9
Apr.	1003.7	28.6	13.4	42.8	13.3	20.4	56.9	97.8
May	1001.3	28.5	10.1	43.3	17.2	28.3	213.4	149.3
June	997.5	28.6	7.0	38.9	19.4	32.5	331.2	396.2
July	997.0	28.8	6.0	36.1	18.9	33.2	325.9	179.6
Aug.	998.7	28.7	5.7	36.1	22.2	33.1	330.2	204.2
Sep.	1002.7	28.6	6.2	36.1	21.1	32.5	286.8	271.0
Oct.	1007.5	26.9	8.2	35.6	16.1	28.0	130.3	278.1
Nov.	1011.1	22.6	10.9	32.8	10.6	20.7	19.6	98.0
Dec.	1013.3	18.7	12.3	30.0	6.7	15.7	2.3	38.6
Annual	1005.4	25.2	10.2	43.3	3.9	23.9	1756.4	396.2

Month	Mean evap. (mm)	Number of days with			Mean cloudi- ness (oktas)	Mean sun- shine (h)	Wind	
		precip.	thunder- storms	fog			prevalent direct.	mean speed (m/sec
Jan.		1.1	0.5	1.3	1.9		360	0.5
Feb.		3	0.8	0.7	2.7		360	0.6
Mar.		3	3	0.4	2.3		090	0.9
Apr.		5	6	0.2	3.5		090	1.2
May		11	13	0	5.5		090	1.3
June		17	14	0	7.3		090	1.0
July		21	8	0	7.7		090	1.0
Aug.		21	8	0	7.5		090	0.9
Sep.		15	11	0	6.6		090	0.7
Oct.		5	3	0.1	3.6		050	0.5
Nov.		0.5	0.4	0.2	2.1		360	0.5
Dec.		0.1	0	0.3	1.6		360	0.5
Annual		103	68	3	4.4		180	0.8

TABLE XXXIII

CLIMATIC TABLE FOR AHMADABAD
Latitude 23°04′N, longitude 72°38′E, elevation 55 m

Month	Mean sta. press. (mbar)	Temperature (°C)				Mean vapour press. (mbar)	Precipitation (mm)	
		mean daily	mean daily range	extremes			mean	24-h max.
				max.	min.			
Jan.	1008.3	20.3	16.8	36.1	3.3	9.3	3.0	30.7
Feb.	1006.4	22.7	16.5	40.6	2.2	10.1	0.3	26.4
Mar.	1004.0	27.1	17.1	43.9	9.4	12.0	0.9	12.2
Apr.	1001.3	31.3	16.7	46.2	12.8	14.9	1.9	21.6
May	997.9	33.5	14.4	47.0	19.4	22.5	4.5	46.2
June	994.1	32.7	10.6	47.2	19.4	28.9	100.0	130.8
July	993.7	29.5	7.5	42.2	21.1	30.9	316.3	414.8
Aug.	995.7	28.2	7.2	38.9	21.7	29.7	213.3	150.6
Sep.	999.1	28.7	8.9	41.7	21.6	27.6	162.8	257.8
Oct.	1003.7	28.4	14.4	42.8	14.4	19.1	13.1	52.8
Nov.	1006.9	24.5	16.9	38.9	9.4	12.7	5.4	53.3
Dec.	1008.5	21.1	17.0	35.6	6.1	10.7	0.7	14.0
Annual	1001.7	27.3	13.7	47.8	2.2	19.1	823.1	414.8

Month	Mean evap. (mm)	Number of days with			Mean cloudi-ness (oktas)	Mean sun-shine (h)	Wind	
		precip.	thunder-storms	fog			prevalent direct.	mean speed (m/sec)
Jan.	4.9	0.5	0.4	0	1.4	9.7	050	1.5
Feb.	7.0	0.1	0.1	0	1.3	10.2	030	1.4
Mar.	9.3	0.1	0.3	0	1.3	9.3	360	2.0
Apr.	11.8	0.5	0.8	0	1.4	10.3	310	2.1
May	13.1	0.5	0.7	0.1	1.5	10.9	270	2.6
June	11.1	7	4	0	4.3	8.4	260	3.0
July	7.2	17	3	0	6.8	5.1	240	3.0
Aug.	5.6	17	1.8	0	6.7	4.4	240	2.3
Sep.	5.7	9	3	0	4.5	7.2	270	1.9
Oct.	6.3	1.5	1.0	0	1.7	9.6	010	1.3
Nov.	5.5	0.8	0.4	0	1.3	9.9	070	1.1
Dec.	5.0	0.3	0.1	0.2	1.3	9.7	060	0.6
Annual	7.7	54	16	0.3	2.8	8.7	180	1.9

TABLE XXXIV

CLIMATIC TABLE FOR DACCA (NARAYANGANJ)
Latitude 23°37′N, longitude 90°30′E, elevation 9 m

Month	Mean sta. press. (mbar)	Temperature (°C)				Mean vapour press. (mbar)	Precipitation (mm)	
		mean daily	mean daily range	extremes			mean	24-h max.
				max.	min.			
Jan.	1014.3	22.0	12.4	31.1	7.2	14.5	8.1	46.2
Feb.	1012.3	21.3	12.6	34.4	6.7	16.0	31.5	73.7
Mar.	1009.1	26.1	12.0	38.9	10.6	20.1	60.7	133.3
Apr.	1005.9	28.4	10.3	40.6	16.7	26.1	137.2	122.4
May	1003.3	28.6	8.5	38.3	14.1	30.1	244.9	129.5
June	999.3	28.7	6.2	36.7	19.4	32.5	314.7	185.2
July	998.9	28.6	5.1	35.0	21.1	32.5	329.4	235.2
Aug.	1000.5	28.5	5.0	35.6	21.7	32.2	336.8	207.3
Sep.	1004.1	28.7	5.5	36.7	21.1	32.2	247.9	177.8
Oct.	1008.9	27.1	7.1	35.0	18.3	28.7	133.9	198.1
Nov.	1012.1	23.9	9.9	34.4	12.2	21.7	24.1	179.8
Dec.	1014.3	20.0	11.6	30.6	8.9	16.6	5.1	44.5
Annual	1006.9	25.8	8.9	40.6	6.7	25.3	1874.3	235.2

Month	Mean evap. (mm)	Number of days with			Mean cloudiness (oktas)	Mean sunshine (h)	Wind	
		precip.	thunderstorms	fog			prevalent direct.	mean speed (m/sec)
Jan.		2	0	10	1.9		310	0.6
Feb.		3	1.7	7	3.1		310	0.8
Mar.		5	4	3	3.8		230	1.4
Apr.		9	9	1.1	4.7		180	2.1
May		15	9	0	6.3		180	2.0
June		21	7	0	8.0		130	2.2
July		26	3	0	8.6		130	2.3
Aug.		25	3	0	8.7		130	2.0
Sep.		19	7	0	7.6		130	1.5
Oct.		8	2	0.3	4.5		090	0.8
Nov.		1.6	0.7	5	2.7		310	0.5
Dec.		0.7	0	11	1.9		310	0.5
Annual		135	46	37.7	5.1		200	3.1

TABLE XXXV

CLIMATIC TABLE FOR SAGAR
Latitude 23°51′N, longitude 78°45′E, elevation 551 m

Month	Mean sta. press. (mbar)	Temperature (°C)				Mean vapour press. (mbar)	Precipitation (mm)	
		mean daily	mean daily range	extremes			mean	24-h max.
				max.	min.			
Jan.	952.4	18.1	12.9	31.7	1.7	9.9	29.6	84.1
Feb.	950.5	20.7	13.5	35.0	1.1	8.9	12.8	43.2
Mar.	948.5	25.6	14.2	41.1	7.2	8.3	9.6	61.0
Apr.	945.7	30.3	14.0	43.9	10.6	9.5	4.2	23.9
May	941.7	33.9	13.7	45.6	16.9	13.0	8.3	58.2
June	938.4	31.3	11.2	45.6	18.3	23.3	145.4	235.5
July	937.7	26.3	6.8	41.1	16.1	29.6	462.6	284.5
Aug.	939.6	25.3	6.2	35.6	19.4	27.9	418.5	274.3
Sep.	942.9	25.3	7.6	38.3	16.7	26.3	228.1	159.3
Oct.	948.5	24.9	11.3	37.8	11.7	17.5	43.9	198.2
Nov.	951.9	21.6	12.2	34.4	6.1	10.8	24.0	105.2
Dec.	952.8	19.1	12.5	31.7	4.4	9.6	6.9	73.9
Annual	945.9	25.2	11.4	45.6	1.1	16.3	1393.9	284.5

Month	Mean evap. (mm)	Number of days with			Mean cloudiness (oktas)	Mean sunshine (h)	Wind	
		precip.	thunder-storms	fog			prevalent direct.	mean speed (m/sec)
Jan.		3	2	1.1	2.5		130	1.9
Feb.		1.9	1.5	0.1	1.7		130	1.9
Mar.		1.6	2	0	1.9		130	2.3
Apr.		1.3	2	0	2.1		270	2.5
May		2	3	0	2.3		270	2.8
June		12	9	0	4.7		270	3.1
July		21	7	0.1	6.7		270	2.9
Aug.		22	7	0.1	6.8		270	2.8
Sep.		15	6	0.1	5.1		270	2.7
Oct.		4	2	0	2.3		230	1.9
Nov.		1.5	0.1	0.1	1.5		130	1.8
Dec.		1.1	0.4	0.4	1.9		130	1.8
Annual		86	42	2	3.3		210	2.3

TABLE XXXVI

CLIMATIC TABLE FOR BARODA
Latitude 22°18′N, longitude 73°15′E, elevation 34 m

Month	Mean sta. press. (mbar)	Temperature (°C)				Mean vapour press. (mbar)	Precipitation (mm)	
		mean daily	mean daily range	extremes			mean	24-h max.
				max.	min.			
Jan.	1010.1	20.5	19.3	35.6	−1.1	11.9	2.0	33.0
Feb.	1008.5	22.5	19.7	41.7	1.7	11.9	2.4	33.0
Mar.	1006.3	26.6	20.0	43.3	6.7	12.9	0.7	12.9
Apr.	1003.8	30.8	18.2	45.9	11.7	16.5	2.9	71.4
May	1000.7	33.4	14.6	46.7	18.9	23.6	4.1	59.7
June	997.3	32.1	10.1	45.6	21.5	29.4	114.8	133.3
July	996.7	28.9	7.0	40.0	21.1	31.5	365.8	247.4
Aug.	998.6	28.1	6.7	37.2	22.8	30.5	263.5	250.7
Sep.	1001.9	28.3	8.5	41.1	18.9	29.4	185.5	372.1
Oct.	1005.9	27.5	15.1	41.7	11.7	23.5	34.5	153.2
Nov.	1008.9	23.9	19.1	38.3	7.2	17.4	9.0	48.3
Dec.	1010.5	21.2	19.6	36.1	3.3	14.1	0.2	9.9
Annual	1004.1	27.0	14.8	46.7	−1.1	21.1	985.4	372.1

Month	Mean evap. (mm)	Number of days with			Mean cloudi-ness (oktas)	Mean sun-shine (h)	Wind	
		precip.	thunder-storms	fog			prevalent direct.	mean speed (m/sec)
Jan.		0.5	0.2	0.2	1.2	9.8	040	1.0
Feb.		0.4	0.2	0.2	1.0	10.3	030	0.9
Mar.		0.3	0.2	0	1.5	9.7	230	1.1
Apr.		0.3	0.4	0	1.0	10.5	250	1.4
May		0.5	0.6	0	1.8	11.2	230	2.4
June		8	2	0	3.8	7.9	230	2.7
July		21	2	0	6.6	4.4	230	2.3
Aug.		18	1.5	0	6.4	4.5	230	1.9
Sep.		11	0.3	0	4.4	6.9	240	1.3
Oct.		2	1.6	0.1	1.7	9.6	170	0.8
Nov.		1.5	0.3	0.2	1.1	10.1	060	0.7
Dec.		0.2	0	0.2	1.1	9.9	040	0.8
Annual		64	9	0.9	2.5	8.7	170	1.4

TABLE XXXVII

CLIMATIC TABLE FOR CHITTAGONG
Latitude 22°21′N, longitude 91°50′E, elevation 29 m

Month	Mean sta. press. (mbar)	Temperature (°C)				Mean vapour press. (mbar)	Precipitation (mm)	
		mean daily	mean daily range	extremes			mean	24-h max.
				max.	min.			
Jan.	1012.1	19.4	13.2	31.7	7.2	15.5	6.1	68.8
Feb.	1010.6	21.4	13.1	33.9	7.8	17.1	27.9	234.9
Mar.	1007.8	25.1	11.0	37.2	10.6	23.1	62.5	177.0
Apr.	1005.3	27.2	8.7	38.9	15.0	28.1	150.6	180.3
May	1002.4	27.9	7.4	36.7	18.3	30.1	264.7	161.8
June	998.7	27.7	5.6	36.7	20.0	31.3	533.2	382.5
July	998.1	27.4	5.2	34.4	19.4	31.3	597.7	268.5
Aug.	999.7	27.3	5.2	33.9	22.2	31.3	518.7	271.0
Sep.	1002.9	27.6	5.9	35.0	21.7	31.3	321.1	259.1
Oct.	1006.9	26.7	7.5	34.4	16.7	29.1	180.1	420.9
Nov.	1009.9	23.6	10.1	33.9	11.1	22.6	55.1	136.7
Dec.	1012.0	20.1	12.1	31.1	8.3	17.7	16.3	116.6
Annual	1005.5	25.1	8.8	38.9	7.2	25.7	2733.8	420.9

Month	Mean evap. (mm)	Number of days with			Mean cloudi-ness (oktas)	Mean sun-shine (h)	Wind	
		precip.	thunder-storms	fog			prevalent direct.	mean speed (m/sec)
Jan.		1.5	0.3	2	1.7		050	1.1
Feb.		3	1.0	3	2.4		050	1.3
Mar.		5	3	2	2.9		130	1.8
Apr.		7	5	1.5	3.7		130	2.3
May		13	11	0.3	5.9		130	2.2
June		22	7	0.1	8.1		130	2.4
July		26	2	0	8.5		130	2.5
Aug.		25	5	0	8.3		130	2.3
Sep.		21	7	0.4	7.2		130	1.6
Oct		8	5	4	4.7		050	1.0
Nov.		2	0.5	2	3.3		050	1.0
Dec.		1.0	0	4	1.9		050	1.0
Annual		135	47	19	4.9		100	1.7

TABLE XXXVIII

<small>CLIMATIC TABLE FOR CALCUTTA (ALIPORE)</small>
Latitude 22°32′N, longitude 88°20′E, elevation 6 m

Month	Mean sta. press. (mbar)	Temperature (°C)				Mean vapour press. (mbar)	Precipitation (mm)	
		mean daily	mean daily range	extremes max.	min.		mean	24-h max.
Jan.	1014.9	20.2	13.2	31.9	6.7	15.1	13.8	46.7
Feb.	1012.2	23.0	13.0	36.7	7.2	16.0	24.2	80.8
Mar.	1009.0	27.9	12.8	41.1	10.0	21.1	26.5	69.9
Apr.	1005.8	30.7	11.3	43.3	16.1	27.5	42.7	107.4
May	1001.7	31.1	9.3	43.7	18.3	31.9	120.6	156.2
June	998.2	30.4	7.4	43.9	21.1	33.1	259.1	303.5
July	998.0	29.1	5.7	36.7	22.8	33.5	300.6	183.6
Aug.	999.6	29.1	5.7	36.1	23.3	33.5	306.3	253.0
Sep.	1003.1	29.2	6.2	36.1	22.2	33.1	289.7	369.1
Oct.	1000.9	27.9	7.9	35.6	17.2	28.7	160.2	206.0
Nov.	1012.5	23.9	11.1	33.9	10.6	20.3	34.9	85.1
Dec.	1014.7	21.6	12.8	30.6	7.2	16.1	3.2	53.1
Annual	1006.5	26.9	9.7	43.9	6.7	25.9	1581.8	369.1

Month	Mean evap. (mm)	Number of days with			Mean cloudiness (oktas)	Mean sunshine (h)	Wind	
		precip.	thunderstorms	fog			prevalent direct.	mean speed (m/sec)
Jan.	2.3	1.4	0.5	8	1.5	8.7	350	0.8
Feb.	3.2	3	1.5	7	1.9	8.9	290	0.9
Mar.	4.7	4	4	4	2.2	8.6	250	1.5
Apr.	5.9	5	6	0.7	2.9	9.3	210	2.0
May	6.1	10	11	0	4.0	8.7	190	2.4
June	5.9	17	13	0	6.2	4.9	180	1.8
July	4.8	23	9	0	6.9	4.2	170	1.7
Aug.	4.1	24	12	0	6.8	4.0	180	1.4
Sep.	3.9	20	16	0.1	6.1	5.1	140	1.2
Oct.	3.6	12	9	0.6	3.9	6.5	030	0.8
Nov.	3.0	1.9	0.5	2	1.9	8.2	350	0.7
Dec.	2.3	0.7	0.9	6	1.4	8.6	350	0.7
Annual	4.1	122	83	28	3.9	7.1	240	1.4

TABLE XXXIX

CLIMATIC TABLE FOR INDORE
Latitude 22°43′N, longitude 75°48′E, elevation 567 m

Month	Mean sta. press. (mbar)	Temperature (°C)				Mean vapour press. (mbar)	Precipitation (mm)	
		mean daily	mean daily range	extremes			mean	24-h max.
				max.	min.			
Jan.	950.1	17.9	16.5	32.2	−1.1	9.7	8.4	30.5
Feb.	948.5	19.9	17.9	36.7	−2.8	8.1	1.1	32.0
Mar.	946.9	24.5	18.4	41.1	5.0	8.5	3.5	19.3
Apr.	944.6	29.1	17.3	44.6	7.8	11.0	3.5	51.1
May	941.5	32.3	15.1	45.6	16.7	15.5	13.2	99.1
June	938.3	30.1	11.3	45.0	18.9	24.1	147.1	127.0
July	937.5	26.1	6.9	38.3	18.9	26.6	316.0	293.4
Aug.	939.3	25.1	6.2	35.0	18.9	25.9	266.5	209.8
Sep.	942.3	25.1	8.3	37.2	16.7	24.5	220.9	136.7
Oct.	947.1	24.1	13.9	37.8	8.9	16.9	48.4	88.4
Nov.	949.9	20.5	16.7	35.0	5.6	11.7	22.1	64.5
Annual	944.8	24.4	13.8	45.6	−2.8	16.0	1053.4	293.4

Month	Mean evap. (mm)	Number of days with			Mean cloudi-ness (oktas)	Mean sun-shine (h)	Wind	
		precip.	thunder-storms	fog			prevalent direct.	mean speed (m/sec)
Jan.		1.9	1.1	0.5	2.1		090	2.8
Feb.		0.6	0.3	0.2	1.5		020	3.0
Mar.		0.9	1.1	0	1.5		010	3.5
Apr.		1.1	1.4	0	1.7		300	4.2
May		2	3	0	1.8		280	6.8
June		11	8	0.3	4.7		270	7.5
July		20	5	0	7.1		270	7.3
Aug.		20	4	0.2	6.9		270	5.9
Sep.		15	5	0.7	5.7		290	5.1
Oct.		3	1.9	0.3	2.4		300	2.7
Nov.		1.9	0.5	0.6	1.7		090	2.0
Dec.		0.9	0.5	0.5	1.7		080	1.9
Annual		78	32	3	3.3		190	4.4

TABLE XL

CLIMATIC TABLE FOR NAGPUR
Latitude 21°06′N, longitude 79°03′E, elevation 310 m

Month	Mean sta. press. (mbar)	Temperature (°C)				Mean vapour press. (mbar)	Precipitation (mm)	
		mean daily	mean daily range	extremes			mean	24-h max.
				max.	min.			
Jan.	979.0	20.7	15.9	35.0	3.9	12.7	15.4	60.3
Feb.	976.9	23.8	17.4	38.9	5.0	11.5	1.9	51.6
Mar.	974.7	27.7	17.3	45.0	8.3	12.1	24.5	45.0
Apr.	971.9	31.8	15.8	46.1	13.9	14.8	20.2	59.4
May	967.3	35.6	14.4	47.8	19.4	15.9	9.9	58.4
June	965.1	32.7	11.5	47.2	20.0	23.1	174.3	315.0
July	965.8	27.6	17.2	40.6	19.4	28.1	351.5	219.2
Aug.	966.7	27.1	6.7	37.8	18.3	28.1	277.1	200.7
Sep.	969.5	27.3	8.4	38.9	18.3	27.1	180.5	182.6
Oct.	974.3	25.9	11.9	38.3	11.6	21.5	61.6	164.6
Nov.	978.7	22.0	15.8	35.6	6.7	14.3	8.7	81.5
Dec.	979.9	20.4	16.6	33.9	5.6	13.0	1.7	45.7

Month	Mean evap. (mm)	Number of days with			Mean cloudi- ness (oktas)	Mean sun- shine (h)	Wind	
		precip.	thunder- storms	fog			prevalent direct.	mean speed (m/sec)
Jan.	4.1	2	0.9	0.2	2.0	9.4	030	
Feb.	5.9	2	1.9	0	1.5	10.0	010	
Mar.	8.3	3	4	0	2.1	9.5	040	
Apr.	10.5	4	5	0	3.1	9.5	290	
May	13.5	3	5	0	3.5	9.8	290	
June	10.3	15	12	0	5.7	6.0	280	
July	5.5	24	8	0	7.2	3.2	260	
Aug.	5.3	21	7	0.1	7.1	3.7	260	
Sep.	5.1	17	7	0.2	6.1	5.3	310	
Oct.	5.0	5	2	0.2	3.8	8.5	010	
Nov.	4.2	1.8	0.6	0	2.1	10.4	030	
Dec.	3.9	0.6	0.5	0.1	1.7	9.6	010	
Annual	6.8	98	54	0.8	3.8	7.9	150	

TABLE XLI

CLIMATIC TABLE FOR BHAVNAGAR (AIRPORT)
Latitude 21°45′N, longitude 72°11′E, elevation 11 m

Month	Mean sta. press. (mbar)	Temperature (°C)				Mean vapour press. (mbar)	Precipitation (mm)	
		mean daily	mean daily range	extremes max.	min.		mean	24-h max.
Jan.	1013.1	19.3	16.5	35.0	0.6	9.3	2.5	43.7
Feb.	1011.7	22.6	15.4	39.4	2.8	9.3	0.5	16.5
Mar.	1009.5	27.1	15.1	43.3	8.3	12.1	3.4	35.8
Apr.	1006.9	30.7	13.7	45.0	12.8	17.7	12.5	306.6
May	1003.7	32.8	13.6	46.7	19.4	24.9	6.2	97.4
June	1000.1	32.3	10.5	45.0	20.6	28.3	13.7	219.7
July	999.6	29.6	7.2	40.0	21.6	29.5	242.8	195.6
Aug.	1001.6	28.5	7.5	38.9	21.7	28.0	152.0	125.0
Sep.	1004.9	28.7	9.0	41.1	20.6	27.7	121.9	242.3
Oct.	1008.5	28.3	11.7	41.1	13.3	20.1	37.8	233.9
Nov.	1001.9	24.8	13.6	38.3	6.1	13.7	7.3	105.7
Dec.	1013.5	21.4	14.4	35.0	5.0	11.1	0.2	25.9
Annual	1007.1	27.2	12.4	46.7	0.6	19.3	600	306.6

Month	Mean evap. (mm)	Number of days with			Mean cloudiness (oktas)	Mean sunshine (h)	Wind	
		precip.	thunder-storms	fog			prevalent direct.	mean speed (m/sec)
Jan.		0.3	0.1	0	1.3		260	2.3
Feb.		0.1	0.1	0.1	1.1		250	2.4
Mar.		0.2	0.2	0	1.1		270	2.7
Apr.		0.4	0.5	0	1.3		260	3.1
May		0.6	0.9	0	1.4		250	3.8
June		7	4	0	4.5		240	4.3
July		15	2	0	6.6		240	4.0
Aug.		12	1.5	0	6.5		240	3.5
Sep.		9	1.9	0	5.0		250	2.8
Oct.		1.3	1.2	0	3.1		260	2.0
Nov.		0.5	0.3	0	1.6		270	1.9
Dec.		0.1	0	0	1.4		270	1.9
Annual		47	13	0.1	2.9		250	2.9

TABLE XLII

CLIMATIC TABLE FOR CUTTACK
Latitude 20°48′N, longitude 85°56′E, elevation 27 m

Month	Mean sta. press. (mbar)	Temperature (°C)				Mean vapour press. (mbar)	Precipitation (mm)	
		mean daily	mean daily range	extremes			mean	24-h max.
				max.	min.			
Jan.	1012.2	22.3	13.2	35.6	7.8	16.5	10.4	61.0
Feb.	1009.7	24.9	13.3	38.9	10.6	18.2	28.5	98.0
Mar.	1006.9	29.0	13.8	42.8	14.4	21.5	19.5	99.1
Apr.	1003.8	31.8	13.0	45.0	17.2	26.9	27.0	94.5
May	999.5	32.9	11.9	47.7	20.6	30.7	71.8	142.7
June	996.2	31.1	9.3	47.2	21.7	31.7	214.6	205.7
July	996.1	28.6	6.0	40.0	21.1	32.1	355.1	210.8
Aug.	997.7	28.6	6.0	37.2	21.7	32.1	364.5	320.8
Sep.	1000.7	28.9	6.7	36.7	21.7	31.9	252.1	249.2
Oct.	1006.3	27.7	8.3	36.7	16.7	28.1	167.6	292.6
Nov.	1010.2	24.5	11.3	35.0	10.6	20.3	41.4	195.6
Dec.	1012.3	21.9	12.9	33.3	8.9	16.6	4.7	54.9
Annual	1004.3	27.7	10.4	47.7	7.8	25.5	1557.2	320.8

Month	Mean evap. (mm)	Number of days with			Mean cloudiness (oktas)	Mean sunshine (h)	Wind	
		precip.	thunderstorms	fog			prevalent direct.	mean speed (m/sec)
Jan.		0.9	0.1	3	1.7		010	0.8
Feb.		1.9	1.0	3	2.1		330	1.0
Mar.		2	2	1.0	2.5		210	1.5
Apr.		3	4	3	2.7		200	2.1
May		6	7	0	4.3		190	2.5
June		14	9	0	5.7		230	2.0
July		21	5	0	6.7		230	1.8
Aug.		21	7	0	6.6		250	1.7
Sep.		17	7	0	5.8		290	1.3
Oct.		10	5	0	4.3		340	1.2
Nov.		2	0.5	3	2.5		010	0.9
Dec.		0.8	0.1	0.5	1.7		350	0.6
Annual		100	48	13	3.9		210	1.4

TABLE XLIII

CLIMATIC TABLE FOR POONA
Latitude 18°32′N, longitude 73°51′E, elevation 559 m

Month	Mean sta. press. (mbar)	Temperature (°C) mean daily	mean daily range	extremes max.	min.	Mean vapour press. (mbar)	Precipitation (mm) mean	24-h max.
Jan.	950.3	21.3	18.7	35.0	1.7	11.5	1.9	22.3
Feb.	949.1	23.1	19.6	38.9	3.9	11.0	0.3	16.3
Mar.	947.7	26.5	19.3	42.8	7.2	11.5	3.1	35.1
Apr.	946.1	29.1	17.3	43.3	10.6	14.7	17.6	51.1
May	944.1	29.9	14.6	43.3	13.9	18.9	34.7	82.5
June	941.9	27.5	8.9	41.7	17.2	23.6	102.8	97.0
July	941.7	24.9	5.8	35.6	18.9	24.7	186.8	130.4
Aug.	942.9	24.6	6.2	35.0	17.2	24.3	106.4	108.7
Sep.	944.9	25.0	8.4	36.1	16.1	23.6	127.3	132.3
Oct.	947.5	25.5	12.5	37.8	11.1	20.4	91.9	149.1
Nov.	950.0	22.9	15.8	36.1	7.2	15.5	37.0	96.8
Dec.	950.7	21.1	18.1	35.0	4.4	12.7	4.9	42.4
Annual	946.4	25.1	13.8	43.3	1.7	18.1	714.7	149.1

Month	Mean evap. (mm)	Number of days with precip.	thunder-storms	fog	Mean cloudi-ness (oktas)	Mean sun-shine (h)	Wind prevalent direct.	mean speed (m/sec)
Jan.		0.5	0.1	0.4	1.7	9.3	240	0.8
Feb.		0.1	0.2	0	1.1	9.9	240	1.0
Mar.		0.9	1.5	0	1.3	9.7	250	1.1
Apr.		3	4	0	2.1	9.5	260	1.5
May		4	5	0	2.5	9.5	270	2.0
June		13	3	0	5.7	5.9	270	2.5
July		24	0.3	0	7.1	3.1	270	2.5
Aug.		23	0.9	0	6.9	3.5	270	2.0
Sep.		14	4	0.1	5.8	5.4	270	1.7
Oct.		8	5	1.5	3.9	7.7	250	0.9
Nov.		3	1.8	0.5	2.7	8.7	170	0.8
Dec.		0.6	0.3	0.2	1.9	9.3	230	0.7
Annual		94	26	3	3.5	7.6	250	1.4

TABLE XLIV

CLIMATIC TABLE FOR BOMBAY
Latitude 18°54′N, longitude 72°49′E, elevation 11 m

Month	Mean sta. press. (mbar)	Temperature (°C)				Mean vapour press. (mbar)	Precipitation (mm)	
		mean daily	mean daily range	extremes			mean	24-h max.
				max.	min.			
Jan.	1011.8	24.3	9.7	33.0	16.1	19.7	2.0	49.3
Feb.	1010.8	24.9	9.2	33.9	17.1	20.5	1.1	41.7
Mar.	1009.1	26.9	8.3	35.0	20.0	23.7	0.4	34.3
Apr.	1007.5	28.7	7.2	34.8	22.9	27.6	2.8	37.3
May	1005.5	30.1	6.4	34.6	24.8	30.2	16.0	126.2
June	1002.5	29.1	5.6	34.4	23.5	31.3	520.3	408.9
July	1002.3	27.5	4.7	31.8	23.1	31.1	709.5	304.8
Aug.	1003.8	27.1	4.7	31.1	23.2	30.1	439.3	287.0
Sep.	1006.0	27.4	5.4	31.7	22.9	29.5	297.0	548.1
Oct.	1008.3	28.3	7.3	35.1	22.3	28.9	88.0	148.6
Nov.	1010.3	27.5	9.5	34.8	20.6	24.5	20.6	122.7
Dec.	1011.7	25.9	10.1	33.7	18.0	21.1	2.2	24.4
Annual	1007.5	27.3	7.4	36.7	15.5	26.5	2099.2	549.1

Month	Mean evap. (mm)	Number of days with			Mean cloudi-ness (oktas)	Mean sun-shine (h)	Wind	
		precip.	thunder-storms	fog			prevalent direct.	mean speed (m/sec)
Jan.	3.7	0.3	0	0	1.2	9.0	080	2.5
Feb.	4.3	0.2	0.1	0	0.9	9.5	050	2.6
Mar.	5.1	0.1	0.1	0	1.1	9.1	070	2.9
Apr.	5.7	0.7	0.4	0	1.7	8.3	010	2.9
May	6.1	2	2	0	2.9	9.3	260	2.8
June	5.5	20	5	0	5.6	5.4	230	3.5
July	3.9	29	1.2	0	7.3	2.2	260	4.0
Aug.	4.0	27	0.8	0	6.9	2.6	260	3.7
Sep.	4.0	21	3	0	5.5	4.9	120	2.8
Oct.	4.2	5	3	0	3.0	8.0	090	2.3
Nov.	3.8	2	0.8	0	2.1	9.2	070	2.3
Dec.	3.4	0.3	0.1	0	1.4	9.6	060	2.3
Annual	4.5	108	17	0	3.3	7.3	130	2.9

TABLE XLV

CLIMATIC TABLE FOR HYDERABAD
Latitude 17°27′N, longitude 78°28′E, elevation 545 m

Month	Mean sta. press. (mbar)	Temperature (°C) mean daily	Temperature (°C) mean daily range	Temperature (°C) extremes max.	Temperature (°C) extremes min.	Mean vapour press. (mbar)	Precipitation (mm) mean	Precipitation (mm) 24-h max.
Jan.	952.5	21.6	14.0	35.0	6.1	15.4	1.7	93.2
Feb.	950.8	23.9	14.5	37.2	8.9	15.3	11.4	42.9
Mar.	949.1	27.4	14.8	42.2	13.2	16.1	13.4	103.1
Apr.	946.9	30.3	13.2	43.3	16.1	19.2	24.1	60.7
May	943.9	32.5	12.5	44.4	19.4	20.8	30.0	65.0
June	942.3	29.1	10.0	43.9	17.8	24.6	107.4	122.7
July	942.3	26.1	7.5	37.2	19.4	25.7	165.0	109.2
Aug.	943.5	25.8	7.4	36.1	19.4	25.2	146.9	190.5
Sep.	945.2	25.7	8.1	36.1	17.8	25.3	163.3	153.2
Oct.	948.6	25.1	10.5	36.7	12.2	22.6	70.8	117.1
Nov.	951.3	22.3	12.7	33.9	7.8	17.9	24.9	95.5
Dec.	952.7	20.6	14.4	33.3	7.2	14.9	5.5	44.5
Annual	947.4	25.9	11.7	44.4	6.1	20.3	764.4	190.5

Month	Mean evap. (mm)	Number of days with precip.	Number of days with thunder-storms	Number of days with fog	Mean cloudi-ness (oktas)	Mean sun-shine (h)	Wind prevalent direct.	Wind mean speed (m/sec)
Jan.	4.5	0.2	0.1	0.8	2.3	10.1	110	2.3
Feb.	6.5	1.2	0.5	0.2	2.1	10.3	090	2.5
Mar.	8.3	2	1.9	0	1.9	9.4	130	2.7
Apr.	9.2	4	5	0	3.1	9.6	220	3.0
May	11.9	4	6	0	3.9	9.1	280	3.4
June	9.0	11	5	0	6.0	7.0	280	6.6
July	7.5	19	1.5	0	7.0	3.6	270	6.1
Aug.	5.8	17	2	0	6.7	4.5	280	5.1
Sep.	5.3	13	4	0	6.1	5.8	300	3.5
Oct.	5.5	6	3	0.1	4.5	7.3	290	2.5
Nov.	4.7	3	0.3	0.2	3.6	8.6	080	2.2
Dec.	4.9	0.9	0	0.8	2.5	9.5	100	2.1
Annual	6.9	81	29	2	4.1	7.9	200	3.5

TABLE XLVI

CLIMATIC TABLE FOR VISAKHAPATNAM
Latitude 17°43′N, longitude 83°14′E, elevation 3 m

Month	Mean sta. press. (mbar)	Temperature (°C)				Mean vapour press. (mbar)	Precipitation (mm)	
		mean daily	mean daily range	extremes			mean	24-h max.
				max.	min.			
Jan.	1014.4	22.6	10.2	33.1	12.8	23.3	7.2	132.1
Feb.	1012.3	24.3	9.9	36.7	13.3	25.3	14.9	64.5
Mar.	1010.3	26.9	8.6	38.3	14.4	27.9	8.7	64.5
Apr.	1007.5	29.3	6.9	40.5	18.3	33.5	12.7	73.9
May	1003.4	30.9	6.2	43.3	20.0	36.7	53.5	145.3
June	1000.0	30.5	6.3	44.4	21.1	36.7	87.8	166.1
July	1000.3	28.9	5.7	38.3	21.3	34.1	121.9	145.0
Aug.	1001.5	29.0	6.0	38.2	21.1	34.5	132.2	121.4
Sep.	1003.9	28.6	6.0	37.8	22.2	33.5	167.3	148.6
Oct.	1008.5	27.7	6.4	36.8	17.8	30.5	259.3	293.3
Nov.	1012.1	25.3	8.1	33.9	16.1	24.1	90.6	270.5
Dec.	1014.1	28.0	9.4	32.8	12.8	22.2	17.5	191.3
Annual	1007.3	27.3	7.5	44.4	12.8	30.2	973.6	293.3

Month	Mean evap. (mm)	Number of days with			Mean cloudiness (oktas)	Mean sunshine (h)	Wind	
		precip.	thunderstorms	fog			prevalent direct.	mean speed (m/sec)
Jan.	5.2	0.9	0.2	0.1	2.0		330	1.7
Feb.	6.3	1.3	0.3	0.1	1.9		290	1.8
Mar.	7.5	0.9	1.1	0	2.1		230	2.7
Apr.	9.0	1.9	4	0	3.5		230	4.0
May	9.1	5	7	0	4.5		220	4.4
June	8.9	9	7	0	6.3		230	3.8
July	6.1	14	5	0	6.9		230	4.6
Aug.	6.2	14	6	0	6.5		240	3.7
Sep.	5.8	15	11	0	6.1		250	2.6
Oct.	5.1	13	7	0	4.9		310	2.2
Nov.	5.7	4	0.7	0	3.6		350	2.2
Dec.	5.4	1.4	0.1	0	2.6		350	2.0
Annual	6.7	80	49	0.2	4.3		270	3.0

TABLE XLVII

CLIMATIC TABLE FOR MASULIPATAM
Latitude 16°11′N, longitude 81°08′E, elevation 3 m

Month	Mean sta. press. (mbar)	Temperature (°C)				Mean vapour press. (mbar)	Precipitation (mm)	
		mean daily	mean daily range	extremes			mean	24-h max.
				max.	min.			
Jan.	1014.1	23.6	8.4	33.3	13.9	22.5	0.9	76.2
Feb.	1012.3	25.2	8.8	37.2	14.4	24.5	11.0	96.5
Mar.	1010.3	27.4	9.0	42.2	16.7	27.5	9.4	150.4
Apr.	1007.7	29.9	8.1	44.4	18.3	30.9	18.4	101.3
May	1003.3	32.3	8.5	47.8	19.4	32.1	35.7	82.5
June	1001.1	31.9	9.0	46.1	20.0	29.3	106.0	133.3
July	1001.1	29.1	6.9	41.7	19.4	30.2	199.3	115.6
Aug.	1003.0	29.0	6.4	38.3	21.7	30.1	151.9	135.6
Sep.	1004.7	28.5	6.1	37.8	20.6	30.8	156.0	116.6
Oct.	1008.3	27.9	5.9	37.2	18.9	30.0	259.0	502.4
Nov.	1011.7	25.5	6.9	34.4	13.9	25.1	111.3	355.6
Dec.	1013.9	23.8	8.0	32.2	14.4	22.1	16.1	159.3
Annual	1007.7	27.9	7.7	47.8	13.9	27.9	1075.0	502.4

Month	Mean evap. (mm)	Number of days with			Mean cloudi- ness (oktas)	Mean sun- shine (h)	Wind	
		precip.	thunder- storms	fog			prevalent direct.	mean speed (m/sec)
Jan.		0.5	0	0.3	1.8		040	2.4
Feb.		1.2	0.1	0.2	2.0		030	2.3
Mar.		0.5	0.4	0.5	1.9		160	2.5
Apr.		1.5	1.3	0	2.3		170	3.2
May		3	2	0	3.3		190	3.8
June		10	4	0	5.1		270	4.0
July		17	3	0	5.9		250	3.5
Aug.		16	4	0	5.7		260	3.1
Sep.		13	5	0	5.5		270	2.4
Oct.		11	7	0	4.7		340	2.1
Nov.		5	1.4	0	3.7		030	2.4
Dec.		1.3	0	0	2.5		030	2.5
Annual		80	28	1.0	3.7		170	2.9

TABLE XLVIII

CLIMATIC TABLE FOR MADRAS (AIRPORT)
Latitude 13°00′N, longitude 80°11′E, elevation 16 m

Month	Mean sta. press. (mbar)	Temperature (°C)				Mean vapour press. (mbar)	Precipitation (mm)	
		mean daily	mean daily range	extremes			mean	24-h max.
				max.	min.			
Jan.	1012.1	24.5	8.5	32.8	13.9	23.3	23.8	212.9
Feb.	1010.6	25.9	9.5	36.7	15.0	23.8	6.8	123.2
Mar.	1008.9	27.9	9.6	40.6	16.7	26.5	15.1	88.1
Apr.	1006.3	25.5	8.9	42.8	20.0	30.3	24.7	96.3
May	1002.7	32.7	9.8	45.0	21.1	30.5	51.7	214.9
June	1001.7	32.5	9.7	43.3	20.6	27.1	52.6	59.2
July	1002.5	30.7	8.9	41.1	21.7	26.9	83.5	116.3
Aug.	1003.0	30.1	8.7	40.0	20.6	27.7	124.3	91.7
Sep.	1004.3	29.7	8.5	38.9	20.6	28.7	118.0	100.3
Oct.	1006.9	28.1	7.4	39.4	16.7	29.3	267.0	233.7
Nov.	1009.5	25.9	6.7	34.4	15.0	25.8	308.7	236.2
Dec.	1011.5	24.6	7.2	32.8	13.9	24.0	139.1	261.6
Annual	1006.7	28.7	8.6	45.0	13.9	27.1	1215.3	261.6

Month	Mean evap. (mm)	Number of days with			Mean cloudi- ness (oktas)	Mean sun- shine (h)	Wind	
		precip.	thunder- storms	fog			prevalent direct.	mean speed (m/sec)
Jan.	4.3	3	0.2	0.8	1.8	8.6	340	2.5
Feb.	4.9	1.4	0.3	0.7	1.1	9.6	300	2.5
Mar.	5.7	1.0	0.7	0.3	0.9	9.8	230	2.8
Apr.	6.6	1.7	2	0.1	1.3	9.2	200	2.9
May	8.2	3	3	0	1.1	8.4	230	3.6
June	8.9	9	5	0	1.6	6.5	250	4.5
July	6.9	14	5	0	1.9	4.8	250	4.0
Aug.	7.0	15	7	0	1.7	5.3	250	3.8
Sep.	6.0	11	8	0	1.6	6.2	250	3.1
Oct.	4.9	14	9	0.4	1.8	6.4	280	2.5
Nov.	4.3	11	3	0.1	2.4	6.9	350	3.2
Dec.	3.7	7	0.5	0.4	2.5	7.6	350	3.5
Annual	5.9	91	44	3	1.7	7.4	270	3.3

TABLE XLIX

<small>CLIMATIC TABLE FOR MANGALORE</small>
Latitude 12°52′N, longitude 74°51′E, elevation 22 m

Month	Mean sta. press. (mbar)	Temperature (°C)				Mean vapour press. (mbar)	Precipitation (mm)	
		mean daily	mean daily range	extremes			mean	24-h max.
				max.	min.			
Jan.	1009.1	26.5	9.7	36.1	16.7	22.5	4.7	40.6
Feb.	1008.5	26.9	8.3	37.8	16.7	25.1	1.9	36.1
Mar.	1007.5	28.1	7.2	37.3	18.3	27.3	8.9	82.8
Apr.	1006.3	29.3	6.3	36.1	20.0	29.1	40.0	117.1
May	1004.8	29.1	6.1	36.7	18.9	30.0	232.7	360.9
June	1004.9	26.7	5.5	34.4	20.0	30.1	981.6	252.0
July	1005.3	26.0	5.0	35.6	20.6	30.0	1058.6	268.2
Aug.	1005.7	26.1	4.9	32.2	20.6	29.7	576.9	232.4
Sep.	1006.5	26.1	5.2	31.7	21.1	29.3	267.0	184.7
Oct.	1007.0	26.8	6.0	34.4	20.0	29.1	205.9	181.6
Nov.	1007.9	27.1	7.9	35.6	17.2	26.3	70.6	112.1
Dec.	1008.9	26.8	9.8	35.6	16.7	22.6	18.2	153.2
Annual	1006.9	27.1	6.8	37.8	16.7	27.6	3467.0	360.9

Month	Mean evap. (mm)	Number of days with			Mean cloudiness (oktas)	Mean sunshine (h)	Wind	
		precip.	thunderstorms	fog			prevalent direct.	mean speed (m/sec)
Jan.		0.3	0.3	0.2	2.2		090	2.4
Feb.		0.1	0.1	0	2.2		080	2.4
Mar.		1.0	1.2	0.3	2.5		080	2.4
Apr.		4	5	0.1	4.1		070	2.5
May		12	8	0	5.5		040	2.7
June		27	5	0	7.1		350	2.6
July		30	1.9	0.2	7.5		240	2.6
Aug.		28	0.8	0.1	7.1		280	2.2
Sep.		22	1.3	0.9	6.3		070	1.9
Oct.		15	7	0.6	5.7		090	2.0
Nov.		6	3	0.4	4.3		090	2.0
Dec.		1.2	0.9	0	2.9		090	2.2
Annual		147	35	3	4.7		130	2.3

TABLE L

CLIMATIC TABLE FOR BANGALORE

Latitude 12°58′N, longitude 77°35′E, elevation 921 m

Month	Mean sta. press. (mbar)	Temperature (°C)				Mean vapour press. (mbar)	Precipitation (mm)	
		mean daily	mean daily range	extremes			mean	24-h max.
				max.	min.			
Jan.	911.5	20.9	11.9	32.2	7.8	13.9	3.3	65.8
Feb.	910.7	23.1	13.2	34.4	9.4	13.1	10.2	67.3
Mar.	909.5	25.7	13.3	37.2	11.1	13.6	6.1	50.8
Apr.	908.3	27.3	12.2	38.3	14.4	17.6	45.7	90.7
May	906.5	26.9	11.6	38.9	16.7	20.1	116.5	153.9
June	905.8	24.3	9.2	37.8	16.7	20.9	80.1	101.6
July	906.0	23.2	8.0	33.3	16.1	21.0	116.6	105.4
Aug.	906.3	23.3	8.1	33.3	14.4	21.0	147.1	162.1
Sept.	907.3	23.3	8.7	33.3	15.0	20.5	142.7	124.7
Oct.	908.5	23.2	8.6	32.2	13.3	20.5	184.9	116.8
Nov.	910.1	21.7	9.1	31.1	10.6	18.3	54.3	114.5
Dec.	911.3	20.5	10.4	31.1	8.9	15.4	16.2	67.3
Annual	908.5	23.6	10.4	38.9	7.8	18.0	923.7	162.1

Month	Mean evap. (mm)	Number of days with			Mean cloudi-ness (oktas)	Mean sun-shine (h)	Wind	
		precip.	thunder-storms	fog			prevalent direct.	mean speed (m/sec)
Jan.	4.7	1.1	0	3	3.3	8.6	090	2.9
Feb.	6.3	0.7	0.5	0.4	2.7	9.2	110	2.7
Mar.	7.7	0.9	1.2	0.2	2.5	9.5	170	2.6
Apr.	7.4	5	7	0.1	4.3	8.6	240	2.5
May	6.6	11	12	0.2	5.3	7.4	270	3.3
June	5.2	13	4	0.1	6.9	4.8	260	4.7
July	4.7	18	2	0.3	7.5	3.0	260	4.9
Aug.	4.1	18	4	0.7	7.3	4.1	260	4.2
Sep.	4.4	15	4	0.6	6.9	5.2	270	3.3
Oct.	4.4	15	7	1.5	6.2	5.6	250	2.2
Nov.	3.9	7	1.3	1.6	5.1	7.2	090	2.3
Dec.	3.8	3	0.1	3	4.1	7.5	080	2.7
Annual	5.3	108	43	12	5.1	6.7	200	3.2

TABLE LI

CLIMATIC TABLE FOR TRINCOMALLEE
Latitude 08°35′N, longitude 81°15′E, elevation 8 m*

Month	Mean sta. press. (mbar)	Temperature (°C)				Mean vapour press. (mbar)	Precipitation (mm)	
		mean daily	mean daily range	extremes max.	min.		mean	24-h max.
Jan.	1011.9	25.5	2.9	28.9	20.0	26.3	211.1	208.5
Feb.	1011.8	26.2	3.8	31.1	18.3	26.1	67.3	108.0
Mar.	1010.4	27.1	4.9	32.8	21.7	28.1	58.4	106.4
Apr.	1008.7	28.5	6.3	36.7	22.2	29.7	54.1	74.4
May	1006.4	29.5	7.2	36.7	21.1	28.5	82.0	271.5
June	1005.7	29.7	7.3	36.1	20.6	26.3	23.6	79.5
July	1006.1	29.5	8.0	36.7	21.7	25.3	43.9	99.8
Aug.	1006.5	29.1	8.1	36.1	21.1	25.9	91.7	107.4
Sep.	1007.3	29.0	8.2	36.7	22.2	26.5	87.1	128.5
Oct.	1008.9	27.6	6.8	36.1	21.7	27.7	242.6	139.2
Nov.	1009.8	26.1	4.7	32.2	20.6	28.1	354.1	200.4
Dec.	1011.2	25.5	3.4	31.7	20.0	27.1	330.2	256.8
Annual	1008.7	27.7	5.9	36.7	18.3	27.1	1646.2	271.5

Month	Mean evap. (mm)	Number of days with			Mean cloudiness (oktas)	Mean sunshine (h)	Wind	
		precip.	thunderstorms	fog			prevalent direct.	mean speed (m/sec)
Jan.		13	0.6	0	6.3		050	4.7
Feb.		5	0.6	0	4.5		050	3.4
Mar.		6	1.6	0	4.2		180	2.5
Apr.		6	4	0	4.7		230	2.8
May		6	3	0	5.5		230	5.5
June		3	1.3	0	6.5		230	7.6
July		3	1.8	0	6.7		230	7.0
Aug.		7	4	0	6.4		230	6.1
Sep.		8	4	0	6.1		230	5.4
Oct.		16	6	0	6.1		230	4.1
Nov.		19	3	0	6.9		230	3.7
Dec.		18	1.0	0	6.7		050	5.1

* 33 m prior to July 1935.

TABLE LII

CLIMATIC TABLE FOR COLOMBO
Latitude 06°54′N, longitude 79°53′E, elevation 8 m

Month	Mean sta. press. (mbar)	Temperature (°C)				Mean vapour press. (mbar)	Precipitation (mm)	
		mean daily	mean daily range	extremes max.	extremes min.		mean	24-h max.
Jan.	1011.1	26.1	7.9	34.4	17.2	25.7	100.6	78.0
Feb.	1011.0	26.3	8.2	33.9	16.7	26.3	66.0	132.6
Mar.	1010.3	27.1	7.6	33.9	17.8	28.0	118.4	89.7
Apr.	1009.2	27.6	6.6	32.8	21.1	29.9	230.4	182.4
May	1008.6	27.8	5.3	32.2	20.6	30.7	393.5	289.6
June	1008.8	27.4	4.4	31.1	21.7	29.9	219.9	91.7
July	1009.2	27.1	4.2	30.6	21.7	29.1	139.5	184.2
Aug.	1009.7	27.1	4.5	30.6	21.7	28.9	102.4	126.0
Sep.	1009.9	27.2	4.8	31.7	21.7	28.7	173.7	153.4
Oct.	1010.3	26.5	5.5	31.7	20.6	28.7	348.2	256.3
Nov.	1010.1	26.3	6.3	32.2	19.4	27.9	332.5	210.3
Dec.	1010.6	26.1	7.3	32.8	17.2	26.1	142.5	113.8
Annual	1009.9	26.9	6.0	34.4	16.7	28.3	2367.5	289.6

Month	Mean evap. (mm)	Number of days with			Mean cloudiness (oktas)	Mean sunshine (h)	Wind	
		precip.	thunderstorms	fog			prevalent direct.	mean speed (m/sec)
Jan.		10	4	0	5.1		050	3.3
Feb.		6	4	0	4.3		050	3.0
Mar.		11	8	0	5.2		050	2.7
Apr.		17	12	0	6.6		230	2.7
May		23	8	0	7.7		230	3.3
June		22	1.9	0	8.3		230	3.5
July		16	1.0	0	8.0		230	3.3
Aug.		14	0.8	0	7.7		230	3.4
Sep.		17	2	0	7.7		230	3.4
Oct.		22	6	0	7.5		230	2.9
Nov.		20	8	0	7.0		050	2.5
Dec.		12	6	0	5.7		050	3.3
Annual		190	61.7	0	6.7		160	3.1

TABLE LIII

CLIMATIC TABLE FOR NOWARA ELIYA
Latitude 06°58′N, longitude 80°46′E, elevation 207 m

Month	Mean sta. press. (mbar)	Temperature (°C)				Mean vapour press. (mbar)	Precipitation (mm)	
		mean daily	mean daily range	extremes			mean	24-h max.
				max.	min.			
Jan.	814.5	13.9	11.7	24.4	−2.8	13.9	176.8	106.4
Feb.	814.8	13.9	14.3	23.9	−2.2	12.5	50.5	120.4
Mar.	814.3	13.7	13.7	25.0	0.0	13.5	103.9	63.5
Apr.	814.1	15.8	12.2	24.4	3.3	15.5	126.5	71.9
May	813.3	16.6	9.4	25.6	0.6	16.6	215.1	189.0
June	813.1	16.0	5.8	23.9	7.2	15.9	264.9	96.3
July	812.9	15.6	6.0	24.4	7.2	15.3	279.9	149.9
Aug.	813.1	15.7	7.0	23.3	6.1	15.5	190.7	64.8
Sep.	813.7	15.6	8.0	23.9	5.0	15.5	209.3	98.0
Oct.	813.9	15.5	8.9	23.9	4.4	15.7	247.9	142.0
Nov.	813.9	15.3	9.1	23.3	0.6	15.5	233.4	75.7
Dec.	814.3	14.5	10.8	23.3	−1.1	14.6	198.4	203.7
Annual	813.9	15.3	9.7	25.6	−2.8	15.0	2297.5	203.7

Month	Mean evap. (mm)	Number of days with			Mean cloudiness (oktas)
		precip.	thunderstorms	fog	
Jan.		14	0.4	0	6.3
Feb.		7	0.6	0	5.1
Mar.		11	2	0	5.7
Apr.		15	4	0	6.5
May		18	1.9	0	7.5
June		25	0.3	0	8.9
July		24	0.4	0	8.7
Aug.		22	1.0	0	8.4
Sep.		20	1.3	0	8.1
Oct.		22	1.4	0	7.9
Nov.		22	1.1	0	7.9
Dec.		17	0.4	0	7.0
Annual		217	14.8	0	7.3

TABLE LIV

CLIMATIC TABLE FOR BIKANER

Latitude 28°00′N, longitude 73°18′E, elevation 224 m

Month	Mean sta. press. (mbar)	Temperature (°C)				Mean vapour press. (mbar)	Precipitation (mm)	
		mean daily	mean daily range	extremes			mean	24-h max.
				max.	min.			
Jan.	990.3	13.7	17.3	31.1	−2.2	7.6	5.6	25.4
Feb.	987.6	17.1	17.9	37.2	−2.2	8.2	7.3	46.7
Mar.	984.7	23.1	17.4	42.8	−0.6	9.1	5.2	43.9
Apr.	980.9	29.2	16.8	47.2	8.3	9.0	4.6	31.0
May	976.0	34.6	14.8	49.4	13.7	13.3	7.5	48.3
June	972.1	35.5	12.4	48.9	17.8	21.1	27.0	110.7
July	971.6	33.5	10.9	47.2	20.6	28.2	86.8	134.1
Aug.	973.8	31.3	9.5	43.3	21.1	27.3	104.5	142.0
Sep.	978.1	30.9	11.4	43.9	19.4	24.3	44.6	165.6
Oct.	984.1	26.9	16.7	42.2	7.8	13.7	5.7	95.8
Nov.	988.7	20.3	20.3	47.2	0.6	10.5	2.6	41.9
Dec.	990.3	15.1	19.1	32.2	−2.8	8.1	2.3	30.0
Annual	981.5	25.9	15.4	49.4	−2.8	15.1	304.7	165.6

Month	Mean evap. (mm)	Number of days with			Mean cloudi-ness (oktas)	Mean sun-shine (h)	Wind	
		precip.	thunder-storms	fog			prevalent direct.	mean speed (m/sec)
Jan.	2.8	1.5	0.2	1.5	1.7		090	1.3
Feb.	4.3	1.3	0.4	0.2	1.5		180	1.4
Mar.	7.0	1.4	0.4	0.1	1.4		110	1.8
Apr.	9.9	0.6	1.2	0.1	1.1		210	2.0
May	12.9	1.3	1.3	0	0.8		220	2.8
June	14.1	3	1.7	0.2	1.3		230	3.7
July	10.9	6	2	0	2.6		220	3.4
Aug.	8.6	8	2	0	2.7		230	3.1
Sep.	7.9	3	1.2	0.1	1.4		250	2.6
Oct.	6.7	0.6	0.4	0	0.4		200	1.4
Nov.	4.3	0.4	0.2	0.4	0.5		140	1.0
Dec.	2.9	0.9	0.1	0.5	1.5		090	1.0
Annual	7.7	28	11	3	1.4		180	2.1

TABLE LV

CLIMATIC TABLE FOR JACOBABAD
Latitude 28°17′N, longitude 68°29′E, elevation 62/61 m

Month	Mean sta. press. (mbar)	Temperature (°C)				Mean vapour press. (mbar)	Precipitation (mm)	
		mean daily	mean daily range	extremes			mean	24-h max.
				max.	min.			
Jan.	1010.3	14.6	16.0	31.7	−3.9	7.1	5.3	27.7
Feb.	1007.5	17.7	16.3	39.4	−1.7	9.1	8.4	25.7
Mar.	1003.7	24.0	17.0	44.4	2.8	10.9	5.6	32.3
Apr.	999.3	29.6	16.8	48.3	8.9	14.9	4.3	48.8
May	994.5	35.2	18.1	52.2	16.1	21.0	3.6	40.6
June	988.9	37.4	16.1	52.8	21.1	28.2	0.6	82.5
July	988.1	35.8	12.8	52.2	21.7	29.9	24.1	84.8
Aug.	990.7	34.1	12.3	47.2	20.0	29.0	22.3	101.6
Sep.	996.1	32.3	14.6	45.0	15.6	26.1	4.3	31.7
Oct.	1002.7	27.6	19.2	44.4	8.3	16.1	0.8	17.3
Nov.	1007.1	21.2	19.3	39.4	2.2	10.3	1.3	46.2
Dec.	1010.6	15.8	17.3	31.7	−0.6	7.9	4.3	47.0
Annual	1000.1	27.1	16.4	52.8	−3.9	17.5	91.4	101.6

Month	Mean evap. (mm)	Number of days with			Mean cloudi-ness (oktas)	Mean sun-shine (h)	Wind	
		precip.	thunder-storms	fog			prevalent direct.	mean speed (m/sec)
Jan.		1.8	0.1	0.1	2.7		310	0.8
Feb.		3	0.6	0.2	2.9		310	0.9
Mar.		2	1.4	0	2.7		310	1.3
Apr.		1.2	0.7	0	1.8		130	1.4
May		0.2	0.9	0.2	0.7		130	1.6
June		0.3	0.4	0	0.7		130	1.8
July		2	1.9	0	2.0		130	1.9
Aug.		1.9	1.6	0	1.7		130	1.5
Sep.		0.1	0.2	0	0.5		130	1.1
Oct.		0	0	0	0.4		310	0.8
Nov.		0	0	0	0.9		310	0.5
Dec.		1.0	0.2	0	2.1		310	0.6
Annual		13	8	0.5	1.6		220	1.2

TABLE LVI

CLIMATIC TABLE FOR NEW DELHI (sfd)
Latitude 28°35′N, longitude 77°12′E, elevation 216 m

Month	Mean sta. press. (mbar)	Temperature (°C)				Mean vapour press. (mbar)	Precipitation (mm)	
		mean daily	mean daily range	extremes			mean	24-h max.
				max.	min.			
Jan.	991.1	14.3	14.0	29.4	−0.6	9.1	24.9	116.8
Feb.	988.5	16.9	13.5	33.3	1.7	9.1	21.8	104.1
Mar.	985.4	22.7	15.1	40.6	4.4	10.0	16.5	62.2
Apr.	981.5	28.6	15.2	45.6	11.7	10.3	6.8	40.9
May	976.5	33.5	13.9	47.2	18.3	13.5	7.9	30.5
June	972.5	34.3	11.2	46.7	18.9	20.7	65.0	235.5
July	972.7	31.3	8.1	45.0	21.7	29.7	211.1	266.2
Aug.	974.8	29.9	7.6	40.0	22.2	29.7	172.9	181.6
Sep.	978.8	29.3	9.5	40.6	17.8	25.4	149.7	176.5
Oct.	985.1	25.9	14.4	39.4	9.4	15.5	31.2	172.7
Nov.	989.1	20.3	16.9	35.0	3.9	9.5	1.2	20.8
Dec.	991.1	15.7	15.4	28.9	1.1	8.9	5.2	53.3
Annual	982.3	25.3	12.9	47.2	−0.6	16.0	714.2	266.2

Month	Mean evap. (mm)	Number of days with			Mean cloudi-ness (oktas)	Mean sun-shine (h)	Wind	
		precip.	thunder-storms	fog			prevalent direct.	mean speed (m/sec)
Jan.	3.3	3	1.5	3	2.9	7.3	270	2.3
Feb.	4.0	3	1.7	0.5	2.5	8.7	300	2.8
Mar.	6.5	3	3	0.1	2.5	7.8	290	3.0
Apr.	10.3	1.3	3	0	1.9	9.2	290	3.0
May	12.1	3	5	0.7	1.5	8.1	270	3.6
June	13.2	5	5	0.2	2.7	6.4	280	4.0
July	8.4	14	7	0.1	5.3	5.6	150	2.9
Aug.	6.9	14	8	0	5.5	6.2	270	2.5
Sep.	5.7	7	4	0	3.1	7.5	270	2.6
Oct.	6.0	2	1.3	0.1	0.9	9.2	270	1.8
Nov.	4.5	0.5	0.4	0.6	0.9	9.5	270	1.8
Dec.	3.3	1.3	0.7	1.7	2.2	8.0	280	2.1
Annual	7.0	57	41	7	2.7	7.8	270	2.7

TABLE LVII

CLIMATIC TABLE FOR YATUNG (CHUMBI)
Latitude 28°55′N, longitude 88°55′E

Month	Mean sta. press. (mbar)	Temperature (°C)				Mean vapour press. (mbar)	Precipitation (mm)	
		mean daily	mean daily range	extremes			mean	24-h max.
				max.	min.			
Jan.	0.2	14.2	19.4		−20.6	3.9	14.2	39.4
Feb.	0.9	13.1	18.3		−15.6	3.7	54.6	164.6
Mar.	4.2	13.4	18.3		−11.7	5.9	52.3	141.0
Apr.	7.8	12.4	22.2		− 6.7	7.7	99.3	85.6
May	10.9	10.3	23.3		− 2.2	10.7	103.1	51.3
June	13.5	8.5	26.7		− 0.6	13.3	145.3	65.0
July	14.7	8.0	25.6		5.0	14.5	162.8	69.1
Aug.	14.3	8.1	28.9		4.4	13.9	155.2	48.5
Sept.	12.8	8.8	23.9		− 0.6	12.4	110.5	120.9
Oct.	8.9	12.3	21.1		− 6.7	8.7	49.3	132.1
Nov.	4.5	13.7	18.9		−10.0	5.8	11.2	71.1
Dec.	1.1	14.3	18.9		−12.2	4.1	9.4	102.1
Annual	7.8	11.4	28.9		−20.6	8.7	967.2	164.6

Month	Mean evap. (mm)	Number of days with			Mean cloudiness (oktas)
		precip.	thunderstorms	fog	
Jan.		4	0	1.8	2.6
Feb.		8	0.1	1.9	3.1
Mar.		11	1.2	3	3.0
Apr.		16	2	1.4	3.4
May		18	1.8	0.2	4.5
June		23	0.7	1.3	5.2
July		23	0.4	1.2	5.4
Aug.		23	0.4	1.1	5.2
Sep.		19	0	1.1	5.3
Oct.		6	0.1	0	3.5
Nov.		1.5	0	0.6	2.2
Dec.		0.6	0	1.1	1.9
Annual		153	8	15	3.8

Reference Index

Geographical Index

Abadan, 225, 231, 234, 236, 238, 240, 244, 245
Abu Kamal, 203, 229, 235, 237, 239, 242, 243, 245, 246, 247, 249
Adalia, 194
Adana, 194, 195, 196, 229, 232, 235, 237, 239, 241, 242, 243, 245, 246, 247
Adriatic Sea, 188
Aegean Sea, 183, 192, 193, 194
Afghanistan, 188
Africa, 73, 185, 190, 192
—, North, 73, 186, 188
Agartala, 270
Agra, 137, 292
Agumbe, 86
Ahmadabad, 156, 270, 300
Ajlum, 208
Akola, 163
Akyab, 11, 12, 17, 18, 20, 43, 268, 269, 270
Aleppo, 203, 229, 232, 235, 237, 239, 241, 242, 243, 245, 246, 247, 248
Allahabad, 98, 270, 296
Alleppay, 88
Alor Star, 11, 17, 18, 19, 35, 62
Alps, 105, 188
Al Wajh, 230, 233, 236, 238, 240
Ambala, 127
Aminidevi, 270
Amman, 209
—, airport, 229, 233, 235, 237, 239, 241, 242, 243, 245, 246, 247, 250
Amritsar, 288
Anantapur, 81
Anatolia, 193, 194
Anatolian Plateau, 184
Andaman Islands, 84, 85, 88, 95, 105
Andes, 26
Andhra Pradesh, 92, 107, 116
Ankara, 195, 196, 229, 232, 235, 237, 239, 241, 242, 243, 246, 247
Annam, 3, 4, 8, 15, 24
Antalya, 195, 196, 229, 232, 235, 237, 239, 241, 242, 243, 245, 246, 247
Aqaba, 209, 210
— airport, 229, 233, 235, 237, 239, 241, 242, 243, 246, 247, 250
— Gulf, 211
Arabia, 73, 101, 189, 191, 192
Arabian Desert, 184

— Gulf, 183, 186, 188, 190, 192, 202, 215, 216, 217, 219, 220, 222, 223, 224, 225
— Peninsula, 184, 185, 187, 189, 190, 191, 192, 215–220, 230, 233, 236, 238, 240, 242, 247, 250, 251
— Sea, 72, 73, 74, 75, 76, 88, 105, 106, 107, 108, 109, 111, 183, 184, 215, 216, 217, 218, 257, 263, 264, 266, 267, 269, 279
Arakan, 109
— Yoma, 70
Arava desert, 213
Aravalli Hills, 70, 88, 95, 116
Arunachal Pradesh, 104
Asia, 5, 66, 71, 101, 183, 185, 186, 187, 220, 221, 226
—, Central Southeast, 19, 20, 21, 22
—, Southeast, 1, 2, 5, 6, 10, 12, 13, 15, 16, 17
—, Southwest, 184–192, 224
— Minor, 186, 190, 193, 196
— — Peninsula, 193
— —Plateau, 184, 188, 200
Asir Mountains, 216, 217
Assam, 73, 74, 75, 80, 84, 85, 90, 92, 101, 103, 116
— Valley, 88, 94, 98
Atlantic Ocean, 188
Atlas Mountains, 188
Australia, 6, 9

Backergunge, 107
Baghdad, 230, 234, 236, 238, 240, 242, 243, 251
Baharaich, 135
Bahrain, 270
Baluchistan, 78, 79, 80, 81, 84, 86, 88, 91, 93, 94, 98, 105
Bandar Anbas, 225
Bandon, 5, 11, 17, 18, 30, 31, 32, 35, 61
Bangalore, 98, 99, 173, 270, 317
Bangkok, 11, 18, 20, 32, 35, 53
Bangladesh, 67, 83, 85, 86, 88, 90, 92, 99, 107, 108, 109, 267, 268, 269
Bannu, 98
Barmer, 143
Baroda, 82, 303
Basra, 221, 230, 234, 236, 238, 240, 242, 243, 252
Battambang, 20, 35, 54
Beer Sheba, 230, 233, 236, 238
Beihan, 230, 233, 236, 238
Beirut, 207, 213, 229, 232, 235, 237, 239, 241, 242, 243, 245, 246, 247, 249
Belgaum, 76

Subject Index